The Breakthrough
BIRD TAXIDERMY
MANUAL

by
the editors of
Breakthrough Magazine:

PUBLISHER and EDITOR
Bob Williamson

AUTHOR
Sallie Dahmes

ART DIRECTOR and MANAGING EDITOR
Ken Edwards

DEDICATION

Within the last 10 years, the art of taxidermy has changed by leaps and bounds. Today, as in days gone by, we have a few very talented and gifted wildlife artists who are working hard to reverse the negative image of taxidermy.

Present day taxidermists such as Henry Wichers Inchumuk, Jim Hall, and Frank Newmyer have contributed enormously to change the status of taxidermy. Because of their gift of portraying wildlife in an artistic and lifelike manner, taxidermy is now being recognized as a valid form of wildlife art rather than merely a craft among sportsmen and collectors of wildlife art.

This book is not just dedicated to the great artists of today; it is also dedicated to taxidermy's great artists of the past, such as Carl Akeley, Leon Pray, and Coloman Jonas, who took those first "giant" steps for taxidermy. These were the early day pioneers of taxidermy.

The "all-important" first steps of these great taxidermists fueled the fire and provided the stimulus and creative spark that inspired the taxidermy "legends" of today. Both our "early" and "modern-day" taxidermy "greats" shared some common characteristics. They exhibited a strong desire to excel, and they were determined to present taxidermy as the beautiful form of art that it is.

We can only imagine how much more our early pioneers could have accomplished with their work if they had access to all of the modern day technology that is readily available today. One thing is "certain" though, even with today's highly advanced technology, these taxidermists would have constantly strived to keep improving their art; just like today's pioneers are doing.

If we asked for some good advice for taxidermists just starting out, both our "early" and "modern-day" pioneers would probably answer something like this: Settle for nothing but "excellence," work hard to achieve "excellence," and remember the "BEST" way to do anything is, as yet, undiscovered!

—Sallie Dahmes

This manual is compiled from information from many sources. The recommendations, procedures and precautions are presented in good faith as the results of experimentation and information from reputable wildlife artists and manufacturers. Due to the fact that we have no control over the actual application of the information provided, the publisher and the authors disclaim any responsibility for the results, including damage by injury, whether or not caused by using the products, techniques, recommendations or suggestions referred to herein. Nor can the publisher or the authors be held responsible for the value of work alleged to be spoiled by the use of a product, technique, recommendation, or suggestion contained herein. It is the user's responsibility to make sure that products and techniques are suitable for their particular requirements.

First Edition published July 1988.
Printed in the United States of America.
Published by *Breakthrough* Publishing Company, 1306 West Spring Street, P. O. Box 967, Monroe, GA 30655.

ISBN 0-925245-08-9
BP1008
The *Breakthrough* Bird Taxidermy Manual
Written by Sallie Dahmes.
Edited by Bob Williamson.
Additional writing by Bryan Bevil, Ken Edwards, Jim Hall, Greg Hildreth, Robert Holshouser and Ed Thompson.
Designed by Ken Edwards.
Principal photography by Bryan Bevil, Ken Edwards, Greg Hildreth and Bob Williamson.
Illustrations and Diagrams by Clairice Mechling.
Typesetting/formatting by Lee Kennedy and Lynn Lloyd.
Production by De Laine Heinlein, Angela Lowe, Gregory Reed and Yvette Rooks.

Turkey mount on cover by Ed Thompson.
Cover photo by Ken Edwards

CONTENTS

The System: An Overview

Birds Are Fun to Hunt— Fun to Mount

The first cool snap of autumn sends waterfowl and upland game bird hunters into a ritualistic frenzy of preparing for their favorite time of the year: *hunting season.* Outboard motors, mudboats, pirogues, ATVs, decoys, hunting dogs, shotguns, and other associated paraphernalia are all put into first class condition in preparation for the upcoming season.

Nothing causes quite as much commotion as the arrival of opening morning. Millions of shotgunners take to the fields, prairies, and swamps in pursuit of good wing shooting, bagging a delightful meal, and bringing home a prize trophy that is suitable for mounting. A brief conversation with any serious sportsman leaves little room for doubt that hunting "birds" is a top priority among most hunters. There is something seriously addictive about bird hunting, and it does not take many trips afield to get "hooked" either.

Just as bird hunting is a favorite pastime with most hunters; bird taxidermy enjoys equal popularity among most taxidermists. It is perhaps the most pleasurable and relaxing form of taxidermy. Although top bird taxidermy work is by no means simple, it is less messy, more artistic and more fun than many other forms of taxidermy.

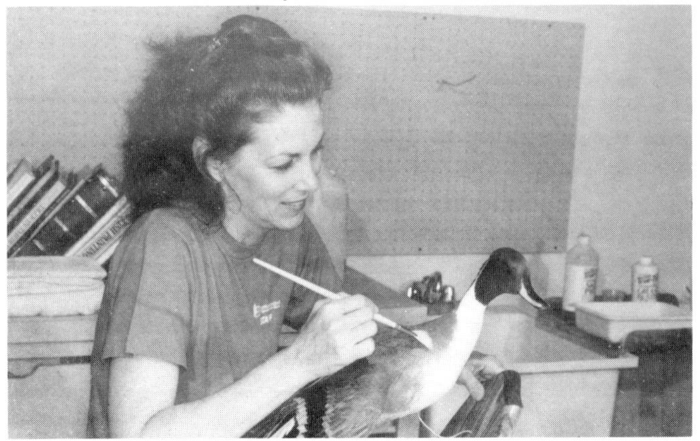

Also, beautiful specimens of birds can easily be obtained by most anyone regardless of what area of the world that they live in. Suffice it to say, that whether a taxidermist mounts only a few birds or hundreds each season, producing "good bird work" is both an enjoyable and a rewarding experience. Quality bird work challenges the taxidermist and offers unlimited avenues for artistic expression, creativity, and imagination. The variety of poses of today's mounts combined with natural and realistic habitat scenes provides the artist with an endless source of new and exciting ideas for their work.

Bird Taxidermy: Use a "System"

Within the last 5 years, bird taxidermy has been transformed from the use of archaic supplies and time-consuming methods of "stuffing" to "state-of-the-art" taxidermy art supplies and techniques. As with other species of wildlife, the WASCO and *Breakthrough* research and development team have carefully examined every method of bird taxidermy that we can find in order to update and develop the "best," most effective, and efficient bird taxidermy "system" ever devised. The suggested techniques and supplies in this book will effectively guide the bird taxidermist through each stage to produce a world class mount. The stages entail: proper field care, skinning, preserving, mounting, finish work and habitat/base work. As with other taxidermy systems that WASCO and *Breakthrough* have developed, the bird taxidermy system is "all" inclusive and encompasses each phase with easy to use "state-of-the-art" supplies and easy to follow "state-of-the-art" techniques.

Quality Mounts Begin With Proper Field Care

Quality mounts begin with proper care in the field.

Taxidermists are limited and can only do so much "repair" work on a bird. We may be able to repair various broken bones, camouflage shot holes and replace or hide missing feathers; but unfortunately, we are not "miracle" workers. In addition to the field care procedures outlined in this book, we recommend that you consider offering The Serious Sportsman Field and Trophy Care book (published by *Breakthrough*) to your customers. This book is designed to help both taxidermists and the sportsmen to implement the proper field care procedures to care for their trophies. By providing this inexpensive book to them, for the first time, taxidermists can easily educate their customers as well as derive some excellent advertising for their studio.

This will benefit both the sportsman customer and the taxidermist as well. There is nothing more frustrating than turning away business or doing a poor taxidermy job on a full plumage bird because it was handled improperly by the customer in the field. This is not only a loss of a beautiful bird but a loss of revenue to the taxidermist.

Small repairs, such as on this Sandhill Crane, can easily be made by taxidermists.

Skin Preparation

Perhaps the most significant milestone in bird taxidermy has been the development of a complete line of products to safely wash, degrease, preserve and tan or leatherize a bird skin and protect it from the shrinkage, grease, discoloration, and ravages of insects that were common in years past.

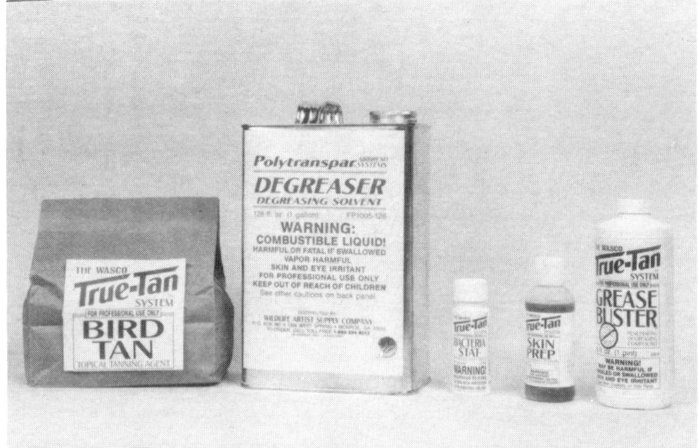

Unlike the poorly preserved mounts from days gone by, today's bird taxidermy procedures allow for longer lasting mounts that can better withstand the test of time.

Taxidermists have always faced the task of cleaning and degreasing birds in such a manner as to avoid damaging delicate tissues and/or feathers. The skin and feathers of birds must be cleaned thoroughly, as fat remaining on the skin or a greasy residue left on the feathers will cause problems and forever look unsightly. This book clearly outlines methods that will safely remove the oils and grease that are present in practically every bird.

Through the years, shrinkage has also been a constant source of aggravation. Over the decades, nearly every conceivable method has been tried at one time or another to keep feathers that were so painstakingly aligned, adjusted, and positioned in their proper tracts and place to stay that way. I have personally carded, wrapped birds with string, groomed and practically watched over every individual feather as it dried, only to turn my back for a day and have the skin shrink slightly and cause the feathers to shift from their desired position to an "undesired" position. The series of steps that we have outlined in this book for cleaning, preserving, and leatherizing bird skins will clearly illustrate how to eliminate such problem areas and pitfalls.

Mannikins and Accessories

Perhaps the best tip that any "would-be" taxidermist could ever receive would be to: *"Always use only the best quality supplies that you can find."* Regardless of all the sales rhetoric, there are very few "actual" bargains in life, and in taxidermy supplies they are almost nonexistent. You simply "must" use good quality supplies to obtain good quality mounts.

We at WASCO/*Breakthrough* did not limit our research to the development of a system for field care and skin preparation alone, but we have utilized our entire resources to offer extremely accurate and easy-to-use bird mannikins, anatomical accessories, references, finishing supplies, etc., of the highest quality (everything necessary for a world class bird mount). The system is so concise that the mannikins are even clearly marked as to where to insert the neck, legs, and wing wires.

With Sportsman Series mannikins, accessories and methods, assembly procedures are cut down to a fraction of the time that it used to take. This allows more time for individual creativity and artistic expression on each mount (the part that taxidermists like the most!).

Wrap-Your-Own Systems

Realizing that many taxidermists will, upon occasion, have custom or unusual mounts that will require techniques and supplies other than the use of a commercially prepared body, or for those few taxidermists who still prefer to use the wrapped body system, we have incorporated a comprehensive "state-of-the-art" system to illustrate the most efficient mounting procedure for the wrapped body techniques to obtain a world class mount.

Miscellaneous Aspects

Because we want this to be the most comprehensive and complete book of bird taxidermy ever written, we have also included a wealth of illustrations, tips and information on a variety of subjects in addition to basic bird taxidermy such as: cast heads, strutting turkey, grooming, finishing techniques, and tips on artificial water scenes. (We have even included some delicious recipes for those who enjoy dining on waterfowl and upland game as much as we do.)

Tips for Success

Incorporating the WASCO bird taxidermy system into your studio, whether using the recommended complete "system," or if following the steps outlined for using the wrapped method, will allow you to produce and display anatomically accurate mounts of world-class quality in less time and with less effort.

The only limiting factors to the degree of success that you, personally, can achieve and enjoy with your work, will be the amount of time and "practice" that you will allot to this art and how dedicated you are to mastering bird taxidermy. Practice and dedication, combined with your own God-given abilities and patience to develop those abilities to their full potential, can make you a World Champion. Our recommendations for success are to follow the techniques, exactly as outlined; *always* use only top quality supplies, and take the time to *practice!* Don't be disheartened by results that are less than what you had hoped for on the first few tries, as with anything worth-while, the necessary skills to do excellent bird taxidermy must be developed over a period of time (patience). Study your reference and work hard to make each mount a little better each time and success is guaranteed!

CHAPTER 1

Showroom: Checking The Mount In

Decorating the Showroom

Nothing makes quite as strong and lasting an impression to a customer as a tastefully decorated showroom. Any "smart" (experienced) sportsman will carefully check out the display mounts in your showroom "before" leaving their "hard-earned" trophies; therefore, it pays to put some careful thought into decoration.

One rule to remember and keep foremost in your mind is the importance of "honestly" displaying mounts in your showroom that reflect the quality of work that is representative of what you regularly produce on a commercial basis. What your customer sees in the showroom should be "very" close to what they should expect to receive back upon completion of their trophy mount. It is misleading to customers to have your showroom full of "competition" quality pieces (or worse, taxidermy work that you did not even do!) and then deliver to them a lesser quality of work. This will certainly mean disappointed customers and a loss of repeat business.

Bob Elzner of Apache Junction, Arizona included an elaborate mural in his showroom.

In addition to your "typical" commercial mounts, it is best to have on display your "top-of-the-line" competition quality pieces as well, in order to offer the customer a choice. Not all customers can afford these more expensive and elegant mounts displayed with glass cases and habitat work, but many of them *can* afford it; and therefore, it makes good business sense to offer both. As an added note, most customers do not realize how far taxidermy work has evolved and are usually quite surprised to see today's unique mounts and the wide variety of poses in which they may display their birds.

This elaborate piece by Frank Newmyer comes complete with a porthole which displays an underwater view of the scene.

Often a customer will prefer a simple base.

You should also exhibit a complete selection of all of the poses that you offer, such as flying, standing, dead game, or swimming. Allow the customer a choice and you will end up with more business, especially if they can actually see a bird already mounted in that pose. It will definitely be to your advantage to allow the customer to actually "see" a bird mounted similarly to their desired pose prior to mounting their trophy.

One excellent showroom scheme is the sportsman "den" atmosphere. This can consist of paneled, rock or brick wall(s) complete with carpet, easy chair, lamp or whatever. Decorate the showroom to simulate a typical living room or den and your customers can relate this directly to their home. By doing this, they can better picture or envision how their mount will look when displayed in such a manner. Unfortunately, many taxidermists display their mounts in cluttered atmospheres, "crammed" together with other mounts on dirty pegboards that have no appeal at all. These taxidermists need to understand that in the customer's mind this is a reflection of their work, and even if the quality is there, they must overcome the negative impression of how it is displayed.

One method that I used when I operated my commercial studio that immensely helped to increase the volume of my bird work, was the introduction of simple habitat materials to the finished pieces in my showroom. This helped to change many an opinion of stubborn wives who initially insisted that the *only* room that they would allow their husbands to display a mount in, would be the garage or possibly a spare room.

Often these same wives would suddenly change their minds when they saw a beautiful wood duck mount standing on a piece of driftwood with a simple base consisting of a few sprigs of cattails and marsh grass. Upon seeing such a piece they would quickly reverse themselves and insist on putting "their" mount right smack dab in the middle of the living room. I love to hear a hunter's wife tell him that he "must" go hunting and bring home another species of bird so *she* can have it mounted "just like that one." (The hunters love to hear this also!)

Remember, it is the "hunter" who brings home the game, *but* often times, it is the "wife" who *ultimately* decides where (or if) the mount will be displayed in their home. A nice variety of full plumage birds on display in a neat, clean, showroom tastefully presented with simple habitat scenes, helps to change the mind of even the most obstinate wife or girlfriend. The net result is that this means extra business for you, so do yourself a favor and take the necessary time and thought to skillfully decorate your showroom.

Checking The Mount In

Upon taking a bird in to mount, first carefully inspect it for missing or damaged feathers, broken bones, or any other unusual characteristics. Many times the hunter is so excited about bagging a good bird, that they often overlook damage done by the force of the shot pellets, "rough-mouthed" retrievers, careless companions, and probably the worst, an over-zealous hunter who has eviscerated the bird in the field. (If done incorrectly, this simple act will often totally ruin a bird for a mount or at the very least prevent the taxidermist from doing a first class job on it.)

A wing in one hand and a bird in the other spells trouble any way you want to look at it.

When a poor specimen is brought in, the customer must be told "as soon as the bird arrives." This is also an excellent time to explain the need for proper care in the field to ensure a quality mount for their next trophy. It is also an excellent time to sell or, if they are a good customer, to "give" them a copy of the *Serious Sportsman Field Care Manual.* Most hunters simply do not know what to do when they bag their trophy, and they really do appreciate the taxidermist providing them with this information. The field care manual saves the taxidermist from having to take the time to personally explain these simple but important procedures. The manual is also a small enough size that it can be taken to the field and by stapling one of your business cards in it they will know exactly

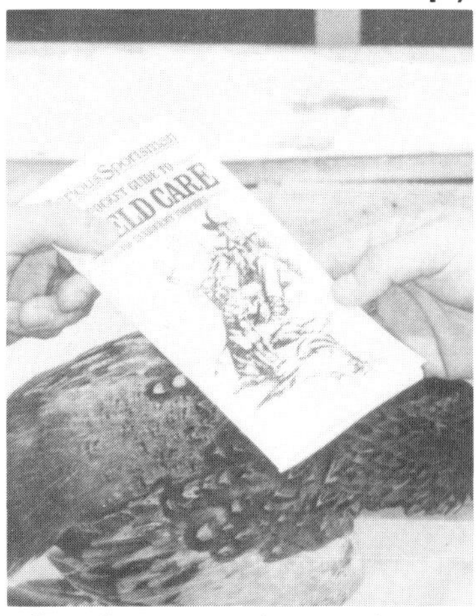

where to bring their trophy after they properly care for it.

In many cases a pose can be selected that will effectively hide problem areas, unless the damage is too severe. We do not recommend mounting birds that are borderline or that are just in too bad of shape. No matter what you try to do to salvage severely damaged birds, it will mean spending considerably "more" time for "less" than optimum results. You will end up being unhappy—and so will the customer.

The only exceptions I make to this policy would be a customer's first bird or rare and unusual birds. One such case I vividly remember was a customer who came into my studio with his young son. The child was very carefully carrying a mutilated green wing teal drake (taken in early teal season, so you can imagine the poor condition of the plumage). Upon careful examination, I found that virtually "every" bone in the bird's

body was shot up and broken, and to make matters worse, there were innumerable feathers missing. I had mixed emotions about whether or not to take the bird in; *but* after seeing the pleading look of the child and the "beaming" look in the father's eyes, I reluctantly decided to mount the bird. However, I made "absolutely certain" to explain to the customer that the poor condition of the bird would mean that they had to expect less than fantastic results (to say the least).

I can assure you that I spent many extra hours trying to salvage that bird. When I did finish, it was "marginal" at best. When the customer and his son came to pick up the mount, the only thing I could think of to explain the condition of the bird and loss of feathers was to say: "Boy, you sure are a good shot; you hit him everywhere!" With that statement, the child proudly exclaimed, "See, Dad, I told you I hit him all three times." I breathed a sigh of relief after seeing the proud excited look in both their eyes.

There are times such as this one when you can justify taking in an occasional bad bird; however, if you do take in a damaged bird, always be sure to explain to the customer the possibility of the bird ending up in a far less than desired mount.

I have found that customers truly appreciate your being honest about the condition of a bird that is unfit for mounting purposes. It may mean that you lose a mount this season, but I can assure you, that these same customers will bring you the very next good bird that they bag. Remember, it is the

finished mount hanging on a customer's wall for many years that is a reflection of you, and your work. Do not be a "short term" business person; be totally honest with your customer even if it means losing a particular job. Choose the wiser course and build a long lasting customer base that will support you and your family for years to come.

As a final note, always remember to log-in anything that is wrong with the bird, *in front of the customer.* If this is not possible, then call them on the telephone and ask them to come by so you can show them the damage. Or at the very least, send them a postcard or letter explaining the damage "before" mounting the bird.

Date _4/23/88_

Dear _Mr. Smith_

Upon further examination of your _pintail_, we discovered that the specimen had experienced considerable damage.

I cannot guarantee an acceptable mount in the pose you originally requested. Please call or come by at your earliest convenience so I can discuss the alternatives with you.

Sincerly,

Martin Lowe

Taxidermy Den
(404) 555-0173

Document your efforts in your business work order log so there will be no misunderstanding. Always make sure that poor condition birds are mounted at the customer's risk, not yours.

Deciding Upon a Pose

Once the condition of the bird has been determined, the next step is to decide upon a suitable pose. I usually explain the different poses and positions that are available with respect to the traits and characteristics of the individual specimen. One way of avoiding any problems with poses is to let the customer choose a pose from one of your display mounts. If you will display mounts in the standing, flying, dead game, and swimming positions, chances are good that the customer will select one of your standard mounts. Often these are the most profitable for you, because the repetition of mounting similar poses repeatedly will build speed and allow you to develop a "fast" system.

Customers who prefer poses other than your standard mounts on display, should always select their poses from your books or reference photographs. (If at all possible, have them do this.) I have found that having my bird taxidermy "Bible," "Prairie Wings," provides a multitude of poses to choose from. I also have my reference photos and duck hunting magazines close at hand and this also helps in selecting a pose. It is far easier to position a bird when using a reference photograph of the desired pose placed directly in front of you when you mount it. This takes all the guesswork out and enables you to do a better job.

Do not force a pose that "you" like upon your customer. Answer questions, offer advice, but let "him" select the pose that "he" wishes. The customer is the one who is paying the bill, and also the one who will be looking at the mount for the next few years, so let him decide. If he is happy, it means repeat business to you as well as all of his friend's business; if he is unhappy, then you lose!

Give your customer a definite price for each possible pose before he decides. Do not be afraid to charge extra for difficult, time-consuming poses or for badly damaged birds. You must be fair to yourself if you plan to stay in business. "Do" try to "sell" your customer on extras such as habitats and/or nice bases. Often the "profit" derived from the extras will actually

outweigh the profit made on the bird and will involve far less work. Additionally, your customer will end up with a much nicer "showpiece" that will be good advertising for you.

Regulations and Record Keeping

After the condition of the bird has been determined and the pose decided upon, the next step is the biggest hassle, *but* is one that is "very" important! That is filling out the proper paper work to encompass and satisfy all of the state and federal regulations.

Remember, it may be a hassle to check hunter's tags and fill in the necessary forms, but I can assure you, that you do not want to have any problems with governmental agencies, "especially" for migratory birds. This chore is quite simple if you simply use and follow the instructions of the "WASCO" business log system. This system is well organized, and the necessary paperwork is not difficult. Simply fill out the necessary information according to the directions and be sure to have the customer sign it. Just remember to handle it promptly and *don't* put it off until tomorrow. All migratory birds must be properly logged in and tagged when they are in your possession.

The proper procedure for state regulations will vary and must be checked out by you through your particular State Game and Fish Commission. Federal regulations are standard and are outlined on the following pages.

Once the paperwork is filled out, it is time to collect the deposit. (I like this part!) The best rule to remember is: "the bigger the deposit, the 'better' when it comes to bird work." The reason is caused by federal and state regulations. Most birds cannot be sold if they are not claimed by the customer. Although many times a taxidermist can use an occasional extra display mount, the large deposit will at least allow you to recover your expense. (Many taxidermist require 100% deposits on birds that cannot be sold.)

50% Deposit Required On ALL Work

Approximate Completion Date

Once the deposit has been collected, give the customer an *"approximate"* completion date and instruct them that *"you"* will notify them when the mount is ready. I generally ask that they *"please"* do not call before the quoted completion date, as it will *not* be ready before then, and it will save us both precious time.

After completing the log-in procedure I politely encourage the customer to look around the showroom. I always have plenty of retail sales items on hand that add up to profit with no labor required on my part. It is surprising just how much these extra sales items add up to in a month. It is no small wonder why every store that you walk in has a wide variety of these items right next to the cash register. Prepare a good selection of retail sales items and you can pay your rent and utilities from the profit.

ARTIST: FRANK NEWMYER

Knowing the Facts:
Migratory Game Bird Laws

Migratory birds are regulated by "federal" laws and regulations. Violations and convictions of the same can result in serious fines, imprisonment, and/or loss of your gun owning privileges for life. Take time to find out how these laws effect you.

Things were getting pretty hectic around Jim's new taxidermy shop. Duck and deer season had only been in for a few weeks yet Jim was overwhelmed with all of the customers that were coming in.

The weather was turning colder and Jim was "dying" to get into the woods and do a little hunting for himself. As he was trying to slip away one afternoon, he noticed two men coming to the door. For an instant Jim thought about running to the door and turning the "open" sign over; however, better logic prevailed and he knew better than to turn away two potential customers.

So instead, Jim warmly welcomed the men and asked if he could help them. They explained that they had been out hunting earlier that morning and had brought down two trophy ducks that were very uncommon for their area. They went on to explain that they had seen some very nice mounts that Jim had mounted for some of their friends and wanted him to perform the same quality work on their birds.

Jim "beamed" at their compliments and quoted the hunters his standard price. After taking their names and addresses, the men told him the position that the birds were to be mounted in and paid their deposit. Jim gave them their receipt and waved them goodbye. Once they were gone Jim quickly put the ducks into his freezer (without taking the time to attach the migratory bird tags to the ducks' legs) and headed for the woods to go hunting himself.

A few weeks later, a federal Fish and Game department officer stopped by unexpectedly. The officer explained that it was just a routine visit. After a few minutes of small talk the law officer asked Jim if he could take a look at his migratory bird permits and records. Nervously Jim produced the records and permits. He asked the officer if there were any problems. After examining the records, the officer told Jim everything seemed in order, but to complete his work he also needed to check the birds in his freezer. There, laying in plain view, were the two ducks Jim had previously taken in.

The officer carefully examined the birds inspecting them for identification tags. He informed Jim that he was in violation of the law and that the birds would have to be confiscated. Further, he was going to have to issue Jim a citation which would involve a fine.

Federal Migratory Game Bird Laws

Beginning taxidermists seem to be overwhelmed by the federal migratory bird laws. Some veteran taxidermists think of these laws as just another example of more bureaucratic red tape from the lawmakers in Washington to plague the small businessmen. But the fact is that these laws were set up to protect our dwindling supply of waterfowl which was almost completely wiped out during the days of market hunting in the late 1800's. During those so-called "glory days," there were little, if any, daily bag limits or restrictions on the taking of waterfowl and game birds.

Fortunately, the decreasing populations of waterfowl throughout America alarmed a variety of concerned individuals. Conservation minded sportsmen banded together with lawmakers to set up regulations and restrictions to help protect this valuable resource. These laws spawned the rules that now apply to all American taxidermists who mount migratory birds for other persons—whether beginner, hobbyist, or full-timer.

"WHAT TYPE OF BIRDS CAN BE LEGALLY MOUNTED WITHOUT A FEDERAL PERMIT?"

A taxidermist may legally mount any migratory game bird that has been lawfully taken *by himself*, for his *personal use only* (not for customers). He can also commercially mount any type of non-migratory game bird (such as quail, grouse, pheasant, turkey, etc.) that has been legally taken.

"WHAT BIRDS ARE CLASSIFIED AS MIGRATORY GAME BIRDS?"

Ducks, geese, swans, and crows are usually the first birds that come to mind, but there are others. Songbirds, such as cardinals and mockingbirds, are classified as migratory birds also. These species cannot be legally taken from the wild without a special permit issued by the U. S. Fish and Wildlife Service (and the state if required). Hawks, owls, and eagles are also considered migratory. But, like the songbirds, they too are protected.

Many taxidermists are asked if they could legally mount hawks, owls, or songbirds. These birds can be legally mounted; however, a scientific permit is required and the birds must be mounted for display only in a public, scientific, or educational institution. Eagles, however, can only be taken, possessed, or transported by public museums, scientific societies or public zoological parks with a valid permit. They cannot be mounted at all unless a person has a valid permit.

We suggest that each taxidermist place a sign in their studio stating that they will not mount these types of birds. This should eliminate the problem of customers asking. However, if a customer asks in "spite" of the sign, quickly tell them "no," *firmly!* Do this without hesitation! Remind them that it is illegal to even have these birds in their possession much less to mount them. Remember it is very common for wildlife, fish and game enforcement officers to work undercover to "set-up" potential taxidermist offenders. Do not fall victim to this type of sting!

"HOW IS A FEDERAL MIGRATORY BIRD PERMIT OBTAINED?"

In order to (legally) mount "common" migratory birds for commercial purposes, a migratory bird permit is required. A federal migratory bird permit application can be received by requesting an application in writing or by phone from your regional office of the United States Department of the Interior, Fish and Wildlife Services (address listed at the end of this chapter.) Once the application has been received be sure to fill it out "completely." (Applications have been sent back to the taxidermist for making such "seemingly" trivial mistakes as not supplying a middle initial for your name.) In order to obtain this permit, a taxidermist must first obtain all state permits that are required. The taxidermist should then provide copies of these when applying for the Federal Migratory Bird Permit. There is no application fee (as of this printing) so all that needs to be done is to fill it out in its entirety and then send it in. If all goes well, the permit will be mailed back to you three to four weeks after the application is received.

"WHAT GUIDELINES MUST BE FOLLOWED?"

Once the permit has been received, an accurate log book must be maintained for all migratory birds received. The records should contain the hunter's name, address, the day the bird was received, species of the bird, number of birds killed that day, signature, and the date that the bird was collected. These records must be kept *"five"* years from the date the permit was obtained.

Every migratory bird must have a properly filled out tag attached to it that includes: the hunter's signature, address, species of bird, date killed, and number of birds taken that day. This tag must be attached to the bird's leg and remain there until the customer picks it up. (The tag can then be temporarily removed during the mounting process but must be returned to the leg when the mount is complete).

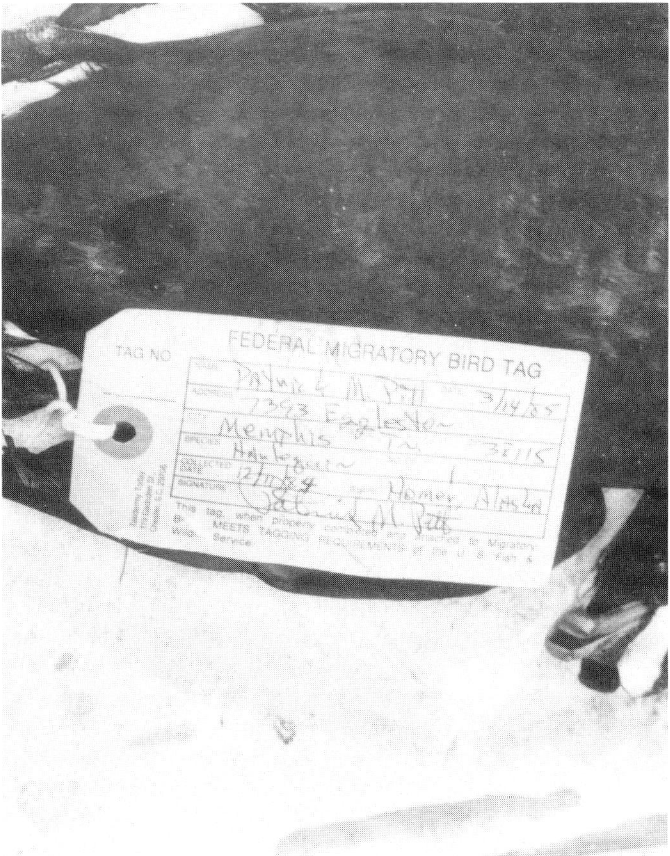

According to current regulations the act of obtaining a federal migratory bird permit gives the Fish & Wildlife Service the right to inspect these records and the birds in your shop to be mounted at any reasonable time without prior notice. (This regulation seems to be somewhat unfair and a direct contradiction to bill of rights provisions for standard search and seizure laws; however, it is the current law and if you want to mount migratory birds you must give up this right.)

To renew the permit you must write for renewal to the Fish & Wildlife service within 30 days of the expiration date on the permit (it will expire after two years from the date it is received).

"CAN UNCLAIMED DUCK OR GEESE MOUNTS BE SOLD?"

The answer to this question is simply no. However, the birds can be placed on display or donated to a public, scientific, or educational institution in order to advertise your business. The best suggestion for compensating your time if you get stuck

with an unclaimed bird is to charge the customer a deposit that is high enough to cover the cost of the materials and time invested.

"WHAT ABOUT MOUNTING PEN-RAISED BIRDS?"

Some taxidermists may have clients who want a mounted duck, but for some reason or other the customers cannot produce a bird. One solution to this problem is for the taxidermist to sell mounts of captive-reared mallards. No permit is required to mount and sell pen-raised mallard drakes and hens. If species other than mallards of captive-reared migratory waterfowl are to be mounted, the taxidermist should have a copy of the permit (Form 3-186 available from the person who sold the birds).

All pen-raised migratory waterfowl must be clearly marked within six weeks of their birth. They should be marked by removing the hind toe on the right foot or by banding one of the legs with a seamless band or by pinoning one (removing the outer section) of the wings. Another method is tattooing a number or letter on the web of one foot. These areas can be painted, repaired, and band removed when the bird is mounted. The bird, however, must be accompanied by the 3-186 form verifying that the bird is captive-reared (excluding captive-raised mallards).

"WHAT ARE THE PENALTIES FOR BREAKING THESE LAWS?"

For those of you wondering how much ol' Jim's fine would have been for having lawfully taken but untagged birds in his possession, it would be considered a misdemeanor and could be punishable by a fine of up to $500 (for each bird) and/or up to six months imprisonment, confiscation of the birds, and possible revocation of his migratory bird permit. If a taxidermist is caught selling wild migratory birds, it would be considered a *felony* and they could be fined up to $2,000 (for each bird) and/or two years in prison (plus loss of owning a gun which means no more hunting). These penalties, fines, and permit revocations are determined according to the seriousness of the offense. Fines for violators of the Endangered Species Act may be up to $20,000 and/or one year in prison. As you can see, the migratory bird laws are something that should not be taken lightly. And, it should be remembered that these laws are here to protect a very valuable resource, so please obey them, for our children's sake.

Addresses of Federal Offices:

United States Department of the Interior Fish & Wildlife Service

P. O. Box 329
Albuquerque, NM 87103

847 Northeast 19th Avenue St. 225
Portland, OR 97232

P. O. Box 129
Newtown Branch
Boston, MA 02258

P. O. Box 45
Federal Building, Fort Snelling
Twin Cities, MN 55111

P. O. Box 25486
Denver, CO 80225

P. O. Box 92597
Anchorage, AK 99509-2597

75 Spring Street SW
Atlanta, GA 30303

CHAPTER 2

The Workshop, Tools, And Supplies

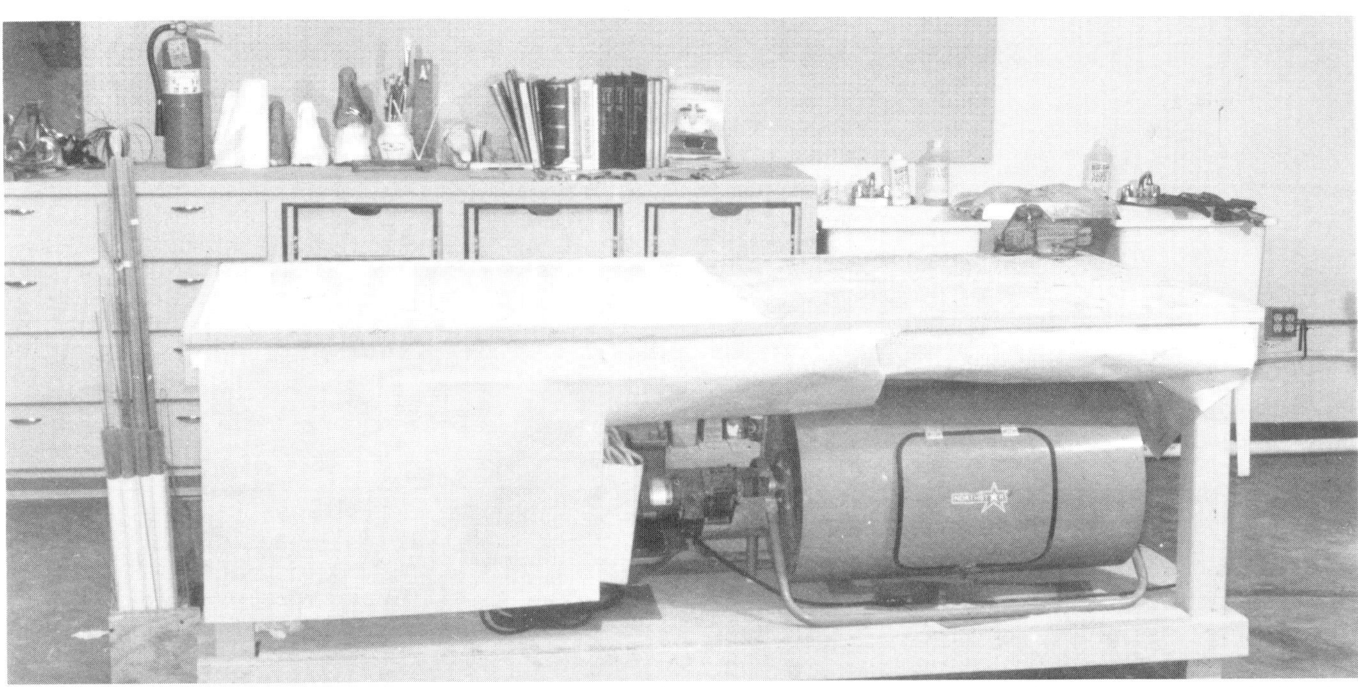

It is "mandatory" to have a clean, well-lighted, and organized workshop if you intend to achieve any degree of success in taxidermy. It is next to "impossible" to produce top quality work in slovenly, dirty surroundings. Likewise, it is frustrating (and a loss of profit) to spend needless, wasted time looking for a certain tool or supply in an unorganized shop.

Organize your tools and develop a strict routine of "cleaning up and putting tools and supplies in their proper places as you go." Due to the very nature of taxidermy, working with dead specimens, skinning, degreasing, etc., it is absolutely necessary that you adhere to this policy. Remember that a dirty, unsanitary, unorganized workshop can change the entire image that your customers have of you and your studio.

Keeping a clean, organized, work area is simply a matter of disciplining yourself to establishing a routine and sticking with it. As soon as you finish each phase of any project, return your tools and supplies to their proper locations and clean up any mess created by that particular phase.

Before I leave at night, I thoroughly clean and sweep my entire work area. If tools and supplies are put in their proper places after use during the day, you will find that it requires a minimal amount of effort to clean up at the end of the day. I can personally attest to the fact that you will be more productive and be able to maintain a better attitude when you can walk into a clean, well organized shop every morning.

It is also more efficient to have specific areas set up for each of the various taxidermy functions. I have specific areas set up for: customer service and "logging" mounts in, skinning and preserving, mounting, drying, and finishing. Organization is the "key" to producing top quality work in a timely and efficient manner. By designating specific areas for specific functions you will organize the tools and supplies needed for these operations and have them within arm's length when you need them. This means extra profit for you as there will be no wasted time running all over the shop looking for them.

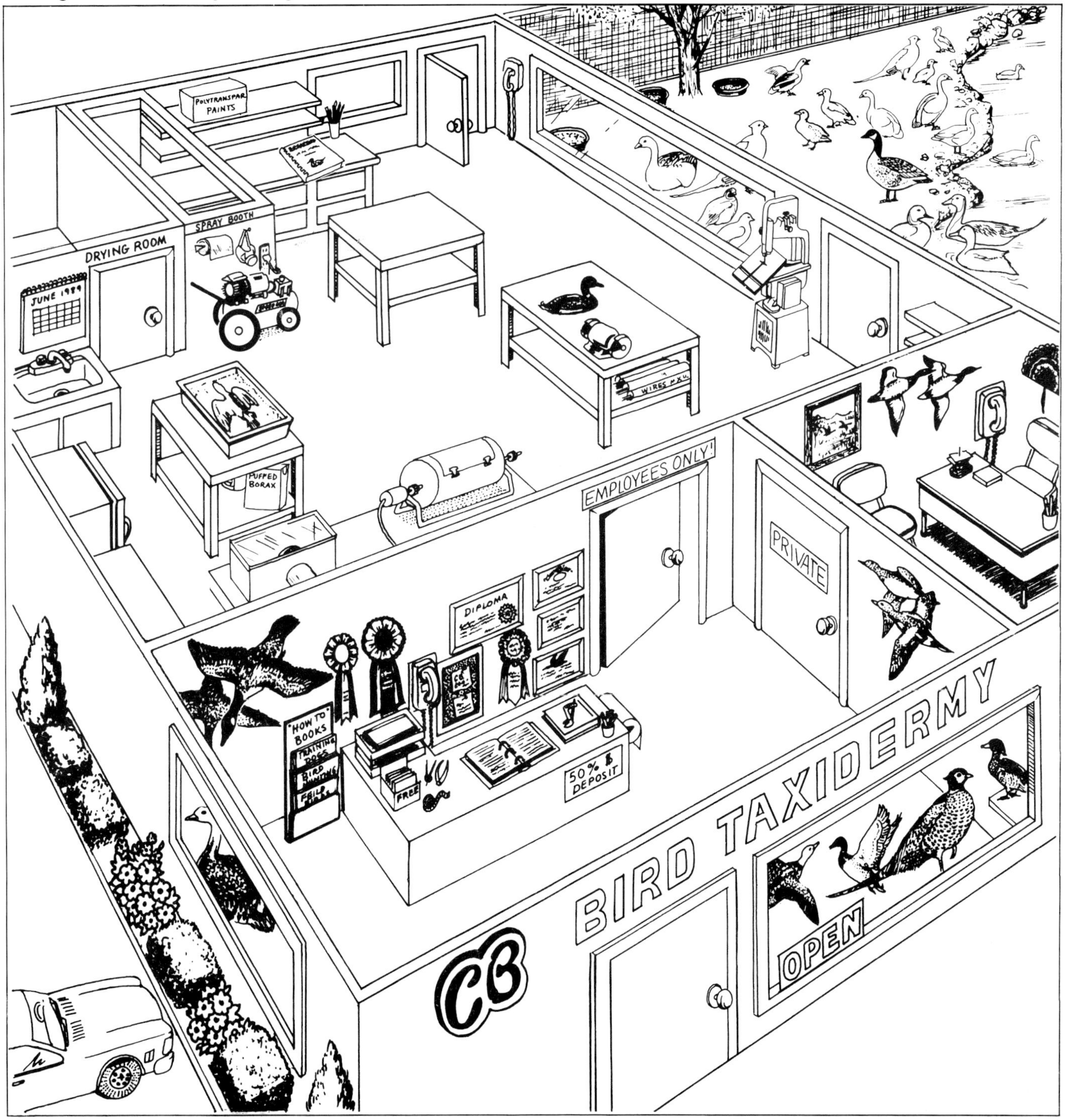

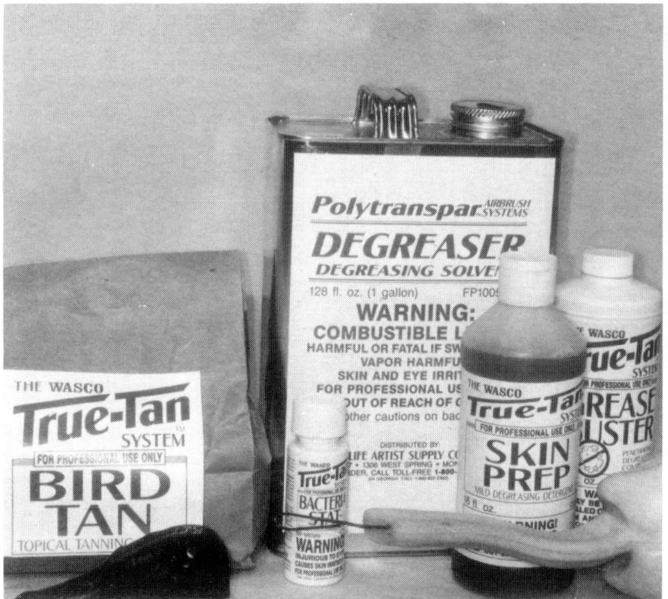

Just as having an organized work area is necessary for producing quality work; it is equally as important to own and use professional tools and "top quality" supplies. With respect to other types of taxidermy, bird taxidermy requires somewhat less in the way of expensive tools and equipment; therefore, it is unnecessary and unwise to try to cut corners when it comes to buying these items. A taxidermist's tools are an integral part of their work, and I firmly believe that you get what you pay for. It may "appear" that you are saving a little money in the beginning, but it is always costly in the long run.

Supplies are even more critical. *It is absolutely essential to obtain and use only the best supplies that you can get your hands on.* This is the "law" of good taxidermy, and with the technology that has been developed in the last few years by WASCO, it is an easy law to adhere to.

I keep a clipboard with a "materials-needed" list within easy reach as I work. Throughout the day as I encounter or think of supplies that I am running low on, I jot them down on my list. I try to order all of the supplies that I need with just one order rather than calling in 3 or 4 different orders. This saves time on the phone and saves money for postage. I schedule my work at least two weeks ahead of time (and sometimes three weeks). During the scheduling process, I check all of the

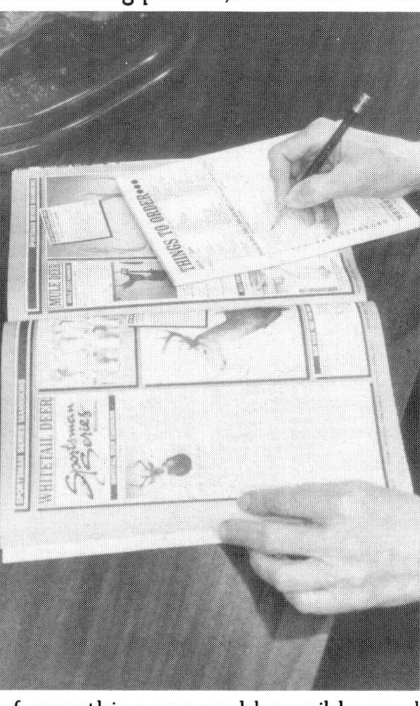

supplies that will be needed for the specimens on my schedule that I intend to mount. Any items that are needed are added to my "daily" list. I then carefully review the catalog for any items that I may have omitted. Also, I write down the correct code numbers of every item that I am ordering on the list. (Giving the correct code numbers to the sales person at the supply company saves time on the telephone and more importantly, helps to prevent shipping errors.)

The following is a list of everything one could possibly need to do "world-class" quality bird work for commercial or competition purposes. The list is separated by function to give a comprehensive picture of the tools and supplies needed for each phase of the skinning and mounting process.

Checking The Mount In

Tools
Calipers (MR130)
Measurite (MR100)
Flexible Tape (GMT)

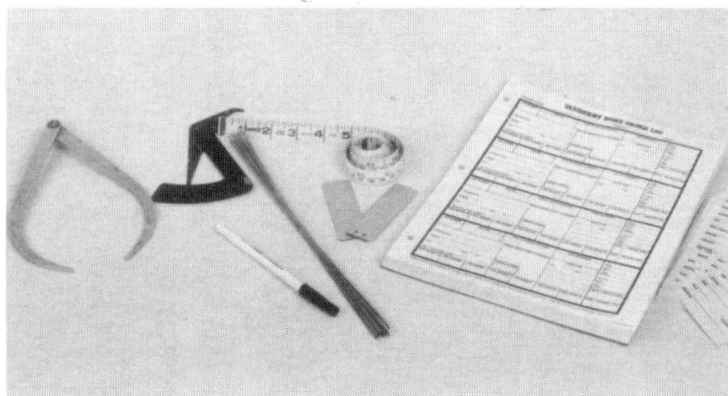

Supplies
Business System (WL100)
Materials List & Clipboard
Work Order Log (WL2)
Metal Identification Tag For Specimen (WL3)
Two Part Manila Identification Tag (WL4)
   For Customer And Specimen
3-Ring Binder
Receipt Book
Ball Point Pen

Serious Sportsman Field Care Book

Advertising Material
Business Card
Brochure/Price List
Tee Shirts, Hats

Miscellaneous
Plastic Bags
Paper Towels
Tape
Twist Ties

Impulse Sales Items
Sportsman-Oriented, High Profit
   Merchandise for Retail Sales

Skinning, Defatting, Cleaning, Tanning

Tools
Scalpel & Extra Blades
Perfect Knife (MR120)
Ultimate Scissors (MR110)
Diagonal Pliers
Light Metal Shears
Chain Hooks
Shallow Tray (Skinning)
Wire Wheel
Apron (APR1)
Basin (Washing)
Tumbler
Hair Dryer

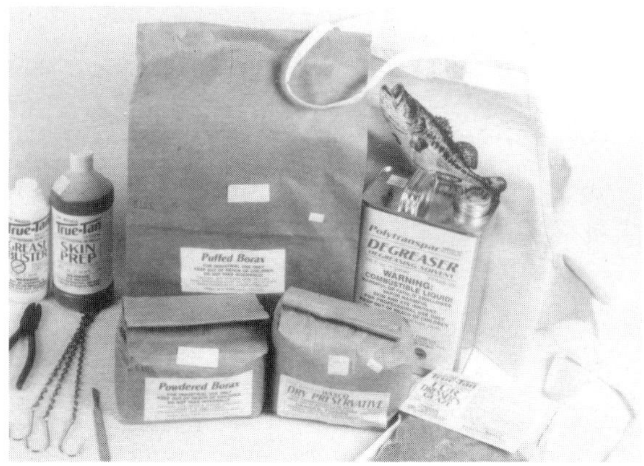

Supplies
Powdered Borax (PB5)
Dry Preservative (DP5)
Skin Prep™ (CT30)
Polytranspar™ Degreaser (FP1005)
Grease-Buster™ (GB16)
True-Tan™ Bird Tan (CT80)
True-Tan™ Conditioner (CT85)
Hardwood Sawdust (CT98)
Puffed Borax (PBX5)
Sawdust Glaze (CT99)

Mounting Tools

Perfect Knife (MR120)
Air Compressor (SAC3/4)
Measurite (MR100)
Ultimate Scissors (MR110)
Diagonal Pliers
Farrier's Rasp (FR)
Needle Nose Pliers
Wire Cutters
Bench Grinder
Foredom Tool (F100)
Screw Gun

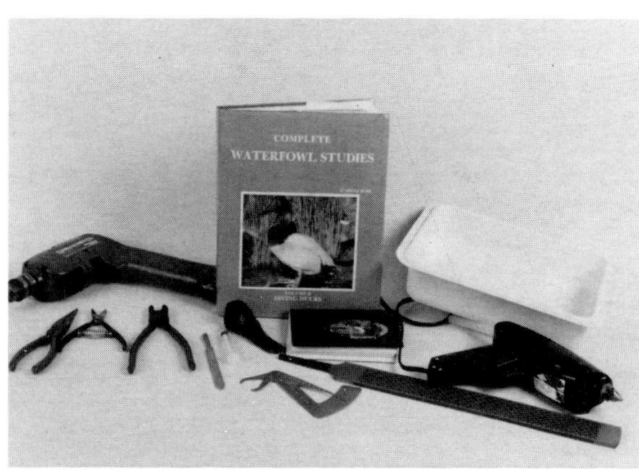

Reference Materials
    (Photos & 3-Dimensional)
Shallow Tray
Syringe (SY)
Jumbo Head Pins (JHP25)
Glue Gun (GG260)
Long Sewing Needle (SN6)
Power Drill
6" Heavy Needle (SN6)
Glover's Needles (SN100)
Lineman's Pliers

Mounting Supplies

Sportsman Series Bird Mounting "System"
Sportsman Series Mannikin
Sportsman Series Accu-Flex Neck
Sportsman Series Artificial Head
    (Optional)
Sandpaper/Screen
Instant Glue (AG7432)
Nu-Glue (EA100)
Annealed Wire
Upholsters Cotton Batting (UBC)
Borax (PB5)
Jumbo Head Pins (JHP25)
Sewing Thread (STS)
Carding Material
5 Minute Epoxy
Paper Clips (JHP25)
WASCO Clay (PC5)
Foot and Leg Injection
Glass Eyes

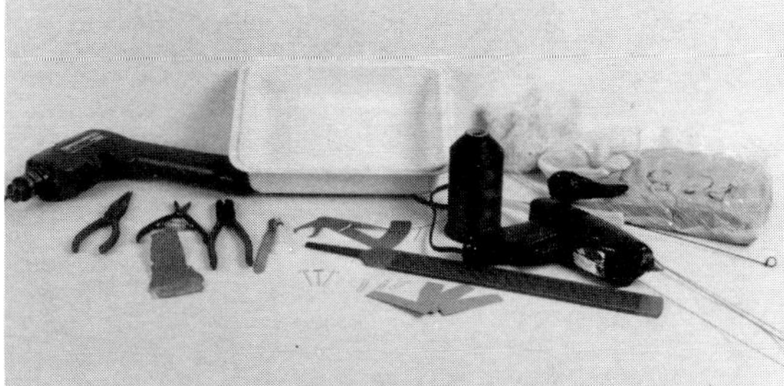

Wrapped Body Method
Paper & Felt Tipped Pen
Fine Excelsior
Jumbo Head Pins (JHP25)
Jeweler's Hemp
Upholsters Cotton Batting (UBC)
Cotton
Foot And Leg Injection
Heavy String
Glass Eyes
Sewing Thread (STS)
WASCO Clay (PC5)
Instant Bonding Glue (AG7432)
Annealed Wire
Borax (PB5)
Carding Material
5 Minute Epoxy
Sportman Series Artificial Head/Bill
    (Optional)

Featherwork Grooming

Tools
Feather Adjuster
Tweezers
Straight Pins
Needle Nose Pliers

Supplies
String or Yarn

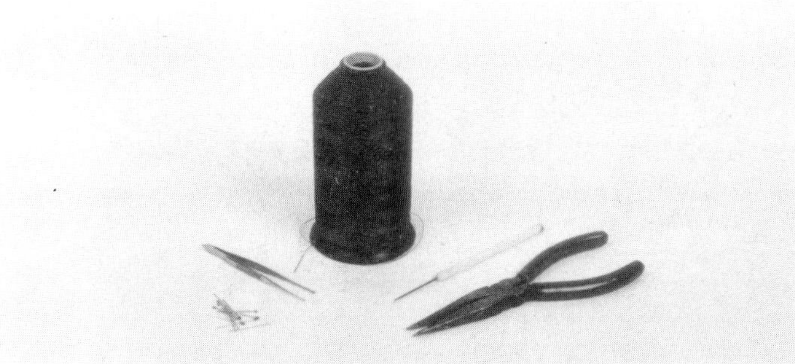

Posing and Attaching To Base

Tools
Power Drill & Bits
Carpenters Saw
Farrier's Rasp (FR)
Hammer
Electric Staple Gun
Wire Cutters
Pliers

Supplies
Instant Glue (AG7432)
5 Minute Epoxy

Finishing: Winsor & Newton and *Polytranspar*™

Tools
Air Compressor (SAC3/4)
Hot Melt Glue Gun (GG260)
VLXL80 Airbrush
  (P43)
Electric Stapler
Screw Gun

Supplies
Winsor & Newton Oil Paints
Polytranspar Airbrush Paints
  (FP or WA)
Feather Duster (FD)
Glue
Eye Cleaner
Silicoil System
Artist's Brushes
Panels
Hangers
Screws
Driftwood
Tuffilm (TF16)

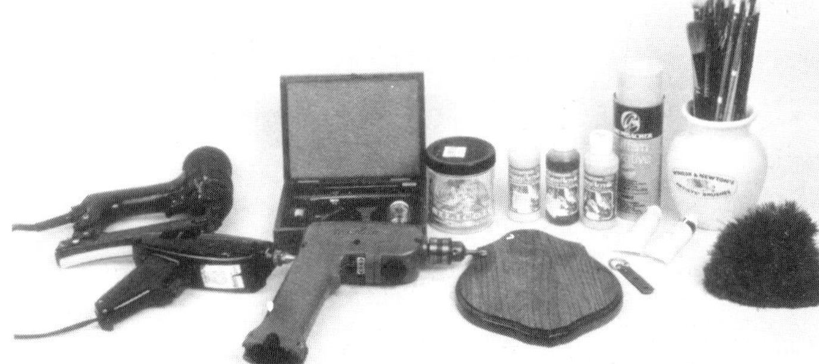

Base Work

Tools
Habitat Manual (BP1007)

Supplies
Cork For Bottom Of Base
Artificial Water (AW100)
Rocks (ARA1)
Foam (UF5)
Artificial Snow (HM600)
Lilly Pads

CHAPTER 3

Reference Material and Posing the Bird

One of the most difficult and frustrating things about writing a book or article is finding the words to adequately explain to the readers the importance of a particular concept or idea. Reference material is an example of just such a topic. If it would be possible, I would enclose a giant poster in each book for every taxidermist to hang in their workshop. The poster would have big, bold letters and would read:

HAVE YOU CHECKED YOUR REFERENCES TODAY?

Obtaining and using good reference materials is the single most valuable resource to any Wildlife Artist!

If you don't have good references, or worse, if you don't use them, what is the point of having a nice studio, organized workshop, and state-of-the-art tools? If you have only a vague idea of what your subject looks like, how are you going to *correctly* recreate it?

Without exception, every successful artist, whether taxidermist, flat artist, carver, or sculptor would not attempt to do a piece of work without his/her reference material *very* close at hand. Using good reference materials is *absolutely* essential if one is going to produce quality work.

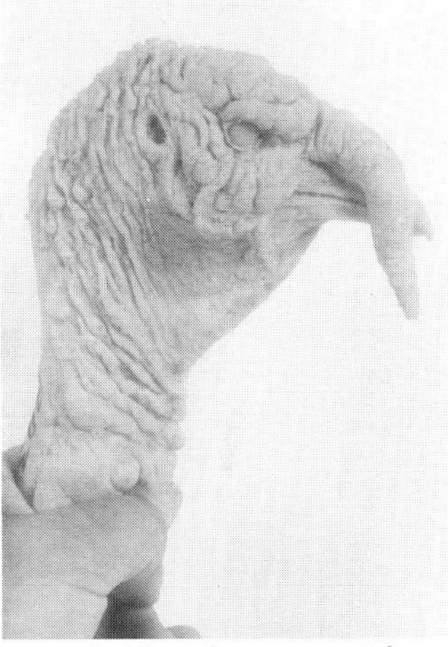

Two-dimensional (left) and three-dimensional (right) reference materials.

Your references should include both two and three-dimensional study aides. The ultimate three-dimensional reference is of course, the live bird. Fortunately, it is relatively easy to have access to live birds. Zoos and parks have excellent facilities for observing and photographing live birds.

Also, building your own aviary is an attractive alternative that you might want to investigate. An aviary takes far less space and expense to build than a whitetail deer pen or similar enclosure.

Owning your own aviary is educational and very enlightening. It is rewarding to observe your own birds daily. Owning an aviary will enable you to study birds at "your" convenience. Times such as when a skin is tumbling or while you are waiting for Ultra Lite filler to set up, etc., are ideal times to walk out to the pen and observe your birds.

Frank Newmyer's aviary provides him with excellent opportunities to obtain good reference photos like this one.

World champion bird taxidermist Frank Newmyer has a glass partition set up in his studio which enables him to observe his birds as he works. Frank states that this also allows the birds to see him continually. The birds get used to his presence and display more natural behavior patterns.

Another good source of "3-D" reference is the wide variety of artificial study bills and feet available for carvers and taxidermists. These are constructed from very accurate molds of bills, legs and feet (before shrinkage) of the actual live or freshly taken bird. Artificial bills and feet can actually be incorporated into the mount or carving or they can be used strictly as reference to correctly recreate this area of the mount when using a natural head.

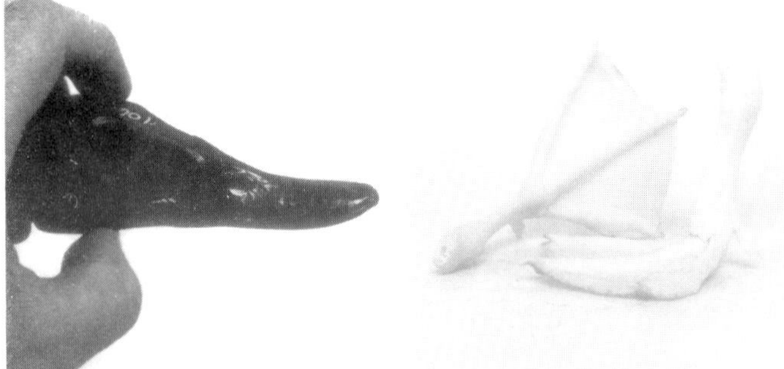

Artificial bird heads and feet offer an excellent source for quality 3-D reference.

One of the best two-dimensional reference sources that bird taxidermists have readily available to them is the huge abundance of photographs available for waterfowl and upland game birds. WASCO recently added an entire line of beautiful close-up photographs of birds including practically every species commonly mounted. Each species is photographed from "top-to-bottom" in this excellent series of photographs. Photographs such as these are worth their weight in gold when finishing a mount or positioning the wings, etc. (see examples below).

Wildlife related magazines are also some of my favorite sources of reference photographs. One in particular that I like is the magazine *Ducks Unlimited*. This magazine is published by the Ducks Unlimited Organization and can be received by joining their ranks. Ducks Unlimited is also a very worthy cause that is totally dedicated to preserving and increasing the numbers of waterfowl. The address for this fine organization is as

follows: Ducks Unlimited, 1 Waterfowl Way, Long Grove, IL 60047. Every issue is full of outstanding photographs.

After I finish reading a magazine, I generally cut out all the pictures, categorize them according to species, and insert them in a binder. Other magazines with good reference photos are: *National Wildlife*, *Audubon*, *Birder's World* and *Ranger Rick*. Additionally, hunting magazines such as *Waterfowler's World*, *American Hunter*, etc., will provide photos that will prove invaluable to you.

Another source of good reference material is the many quality reference books for carvers that are on the market. Because of the nature of the extreme detail work performed by carvers, many of these books have excellent close-up shots of live birds. One in particular sold by WASCO is "Complete Waterfowl Studies" by Bruce Burk. This three volume set covers all North American waterfowl and geese with multiple views.

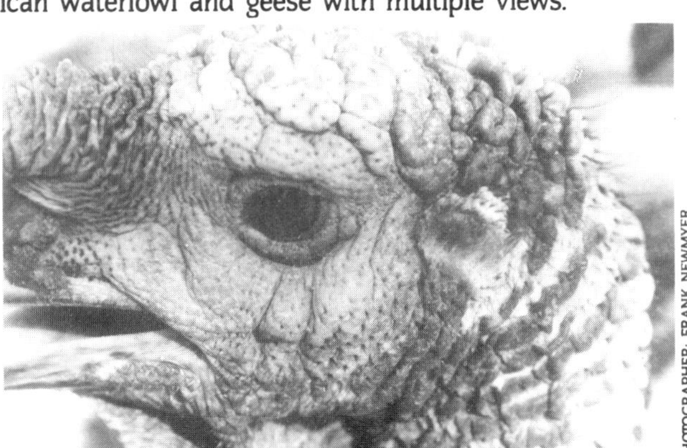

Beautiful color reference photos such as these are available in sets for each specie from Wildlife Artist Supply Company.

Wildlife art magazines such as *Breakthrough* regularly publish color reference photos of different species. Some of *Breakthrough's* "Reference Photos of the Month" are reprinted below.

American goldeneye.

Northern shoveler.

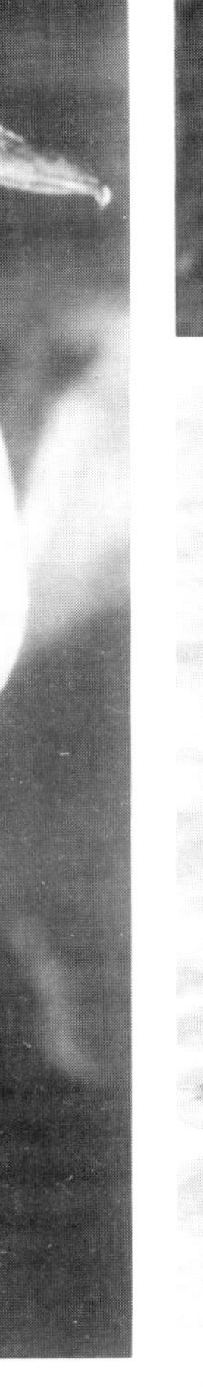

Fulvous whistling duck.

Lesser snow goose.

Mallard hen.

Pintail drake.

Cinnamon teal.

My favorite book, and I might add, one that I would simply not attempt to pose a flying bird without, is "Prairie Wings." "Prairie Wings" is a comprehensive study of birds in flight. I urge everyone who is serious about their bird taxidermy work to purchase a copy of this book. It will be one of the best investments that you will ever make. I feel so strongly about this book, because it not only takes the guess-

work out of posing a flying bird, but it also provides me with such a wide variety of ideas and different poses to try.

Finally, do not impose limits upon yourself when it comes to reference. Study other artist's work, but do not "copy" it. Study the ideas and styles of other wildlife artists and you will be surprised at the "new" ideas of your own that will be stimulated by them. Never use other artist's mounts, prints, and/or paintings as reference. These are merely their "impressions." Always use actual birds or photos for reference. You do not want to be guilty of "their" mistakes.

Interpreting Reference

Have you ever seen a wildlife painting that had numerous objects or animals in it that blended in so well as to appear "hidden" in the painting? The caption under the painting would read "Can you find all of the hidden creatures?"

There are seven herons perched in these trees. Can you find them?

Most people would not be able to identify all of the objects in the painting, even if the artist would list each item individually to make it easier for viewers to find. The reason for this is, that it is all but impossible to see all of the hidden creatures if you concentrate on viewing the "entire" painting at once. However, if you carefully examine the painting by viewing it "one small section at a time" and if you carefully study the very "small details," rather than broadly observing the entire painting at once; then the hidden objects will become quite obvious. After finally locating them all and standing back to observe the painting, it will make you wonder why it was so difficult to spot the hidden objects when you looked at the painting in the first place.

Interpreting references is just like finding the hidden objects in such a painting. Instead of looking at the "entire" specimen and missing so much detail, you must *train* yourself to look "very" closely at *each little section* of the reference material.

If, for example, you were to look at a close-up of a duck's eye, what would you see? An entire eye? Or would you notice the color of the eye? The shape and contour of the upper lid? How the upper and lower lids meet in the front corner of the eye? The position of the nictitating membrane? Would you notice the seating depth of the eye? The cant of the eyeball?

You must study every little "speck" of the photograph in order to fully understand the significance of the picture. An entire session could easily be spent just studying a particular aspect of the eye, especially when the bird is exhibiting different moods (sleeping, alert, resting, swimming).

It is extremely important to study your references *every time* that you mount a bird. It would be very helpful if you were to spend a portion of each day studying your references. Study the bill one week, the legs the next, then the wings, and so forth. Every time that I pick up a photograph, even if I have looked at it a hundred times, I will see something new or observe it in a different perspective. The more you learn about a particular subject the more you will see in your reference material. I can assure you that I have, on many occasions, picked up one of my reference pictures and noticed something new. I have thought to myself that I must be blind not to have noticed that particular little detail before. That is just one of our basic human weaknesses and frailties. It is exactly the reason why it is so necessary to keep those references in plain view when working on a mount.

Our job is to try to recreate "exactly" a live specimen with a mounted bird. This must be done to the best of our abilities, and it is simply impossible to do that job without good reference materials. The more that you know and understand about the species that you intend to recreate, the better you will be able to interpret "what" you are trying to duplicate.

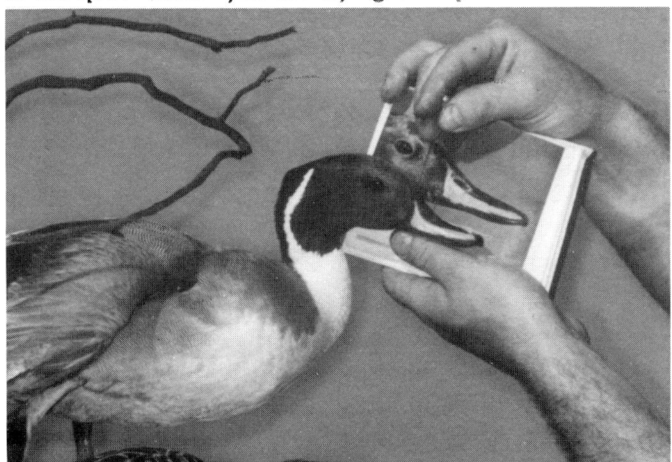

Photographs

I enjoy going to the zoo and photographing live birds. This gives me the opportunity to observe the birds and this also gives me a chance to obtain new ideas for poses. With luck, I can capture unique poses with my camera. It is also advantageous to take pictures at many different angles. This gives you a better idea of balance when trying to pose the bird. The best camera and lens I have found for photographing birds is a 35mm.

I prefer one that is totally automatic (auto shutter speed, focus, etc.). I usually shoot color print film as I don't like to wait and these can be developed in one hour in most places. I also shoot quite a few photos and it is less expensive to have color prints processed.

Death Masks and Study Casts

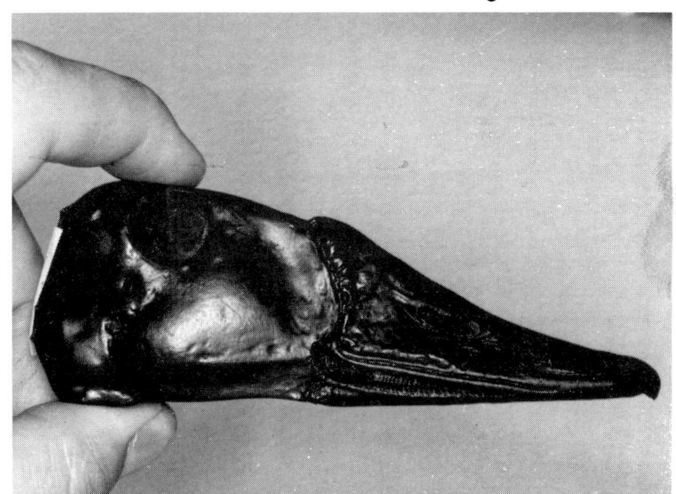

This three-dimensional aspect of taxidermy is not limited to mammal taxidermy alone. The head of a duck, for example, is very complex and not just a round shape with two eyes.

If you carefully examine the head of a bird, you will find that there are many different subtle contours and curves that must be shaped and put back into the mounted head. Study casts such as these are quite valuable when setting the eyes and shaping the head. Also the bills and

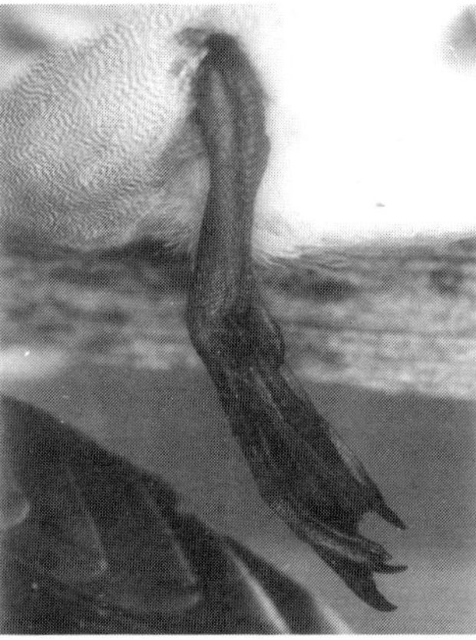

feet on many species will shrink dramatically and this type of reference helps the artist to restore the mount to its original proportions.

Posing the Bird

Choosing a pose involves more than just deciding upon an attractive way to display a customer's bird. Many other factors should be taken into consideration, such as where the bird will be displayed, condition of the bird, and characteristics of the varied species of individual birds.

The customer should have a general idea of where they would like to place the mount in their office or home; however, the taxidermist should make the customer aware of such factors as: lighting, consideration of other mounts in the room, height that the mount is to be hung, if it is to be a table mount, and the type of base, etc. It is essential to find out all the pertinent information from the customer in order to meet and satisfy their expectations.

If a customer decides upon a flying pose with the wings spread, but some of the primaries are damaged or missing, a better choice would be to put the bird in a standing or swimming position, or perhaps a flying pose with the wings slightly closed. In extreme cases, I dig into and use my supply of extra feathers to repair such mounts. (These feathers are collected and put aside when I pluck my personal birds that I am going to put in the freezer for a gourmet meal.) Repairs such as these are made by first removing the damaged feather, then inserting the replacement feather with a spot of Instant Bonding Glue on the base of the feather. It makes for an easy way to solve such problems.

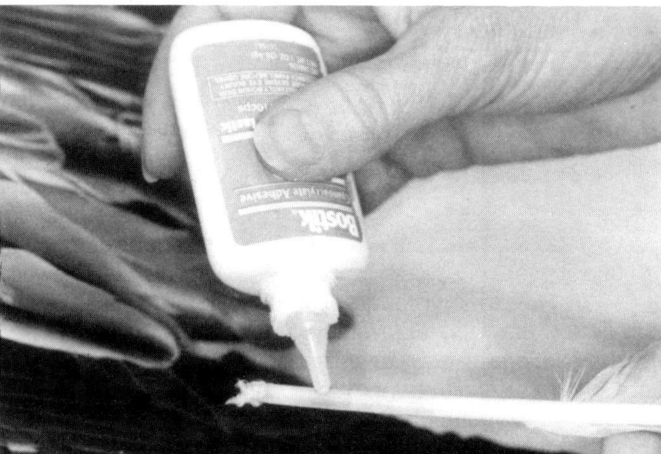

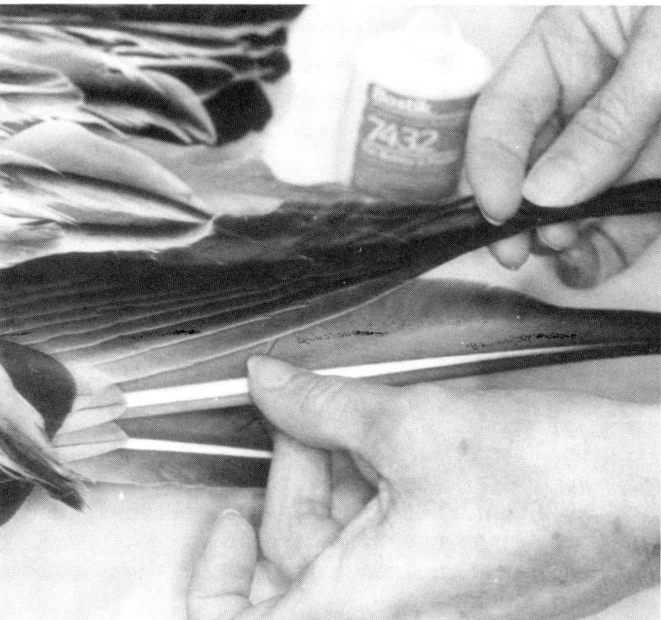

I feel that the most important aspect of posing a bird is understanding the basic characteristics of the individual species of that bird. Birds that are noted for their particular ability to fly or swim should be portrayed in a like manner. The individual characteristics of diving ducks vs. puddle ducks, for instance, play an important role as to the choice of a certain pose. I would not want to portray a coot or canvasback blasting out of the water in the same fashion that I would a wood duck and vice versa.

The shape of the body and anatomical features will vary within certain species and is determined by the manner in which these birds exist and function within their environment.

It is important that you are aware of this aspect of bird taxidermy and have ample references and a thorough knowledge of your subject available to you to substantiate your selection of a certain pose.

Harlequin scene mounted by Frank Newmyer.

Understanding the purpose and function of certain parts of a bird's anatomy greatly enhances your ability to add expression and uniqueness to your mounts. Birds use their wings, feet, and tail to pitch, hover, bank, climb-vertically, drop-in or maneuver through a maze of tree limbs. A complete understanding of the versatility of a particular bird's command of flight will be an important tool in your work.

Not only is is important to choose a position that accurately represents the individual characteristics of the bird, but it is equally as important to display the bird in it's own natural habitat. It would be detracting, for example, to mount a Harlequin, a duck associated with cold, ice, and snow, in a South Louisiana swamp habitat. You should always choose a base and habitat materials that will compliment the bird and not detract from it or completely overpower the mount.

In essence, we want to accurately display the bird in a manner that stays as close to nature as possible yet allows room for artistic and creative expression (see example below).

1985 Carl E. Akeley Award winner mounted by Sallie Dahmes

CHAPTER 4

Anatomy

Do Not Skip This Chapter!

Your first inclination might be to quickly skim over this chapter because the contents are somewhat boring. *But,* let me assure you, that understanding the skeletal, musculature, and outside topography of a bird is "vitally" important to producing quality bird work. In fact, you cannot produce excellent quality bird taxidermy mounts without a "thorough" understanding of the contents of this chapter. *If a taxidermist does not know where the parts belong, how can they reasonably expect to put a specimen back together?*

Just as "external" or "outer body" reference photographs are necessary tools to the life-like recreation of bird mounts, "internal" or "inner body" reference materials are equally as important (if not "more" important) to the taxidermist. In a nutshell, the skin is mounted over the inner-body, and what is underneath the skin determines whether the mount will be accurate or not.

A basic knowledge of how the skeletal system works and ties in with the musculature will prove to be valuable indeed, when you are involved with the assembly procedures. Tracking and aligning the feathers and skin will be easy instead of confusing when you have a thorough understanding of how the feathers and skin relate to each other. If the bird was not assembled properly, no matter how much pinning, carding, "stuffing," and wrapping is done, the feathers simply *will not* go back into their proper positions. Remember that the skin (and feathers) of every bird has been positioned over the exact anatomy of that particular bird's body for the entire length of its life. The skin and feathers will not fit naturally over a different anatomy; it is as simple as that. Consider, for example, if you mounted a human being's arm (something that everyone is intimately familiar with). Suppose the taxidermist that mounted the human arm put the elbow where the biceps are supposed to go. It would look unnatural and you would easily see

it, because you know what a human being's arm is supposed to look like, and you know where the location of the elbow is. It is no different with birds. If you know the anatomy, it is not difficult at all to position the feather tracts, etc. back to where they belong.

Skeletal Anatomy

Below is a skeletal drawing of a typical bird. Notice and carefully study the attachment points of the head, neck, wings, legs and tail in this drawing. Notice where each joint of the wing, neck and leg bends (and where they do not bend) and note the position of each joint with respect to the body.

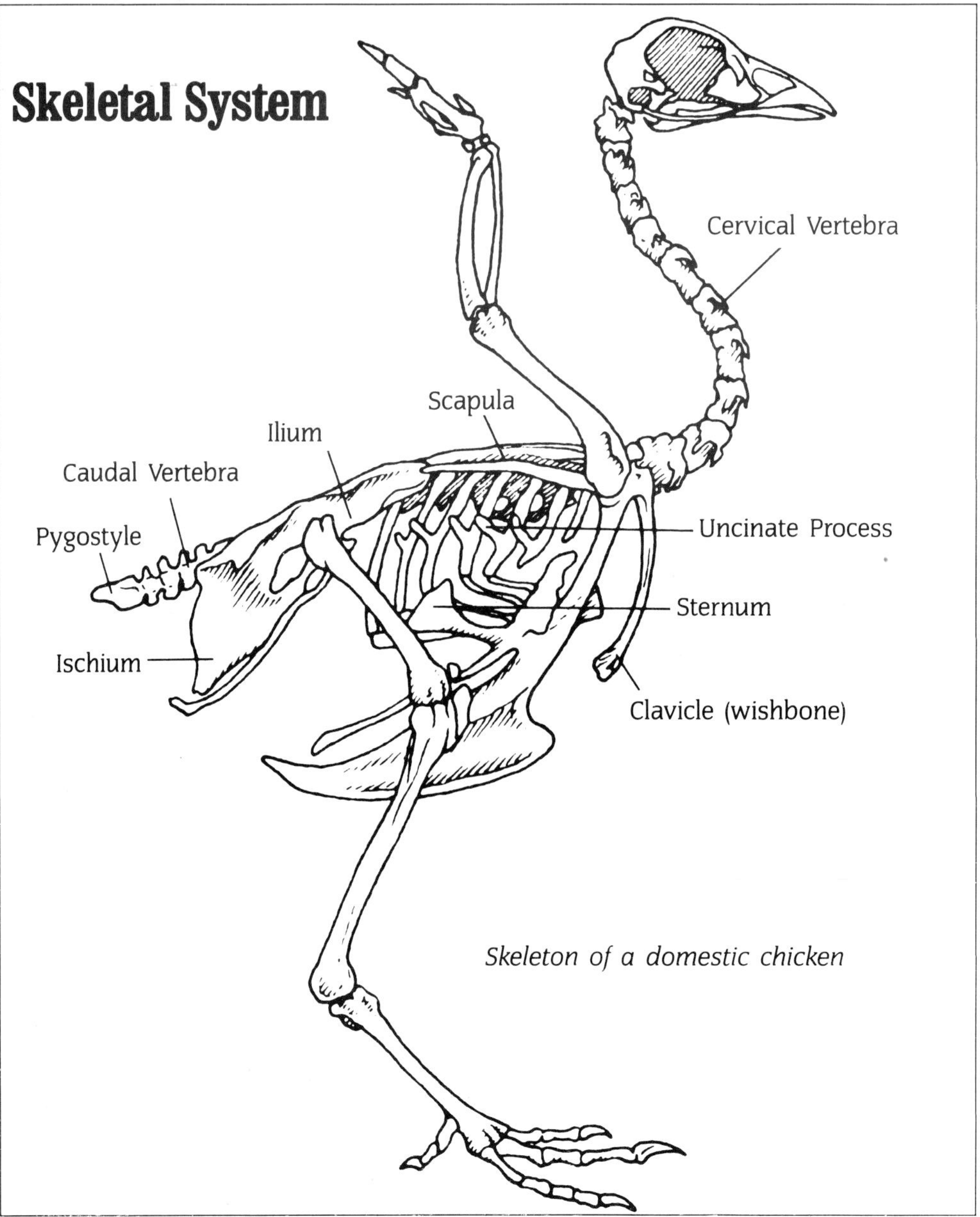

Skull of a Crow

Skull

Leg

Wing

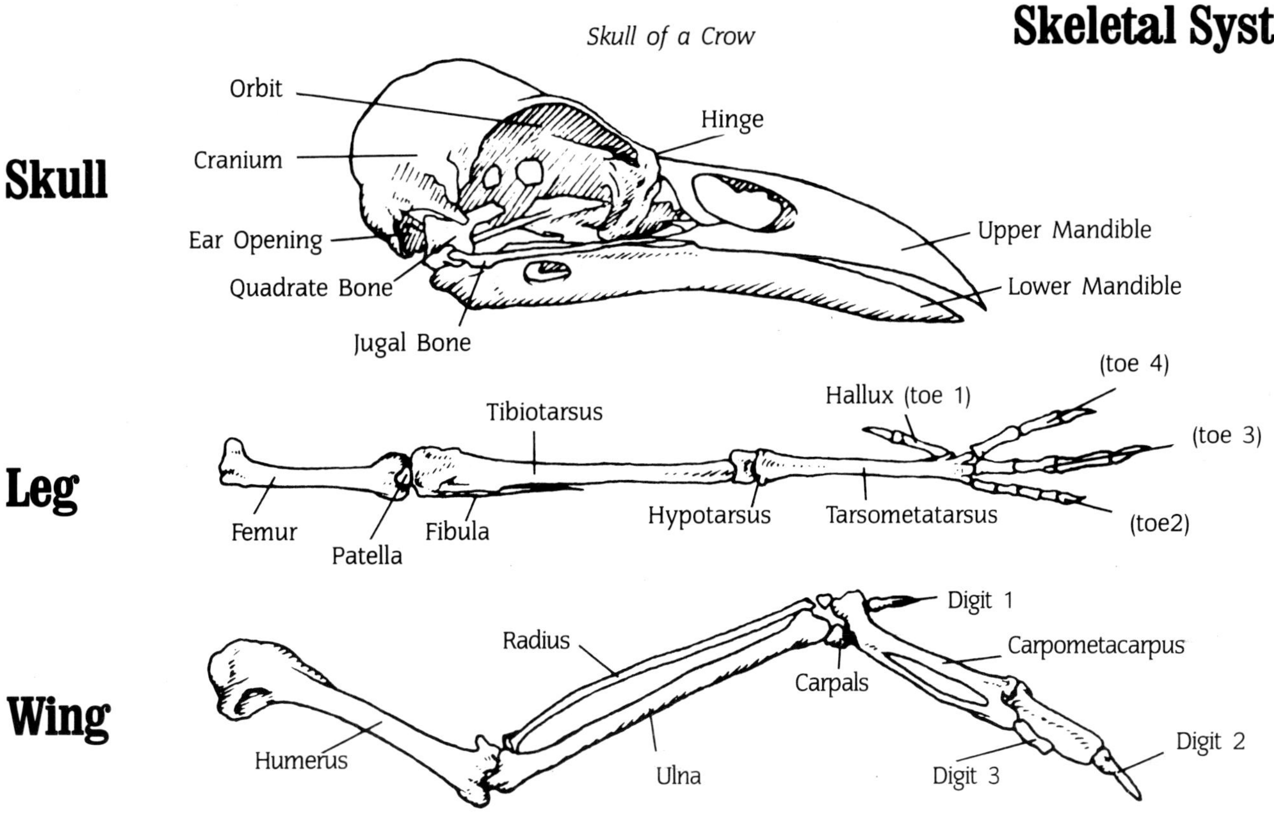

Wings & Legs

Understanding the skeletal anatomy of the wings and legs will make inserting and installing the wires in their proper position much easier. Spend some time studying their composition.

Bill

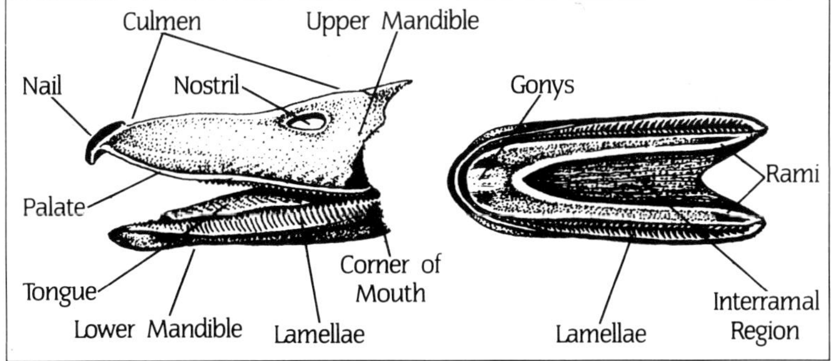

A bird's bill is comprised of several different parts. Study each specimen and get to know your subject both inside and outside. The more that you learn the better your work will become.

Definition of Terms

Listed are the scientific terms used to identify all of the parts of a bird. It is not essential to memorize these, but it certainly will not hurt.

Claw—A sharp, curved nail located on each toe (digit) of a bird. The middle claw has a comb-like serration on the inner edge that is used when the bird preens.

Corner of Mouth—Soft, fleshy area located at the point where the end of the upper and lower mandible meet.

Culmen—The surface area located on the upper most part of the bill running from the tip of the nail to the base of the bill.

Eye Ring—Or eyelid which is comprised of a fold of skin which completely surrounds the eye. In some species this tissue is thick and fleshy.

Gape—The width of the space between the open mandibles.

Gonys—Area on the tip of the lower mandible. Very similar to the nail on the upper mandible.

Heel—The joint located between the tibia and the tarsus on waterfowl and upland game, this joint always bends backward.

Interramal Region—Triangle area of skin located on the lower mandible.

Iris—The pigmented or colored, round, contractile membrane of the eye.

Joints—A point of contact between movable parts such as bones.

Lamellae—Serrations or grooves on the inside edges of the mandibles.

Lobe—A rounded projection of tissue located on the 1st and 2nd digits (toes).

Mandible—Either jaw (upper or lower) of a beaked animal.

Nail—The area on the tip end of the upper mandible.

Nictitating Membrane—The "third eyelid." This transparent membrane is located in the front corner of the eye.

Nostril—Opening located on each side of the upper mandible.

Palate—Roof of the mouth.

Pupil—The black circular aperture located in the center of the iris.

Rami—The lower jaw or mandible.

Scales—The protective covering of the tarsus, digits, lobes and webs.

Tarsus—The area between the tibia and the digits (toes).

Tibia—The bone located between the femur and the tarsus.

Toes—Digits or phallanges.

Tongue—Fleshy organ which fills the lower mandible.

Upper Mandible—Top half of the bill.

Web—Area of skin located between the toes on waterfowl.

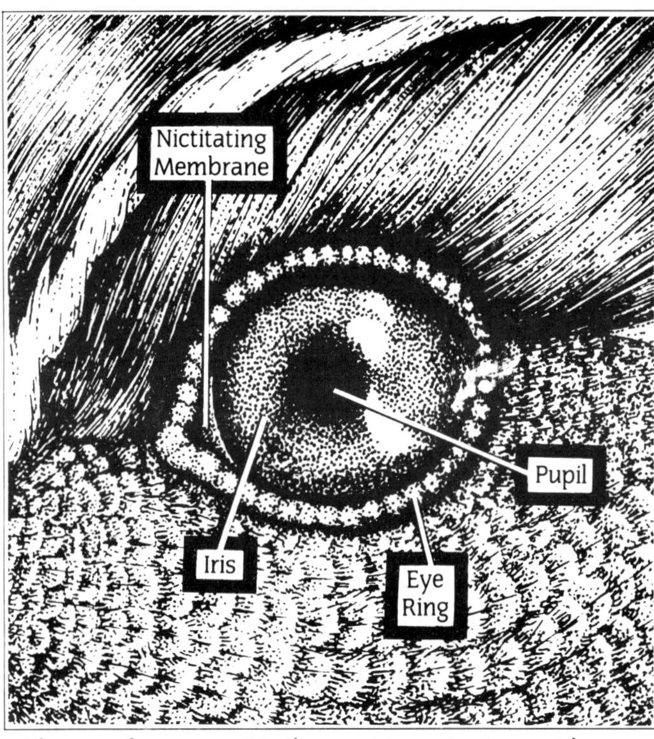

The eye of any mount is the most important aspect because it is the primary focal point. Just about everyone looks at the eyes first. Again, learning as much about your subject as possible will be your ticket to excellent work. Study the "fine" details.

Bird eyes are "fixed" in the sockets and generally they cannot roll their eyes as humans can. Instead, birds turn their neck to see objects that are at a different angle. The neck has several vertebra which look and act something like a string of beads when the head is turned. Bird eyes are also supported by a sclerotic ring which is a series of thin bony plates, which will vary in size and shape depending upon the bird species. All birds have both binocular (able to focus on a subject with both eyes simultaneously) and monocular vision (meaning each eye focusing on a different subject independently of each other). In the retina of each eye, the duck has a fovea located in the center of the retina enabling it to see straight out from the side of the head, thus enabling a duck to see entirely different objects with each eye.

All birds have two eyelids, upper and lower, which are folds of skin. Birds also have a nictitating membrane, or third eyelid, which is a transparent skin that is drawn across the eye from front to back to clean the eye. When observing waterfowl from a "front-on" view, the top of the eye is slanted out farther than the bottom of the eye, which simply follows the contour of the bird's skull. This enables the birds to see front-on while feeding.

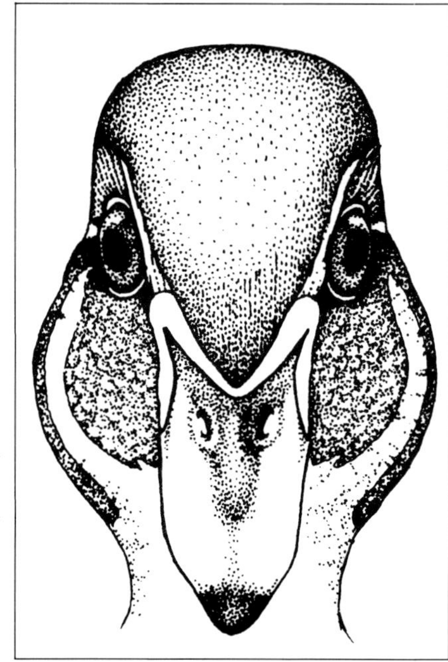

Musculature

The musculature of a bird is extremely important in bird taxidermy because the skin actually fits over it. Sportsman Series mannikins are accurate representations of this anatomy; however, fabricated mannikins will require that "you," individually know what the anatomy is.

A comparison of a plucked duck carcass (above) and a Sportsman Series mannikin (below) clearly illustrates the anatomical accuracy of the mannikin.

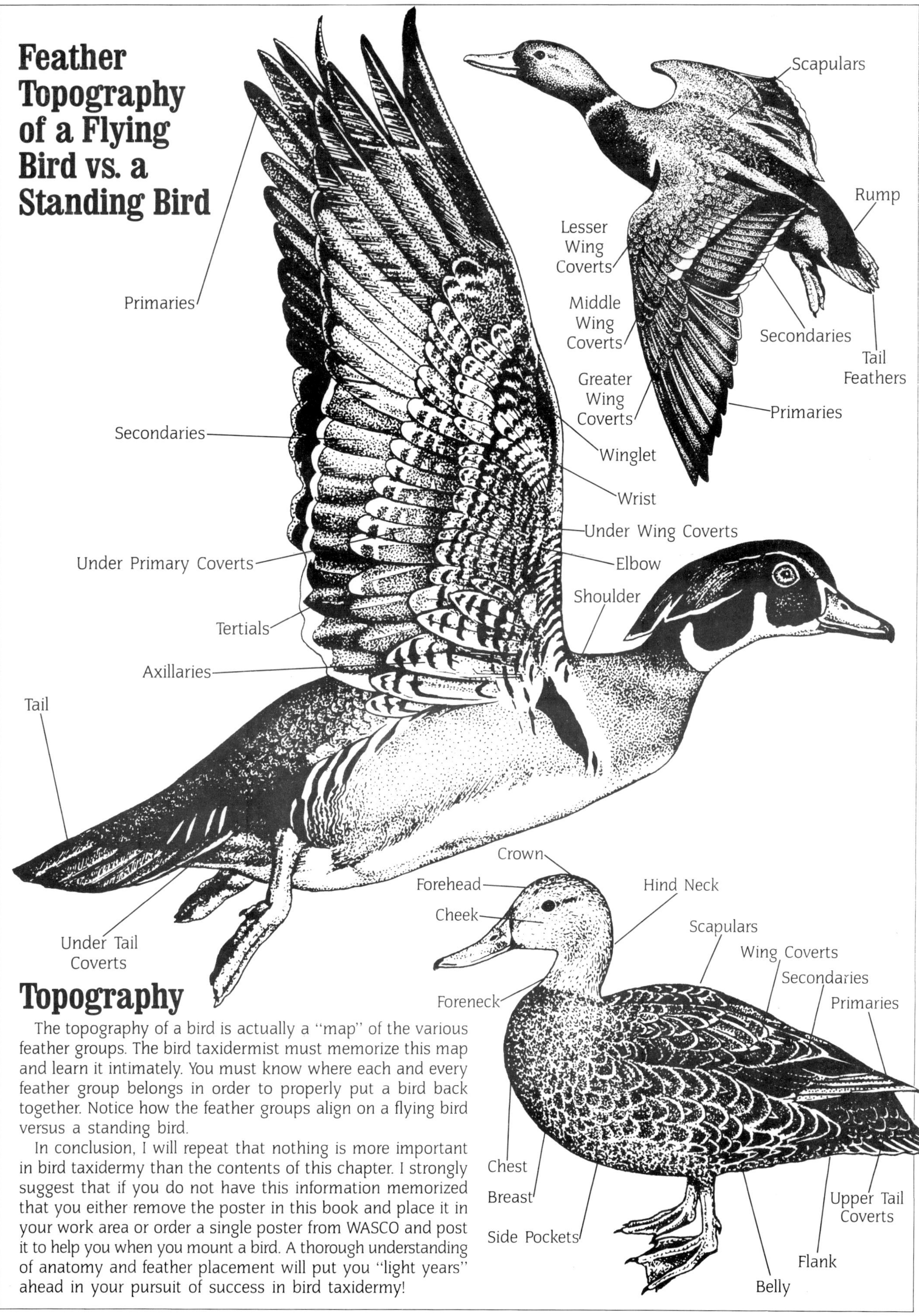

Topography

The topography of a bird is actually a "map" of the various feather groups. The bird taxidermist must memorize this map and learn it intimately. You must know where each and every feather group belongs in order to properly put a bird back together. Notice how the feather groups align on a flying bird versus a standing bird.

In conclusion, I will repeat that nothing is more important in bird taxidermy than the contents of this chapter. I strongly suggest that if you do not have this information memorized that you either remove the poster in this book and place it in your work area or order a single poster from WASCO and post it to help you when you mount a bird. A thorough understanding of anatomy and feather placement will put you "light years" ahead in your pursuit of success in bird taxidermy!

CHAPTER 5

Field Care

Customer Satisfaction Begins in the Field

Proper field care is a very important aspect of bird taxidermy that I "always" stress to each customer. This does require some of my time, but I can assure you that it is time well spent. A properly "cared-for" bird will make the mounting procedure less complicated, take less time, and will definitely ensure a better looking mount. Since I save time when I receive a properly prepared specimen to mount, I figure that it is worth it to spend some extra time educating my customers about this important subject.

I prefer to give field care instructions both verbally and in writing. WASCO sells an inexpensive field care manual that can be stamped with your taxidermy studio's company name, address, and the phone number on it, or one of your business cards can be attached to the inside cover. These inexpensive manuals can be provided as free gifts to "special" customers. They can be used not only for field care instructions, but as good advertising tools as well. They can also be sold in your shop as an extra profit item. Of course, the "big" profit received from these well-written books will come to you when your customer implements the proper field care procedures which will help you when you mount their next trophy. An even better approach may be to sell, rent, loan or give away the field care instructional video tape in addition to the book. It costs quite a bit more but there is nothing like "seeing" the procedures performed.

"All" customers are anxious to receive top quality work on their trophies. You can rest assured that they are really appreciative of your taking the necessary time to explain to them the proper procedures that they will need to follow while in the field.

Proper field care for birds is just as "important," but is generally less complicated than many other species such as big game animals. This is because there is usually no skinning necessary on birds. The major concerns with birds are: rough handling of the specimens, excess body heat, and blood stains.

Many sportsmen use retrievers to locate and recover downed birds. Unfortunately for the taxidermist, many dogs are quite rough with birds. I would like to emphatically stress that, if at all possible, the "hunter" should personally retrieve the bird to avoid any ruffled or broken feathers. If the bird is alive, then the hunter should grasp it tightly across the breast, behind the wings, and squeeze firmly for one or two minutes until it succumbs.

Do not wring the neck of a bird that is to be mounted.

Other than "rough-mouthed" retrievers, "neck-wringing" probably comes in second for ruining the most birds for mounting purposes. Also "field dressing" by overzealous hunters makes a strong third place showing. Tell your friends and/or guides not to do it!

Do not carry birds by their heads; instead, carry them by their feet. Treat the birds gently as you would a "fine" garment.

I like to encourage hunters to go into the field prepared to properly handle a mountable specimen. Drop a mesh game bag (or pillowcase), cotton balls, paper towels, and wrapping material into your shell bucket the next time that you go hunting.

These protective materials will prove to be just as valuable as that "extra" box of shells that was placed in your shell bucket in anticipation of a good morning of shooting.

Most gunshot birds will drain fluids from shot holes or body orifices. Carefully placing a small wad of cotton into the nostril openings, mouth, vent, and shot holes will protect the plumage from these fluids. Once the cotton wads have been put into place, the bird should be either placed in a cooler with ice or at least set aside where air can circulate around it to allow the body heat to escape as rapidly as possible.

Never place a bird in direct sunlight! Never place the bird in a tightly enclosed plastic bag or rubber lined game pouch with other warm birds. Birds can spoil quickly when exposed to excessive heat or direct sunlight. (Even if the outside temperature is fairly cold, direct sunlight can heat a bird up to the point that it will be spoiled very quickly.) The insulation value of their feathers compounds this problem and requires that a bird must be allowed to cool to prevent spoilage. However, this does not mean that the bird should be "gutted." By the time the hunt is over, the decoys have been picked up, and all of the hunting paraphernalia is loaded, the bird should be cooled enough and be ready for transporting out of the field.

Several different methods of wrapping and transporting birds will work quite well to protect the bird. Using a mesh game bag or inserting and carrying the bird in a loose sling fashioned out of a hunting shirt works well. A loose fitting nylon stocking slipped over the bird is a handy way to protect the plumage, as it helps to keep the feathers in place. Also, simply shaping a piece of heavy paper into a cone and taping it and then placing the bird, head down, into the base of the cone works quite well. I am sure there are many other good methods of taking birds out of the field; however, the important thing is to stress "gentle" handling of the bird and for the hunter to develop an awareness of the delicate nature of the plumage. As odd as this may sound, I once actually had a customer that thought that a taxidermist plucks the feathers and then glues them back onto a piece of styrofoam in order to mount a bird. It is no small wonder that he was not worried about a few missing feathers on his badly mangled trophy. Here again, a few minutes spent with such a customer can make a big difference in the condition of the next bird that you receive from them.

After the hunter returns home from the hunt, the bird should be placed in the freezer as soon as possible. The best method that I have found to protect a bird in the freezer is to first tape wet cotton around the feet and head areas with masking tape (be "very" careful not to get any tape on any of the feathers)

and then carefully place the bird into a heavy duty Zip-Lock freezer bag. Do not bend the tail of birds with long tail feathers such as pheasants; instead, use a bag that is big enough to avoid bending the tail or cut a hole in it to allow the tail to exit the bag. Next, either secure the bird to a flat piece of cardboard with freezer tape or make a cardboard tube. The tube can be made by rolling up and taping a section of cardboard the same diameter as the bird and long enough to completely encase the tail.

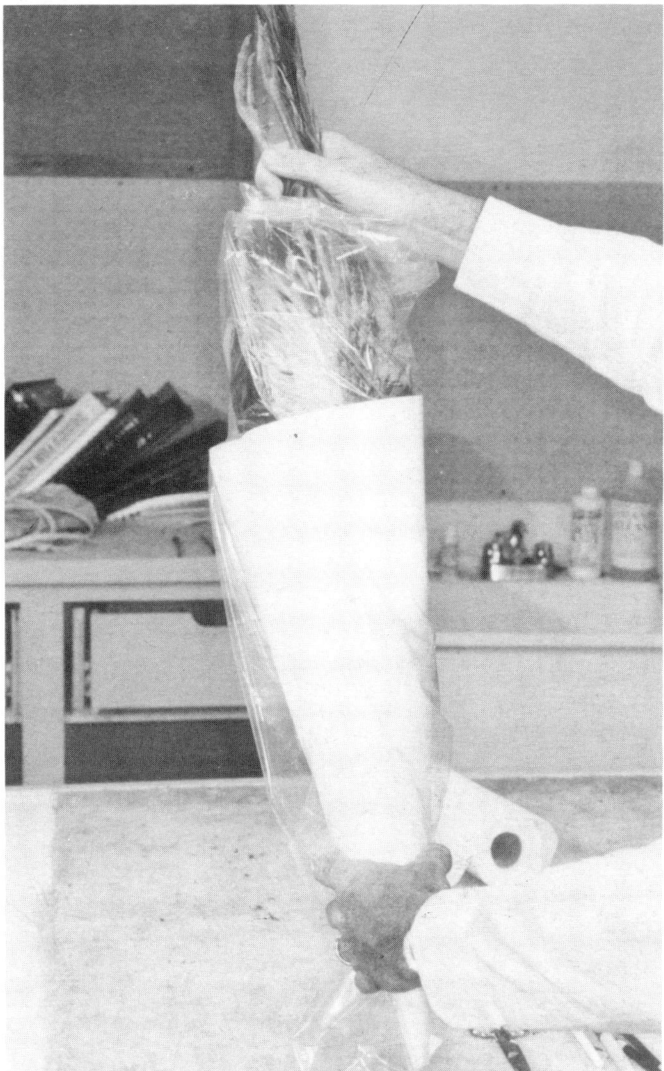

The tube and the bird are then placed into another plastic bag and frozen.

Placing the bird in airtight, "double-plastic" bags will prevent dehydration while frozen. Birds may be kept for months using this method without any damaging effects to delicate tissues and feathers, and although I certainly "do not" advise it, I have actually kept birds stored in such a manner for several years with only a slight problem with freezer burn on the tips of the feet.

One of the most important things to remember is to *never* wrap the bird in newspaper as the paper draws the moisture from the skin and will readily "cause" severe freezer burn with damage to the feet and the delicate tissue of the head area. Always encase the delicate feathers in tubes or double pieces of cardboard in addition to placing the bird in a plastic bag to avoid damage from other items banging around in the freezer with your bird. The cardboard protects the neck and feet and also prevents the very common problem of damage to the tail feathers especially on birds such as pheasants, pintails and old squaws.

I have a few customers who are very dedicated bird hunters and who go to a great amount of expense and difficulty to travel long distances to remote areas to collect specimens. Due to the fact that these hunters are often in isolated areas with no refrigeration, it poses particular problems for the hunters. Usually the birds cannot be refrigerated or frozen and must be skinned by the hunter.

I like to have my customers come into my shop prior to leaving on such a trip and spend some time observing the skinning and defatting procedures. Once they obtain a basic understanding of the fundamentals of skinning a bird, it should pose no special problems to the hunter while in the field.

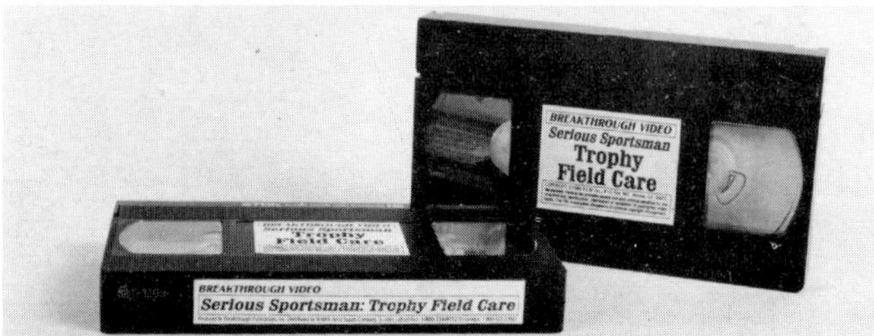

A Serious Sportsman field care videotape can be "given," sold, rented, or loaned to the customer if a visit to your shop is impractical for either of you.

At this point it should be emphasized to the customer that if they are going to perform the skinning operation, it is imperative that accurate measurements be taken "prior" to skinning the bird. Additionally, accurate sketches need to be drawn (or traced) of the body. These will prove to be invaluable to you when you recreate the inner body. The customer should also take color photos and/or make notes of the color of the eyes, bill, and feet if at all possible. Instruct your customers that "any" and all information that can be passed on to the taxidermist will definitely aide in producing a fine mount.

After the preliminary measurements, photos, and sketches, are completed, the bird should be skinned. Having the field care book by their side while they are skinning should help the customer get through the fairly simple process quite easily if they will just take it "slow." After the skin is removed and the meat is cleaned from the wings and leg bones of such a specimen, then WASCO Salt (or non-iodized equivalent) should be generously applied to the skin. Particular attention should be given to the wings, bill, and caudal areas.

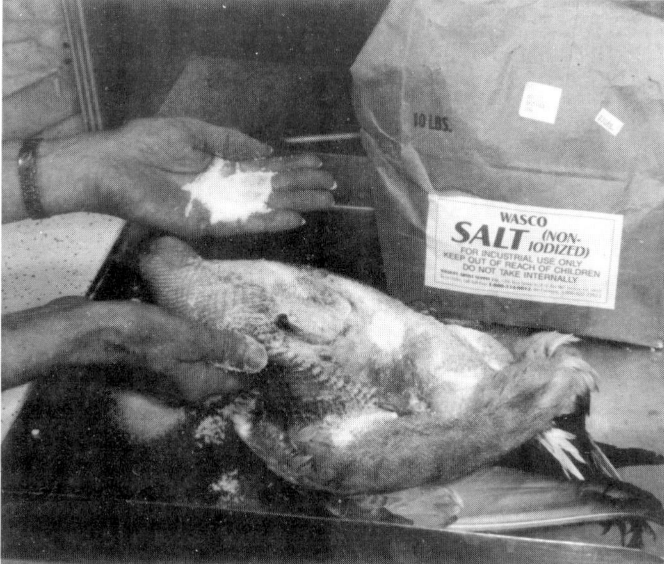

Next, the skinned and salted bird should be placed in a paper bag and kept in a cool area out of direct sunlight. If these steps are followed, the bird will keep well for several days.

When the customer returns from their trip, the salted bird should be taken to the taxidermist just as soon as possible. The taxidermist should not delay in rehydrating and properly preserving the bird skin before freezing it in double plastic bags.

Proper field care consists of basic and simple, "common sense" steps, that if followed by the hunter/customer, will practically "guarantee" a first-class mount to be enjoyed for many years to come.

CHAPTER 6

Skinning, Defatting, Washing and Tanning

Many taxidermists view the skinning, preserving, and leatherizing procedures of a bird as insignificant when compared to the actual mounting and posing process; however, the skinning, preserving, and leatherizing of a bird skin (as well as any species in taxidermy) is "very" important and has a dramatic impact on the final outcome of the mounted bird.

This phase of taxidermy does not necessarily challenge one's creative abilities, but it does require exactness and paying close attention to detail. Without a properly prepared skin it would be difficult, if not impossible, to produce a *lasting*, quality mount. In essence, the bird skin is your medium as an artist, and unless it is properly prepared, your chances of obtaining a good mount are nonexistent.

Thawing the Frozen Bird

The importance of proper handling begins as soon as the bird is removed from the freezer. Do not try to rush the thawing time. Never place a bird by a heat source or place a bird in hot water to thaw it. Exposing the bird to a heat source will increase the risk of slippage and damage to delicate tissues a thousandfold. It is best to remove the bird from the freezer the night before you intend to skin it in order to allow adequate time for the bird to thaw.

As soon as the bird is taken from the freezer, remove the plastic bags and cotton wrapping and place the bird in a tray of WASCO granulated borax to thaw. The WASCO granulated borax will absorb moisture as the bird thaws. It is important to always keep the skin and feathers moisture free. Damp, blood and "fat-soiled" skin and feathers will quickly promote bacteria growth; especially if the temperature is slightly elevated. This will result in only one thing: *feather slippage.*

Large birds such as geese and turkeys will require considerably more time to thaw than smaller birds. The head and neck will thaw out before the body does. Once they have thawed, place the large bird in a refrigerator to finish the thawing process. This will prevent the head and neck from slipping or spoiling before the body is thawed out enough to skin.

Tools and Logging-In

Assemble "all" necessary tools and supplies for the skinning process before you begin. A little time spent organizing yourself and your work area will eliminate much wasted time spent looking for tools and supplies.

If you have the room, it is best to set up a specific area for each function, i.e., one area for skinning and defatting, one area for mounting, one area for finishing, etc. Locate all the tools and supplies needed for that function and leave them there. At the very least, assemble all your tools for that function into different boxes so that they will be organized and readily available.

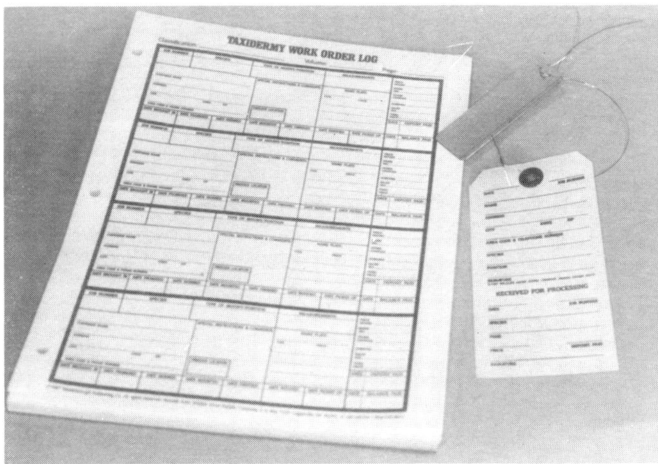

Before beginning the skinning procedure you should have your business work log close at hand. You will need to note in the work log the desired pose, type of mannikin to be used, type of head and bill needed, glass eye to be used, reference notations, and measurements, etc.

These factors will be used to determine how you will handle the skinning of the bird. For instance, if the bird is to be mounted using a Sportsman Series mannikin, the taxidermist would sever the leg between the tibia and the femur (at the point shown in the photo below), because the femur is already sculpted onto the mannikin.

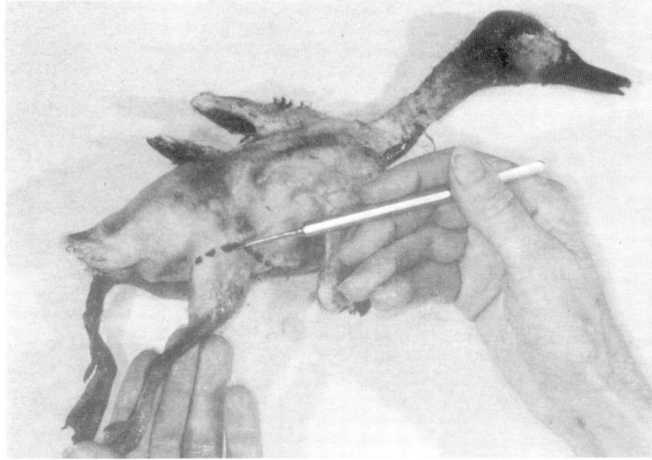

If the taxidermist wanted to use the wrapped body method for a walking or dead game mount, they would need to keep the femur attached because they would need to incorporate it and the thigh muscles on the body. Therefore, they would remove the leg above the head of the femur (at the point indicated in the photo below).

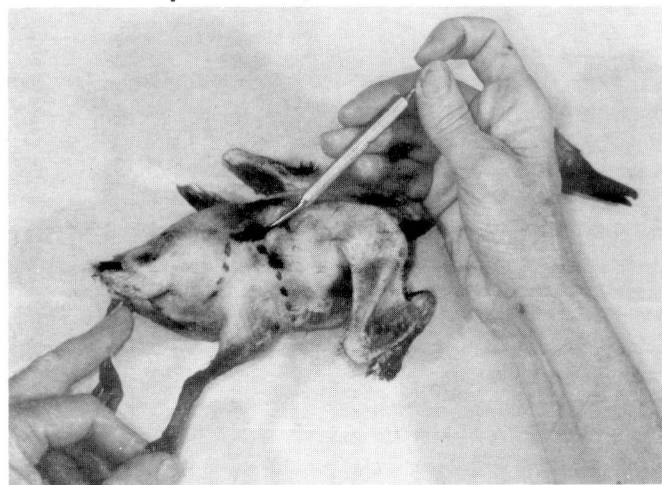

Again, this is dependent upon the pose; some poses can incorporate the femur right into the wrapped body. All these decisions must be made before the actual skinning is started.

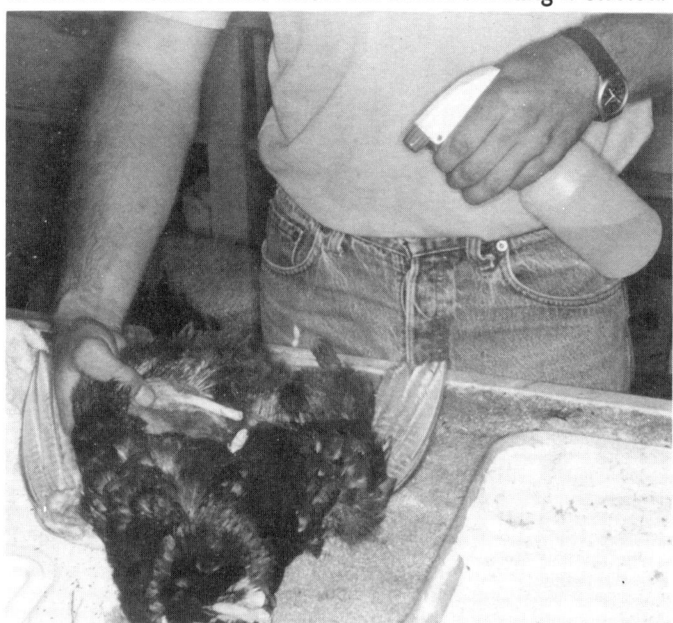

Bird skins have a tendency to dry out fairly quickly because they are so thin. A spray bottle with Bacteria Stat treated water can be used to occasionally *lightly "mist"* the inside of the skin to help keep it from drying out too quickly, especially in very arid climates.

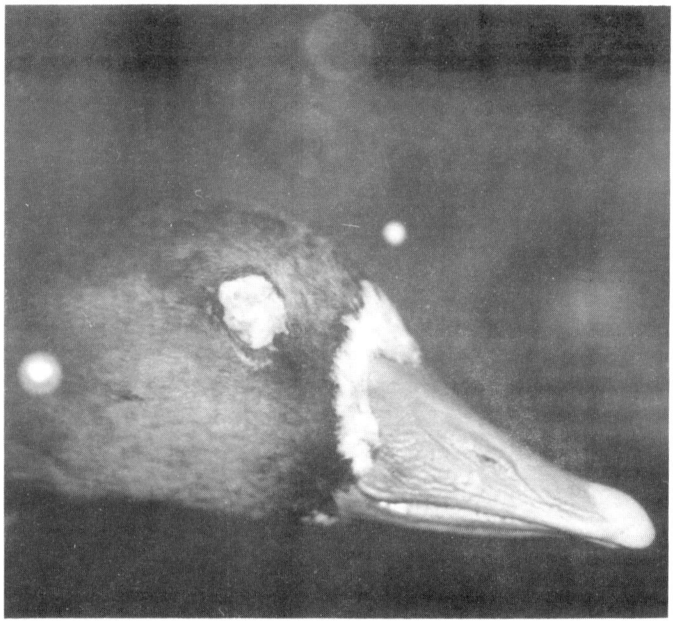

Also, slightly dampened cotton can be put into the head area to prevent it from drying out too fast. It is very important to proceed completely through the skinning process, from "start-to-finish." If you must stop working on the bird, be sure to wrap it in a towel and place it back in the refrigerator. If there is going to be a "long" delay, put the bird back in a plastic bag and wrap the feet and head with damp cotton *before* refreezing it.

Skinning

Once the bird is thawed, lightly massage the abdominal section and gently work the legs and wings back and forth. This action will loosen the joints and skin and will decrease the risk of tearing the skin. Once the skin can be easily moved with your fingers, the bird is thawed and ready to skin.

Incisions

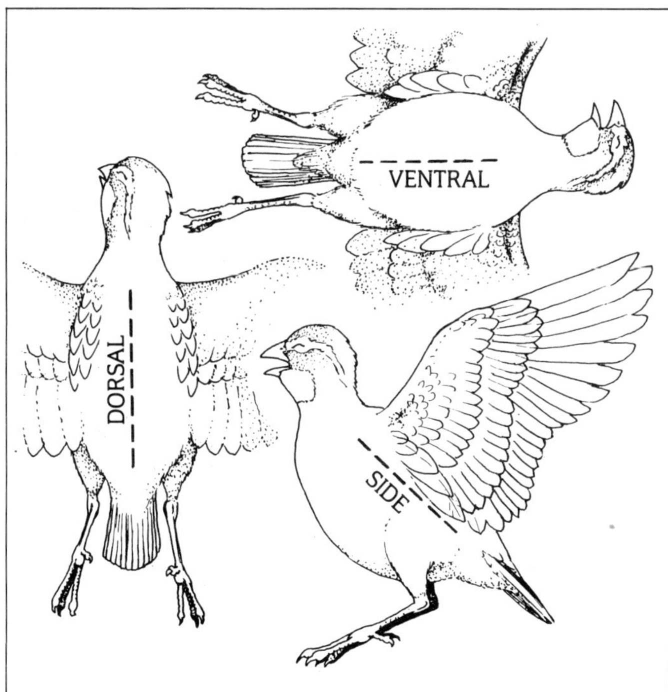

There are several different methods commonly used when skinning a bird: the ventral or belly incision, the dorsal or back incision, and the side incision.

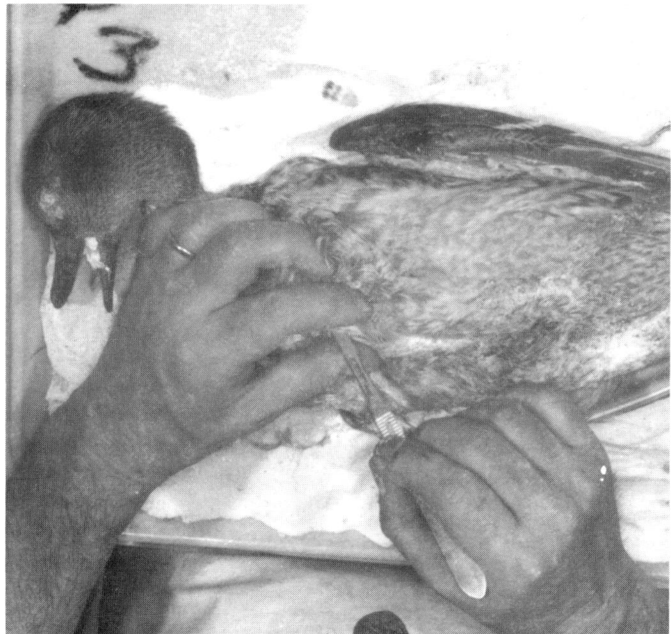

The ventral incision is the easiest and most common method used by taxidermists. The reason is that "most" birds have a natural area without feathers (called the apterium split) in the area of their breast where the incision is made. This "bare spot" makes a "perfect" place for making a good clean incision without cutting feathers.

If you prefer the dorsal or side incision, the same essential skinning procedures as outlined below are used; however, the incision will be made as per the illustration. The bird body, will be skinned and removed in the normal manner except through the side or dorsal incision. One word of caution will be that the incision will need to be made much more carefully because the feathers will need to be pushed aside before cutting. The same holds true for birds without the bare spot on their breast. Do not cut the feathers, cut *around* them.

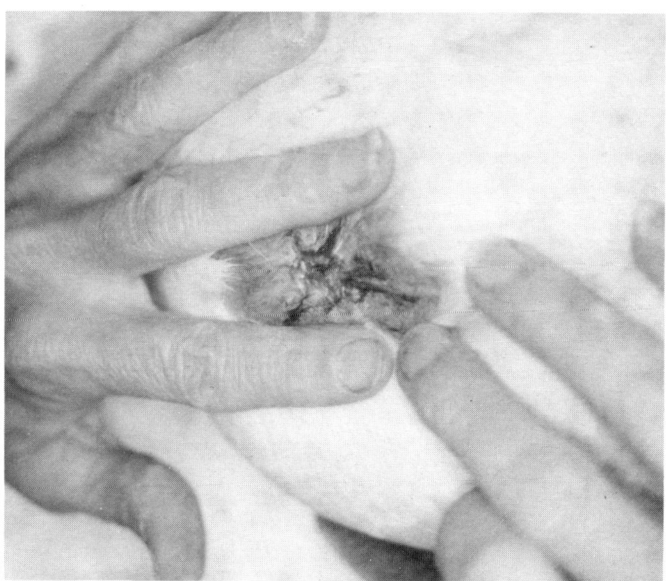

To begin the process, lay the bird on its back in a tray of WASCO granulated borax. Separate the feathers located in the breast area with a piece of Bacteria Stat water-dampened cotton to expose the bare section of the skin on the breast.

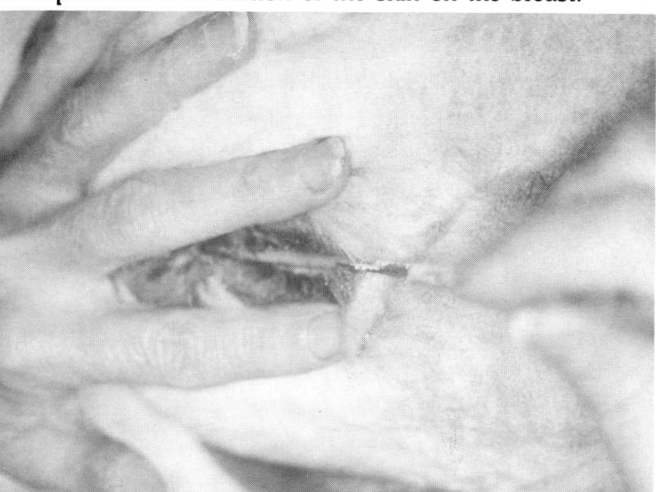

Now make a "clean," straight incision, without cutting through any breast feathers. Use a scalpel to make the incision, and, cut "just" deep enough to penetrate through the skin.

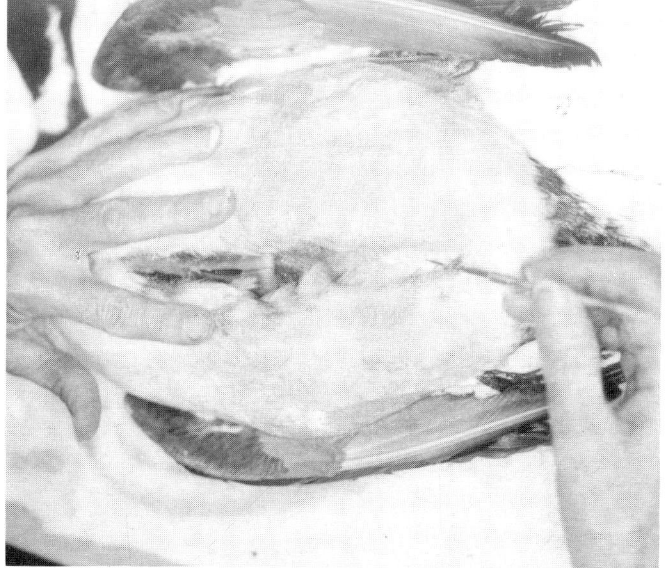

Always make a nice, straight and even cut; this will result in a much neater sewn incision.

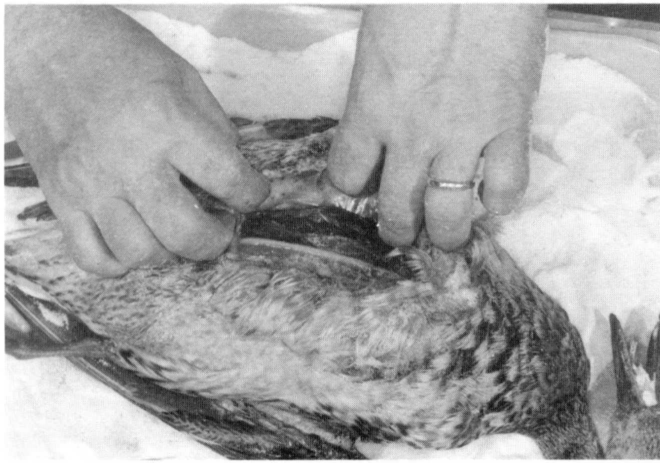

Once the cut is made, use your fingers to lift up and separate the skin from the meat. Gently peel the skin up and away from the sides of the body and gently work towards the vent area. Use your scalpel "only" when needed to cut stubborn tissue; your fingers can be used for 98% of the separation.

Lightly sprinkle WASCO granulated borax on the inside of the skin to absorb body fluids and dip your fingers in borax "often" to increase your grip. In addition to the absorption properties, the borax also helps to keep the "feathers" free from grease and blood. While it is true that the skin will be washed later, it is also true that the cleaner that you can keep the feathers, the better!

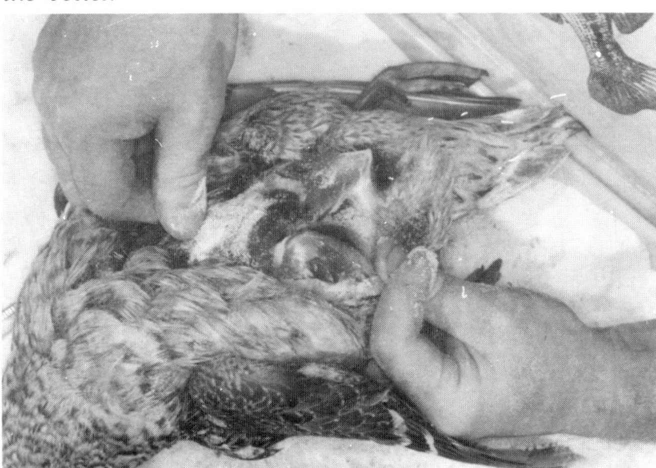

Now that the skin has been worked free from the sides of the body and the legs are exposed, gently bend the leg and push the knee towards the incision. With the knee exposed, hold the tibia to steady the leg and cut the knee joint with diagonal pliers.

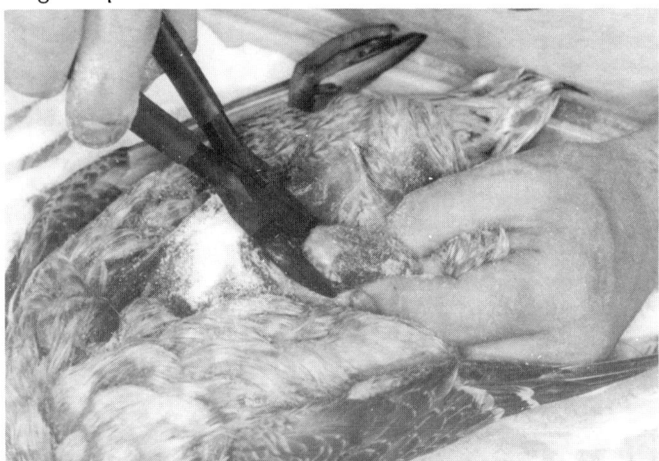

(NOTE: Cut the leg free at the femur if a wrapped body extended leg pose is used.)

With the leg severed, continue to free the skin from the thigh meat. Once both legs are detached and the skin is free from the thigh, proceed to skinning the anal area.

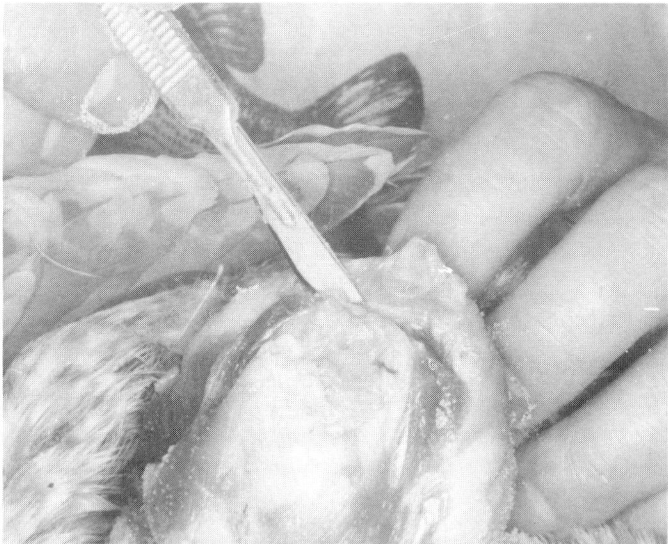

Carefully cut the anus or vent loose by *cutting the anal passage; do "not" cut through the vent itself.* Remember to use granulated borax to soak up the body fluids.

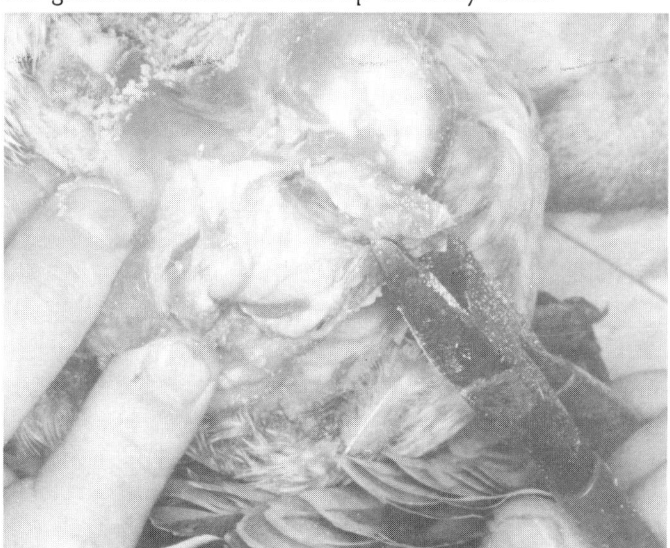

Carefully peel the skin from around the tail section until the tailbone or caudal vertebra is exposed. The caudal vertebrae are located in the middle of the meat and cartilage that holds the tail quills (pygostyle). Cut the vertebra with the diagonal pliers; this will release the tail from the body. Now continue to peel the skin over the rump and down the back.

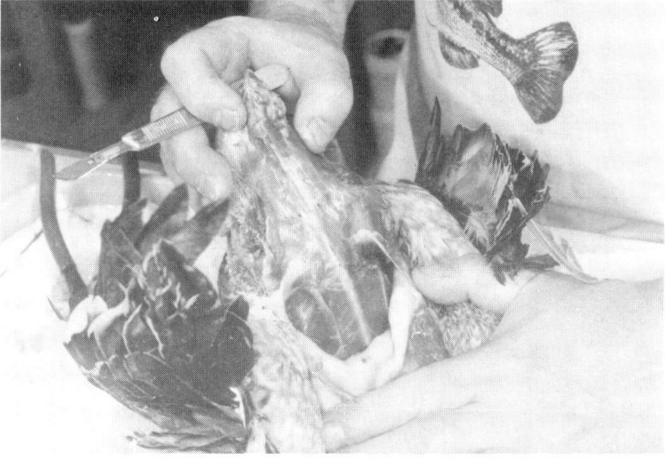

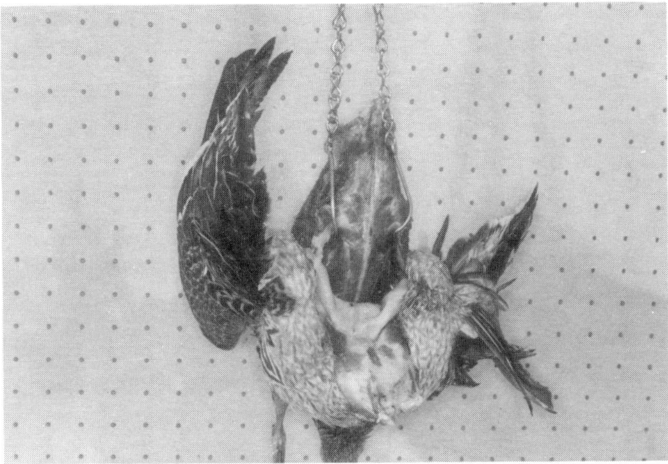

The carcass can then be attached to the chain hooks of the bird skinning gambrel to facilitate the skinning process. One hook is attached to the top of the breast and the two remaining hooks are attached to either side of the rump area. Attaching the carcass to the chain hooks in this manner holds the bird up where you have access to it and eliminates trying to hold the bird with one hand while skinning with the other.

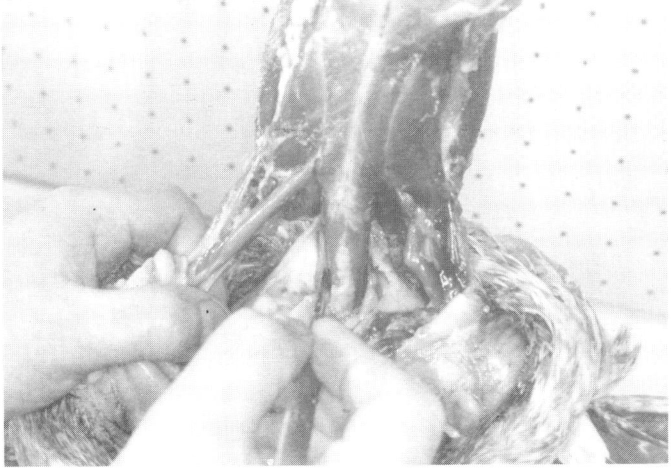

Now continue skinning down the back until the wings are encountered. "Care" should be taken when skinning down the back, as the skin is very thin in this area. Use your fingers to push the skin loose and use your scalpel *sparingly*. Go very slowly until you develop a "feel" for the bird skin.

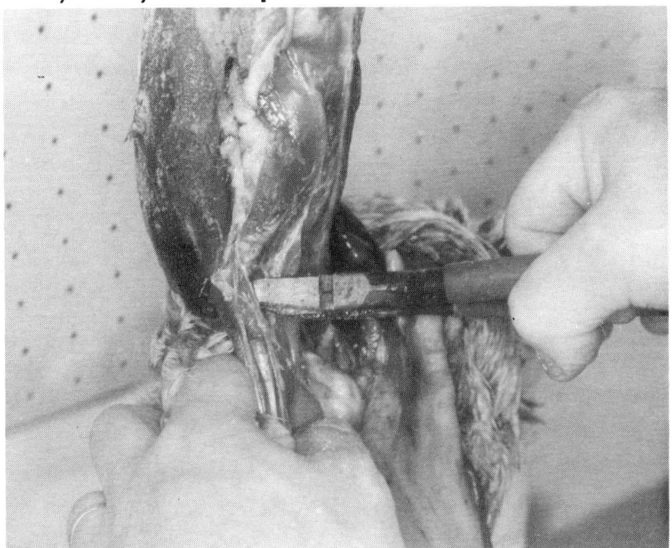

Partially skin down each wing bone and expose the top part of the humerus.

With a sharp pair of diagonal pliers, cut the humerus close to the joint of the shoulder. Care should be taken when cutting the humerus as it has a tendency to fragment and splinter when cut.

With both wings free, peel the skin over the neck to the base of the skull. This is easier if you dip your fingers in the borax first.

With certain birds, the skin can be peeled completely over the head all of the way to the edge of the bill. However, on many other birds, especially waterfowl, the skin will not invert over the head because the head is too big for the small neck skin to fit over.

If this is the case, sever the neck at the very base of the skull with diagonal pliers. A decision must now be determined as to whether or not to remove the natural skull or leave it attached.

If an artificial head and bill is to be used, the skin can be quickly and easily completely removed from the skull. If the natural skull is used, the skin must be left attached.

Natural Skull Incisions

If the skull is to be used, there are three different incisions that can be made on the head. The choice of the type of incision is dependent upon the pose and the type and length of the feathers on the head. Basically, an incision should be chosen that is easy to camouflage. For instance, on hooded birds, such as wood ducks, it might be advantageous to make a cut on the back of the head as the hood feathers will easily cover the incision. However, a bird with short feathers on the back of the head might require a different incision to more easily "hide" the stitches.

For a wall-mount, either standing or flying, an incision made on the side of the head from the ear opening extending approximately 1" down the side of the neck (make the incision "just" big enough to allow the skull to fit through) is probably the easiest incision to hide. For standing bird table mounts, an incision made under the head starting midway between the mandibles and down the throat, can be partially hidden by the natural curve of the neck. (Viewers would have to get below the table mount to be able to even see it.) For birds positioned in a flying, coming-in pose or birds with long crown feathers, an incision in the back of the head probably would work the best. Remember to choose the type of incision that will work best for your mount.

Skinning the Head

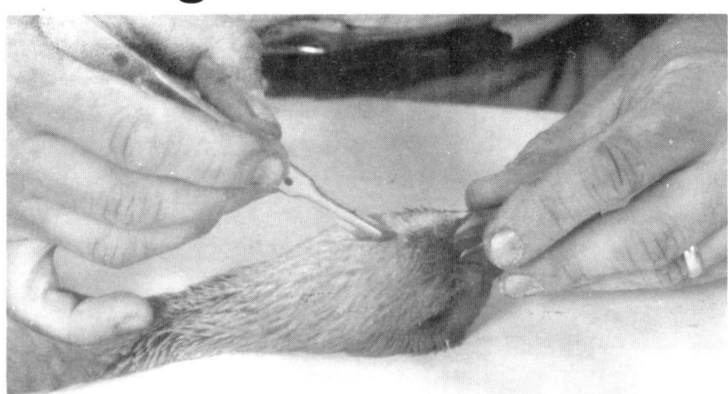

To make the incision, carefully part the feathers and slowly make an incision that is "just" large enough to allow the skull to fit through. Try to make the incision straight and clean. Do not make a ragged incision, as it will be too hard to sew up and hide the stitches later.

Once the incision is made, gently push the skull through the opening in the skin. (Again use borax or dry preservative to absorb any excess fluids and to give you a better grip.) Peel the skin to the base of the ears and carefully cut the ear canal as close to the skull as possible.

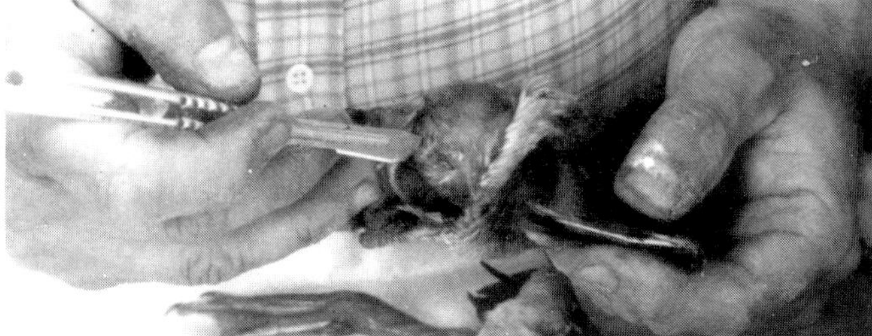

Actually put the knife point down inside the skull where the ear enters it and lift it out. It is a common "beginners mistake" to cut a big hole in this area! Do not let this happen! Carefully work the ears out and sever the ear canal deep inside the skull.

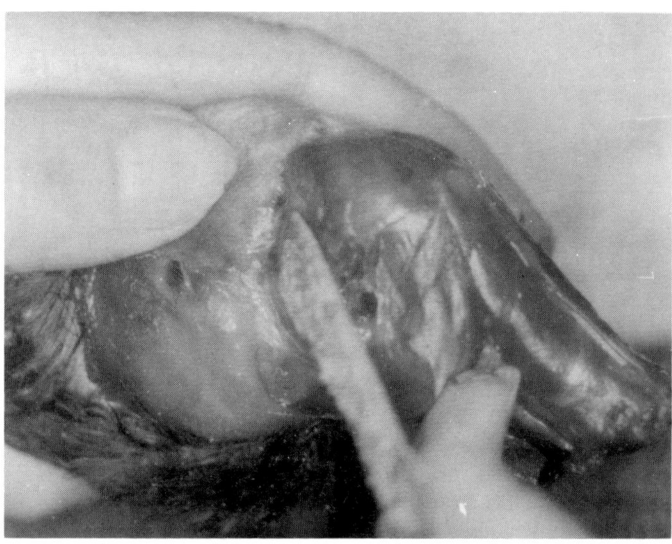

Next, the eyes will be encountered. You will usually be able to see them through the skin as you skin towards the bill. They will appear to be a dark blue color. Be "very" careful in this area. The eyes must be skinned without cutting through the eye membrane.

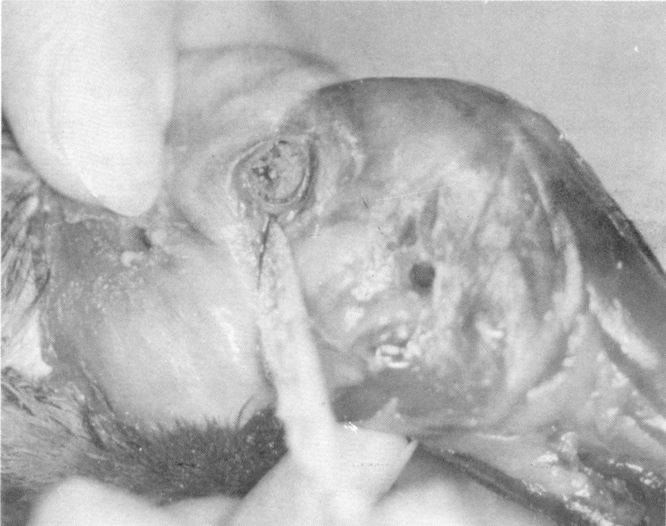

Go slow in this area. Once past the eyes, continue skinning to the very edge of the bill (where the feathers stop).

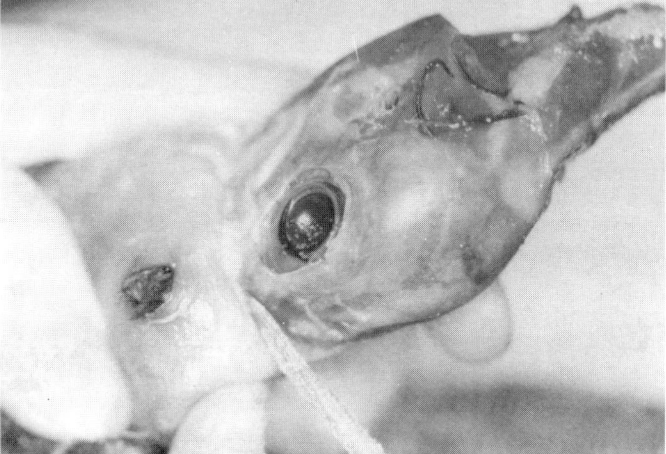

Be sure and skin as close to the edge of the bill as possible without actually severing the skin. Many taxidermists who use the natural skull do not skin it far enough! This will cause distortion and shrinkage and will attract insects. Once the edge of the bill has been reached, the skinning procedure is completed and the skull is ready to be cleaned.

Artificial Head Incision

If an artificial head is to be used, the process is greatly simplified. In fact, it is not necessary to make a neck or head incision at all, nor is it necessary to clean the skull. (A tough job!) It is also much easier to clean and tumble a bird skin with the skull and bill removed, and finally, you have the distinct advantage of being able to completely paint the artificial bill separately before mounting the bird and not have to worry about getting paint on the feathers.

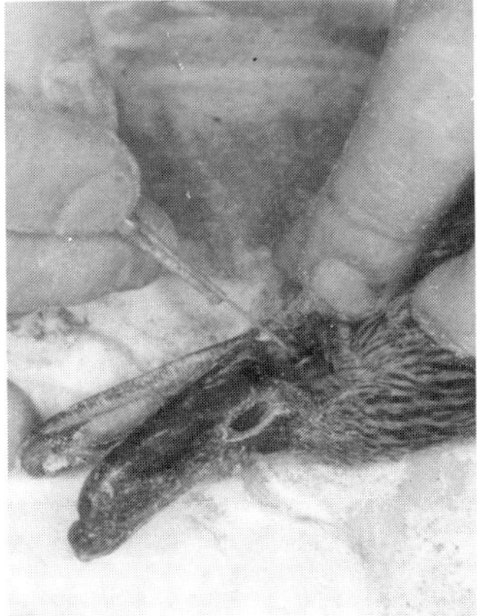 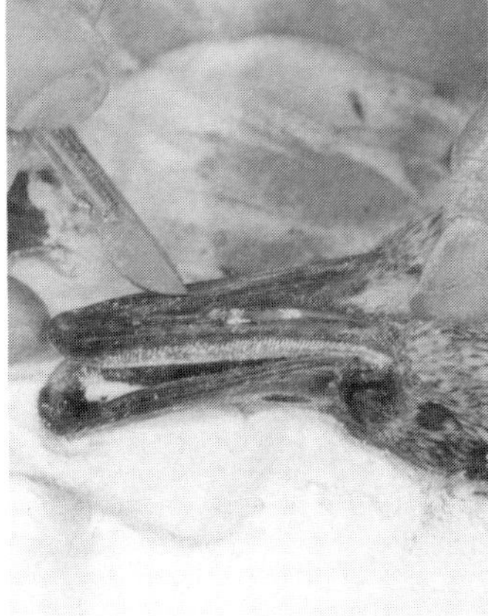

To begin the process, simply make an incision separating the skin from around the edge of the entire bill. Keep the incision as "close" to the bill as possible!

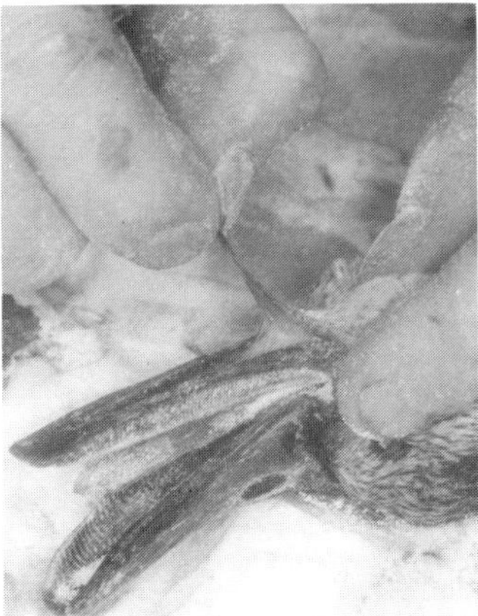 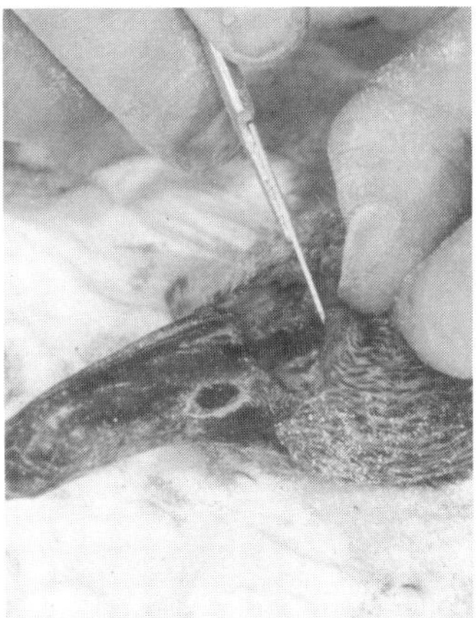

Now peel the skin back to the eyes. Next, remove the lower mandible by grasping the hinge on one side of the lower mandible with pliers and gently, but firmly, pull it out.

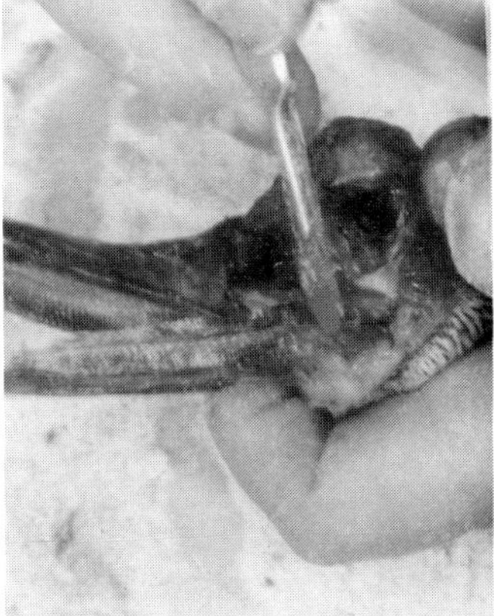 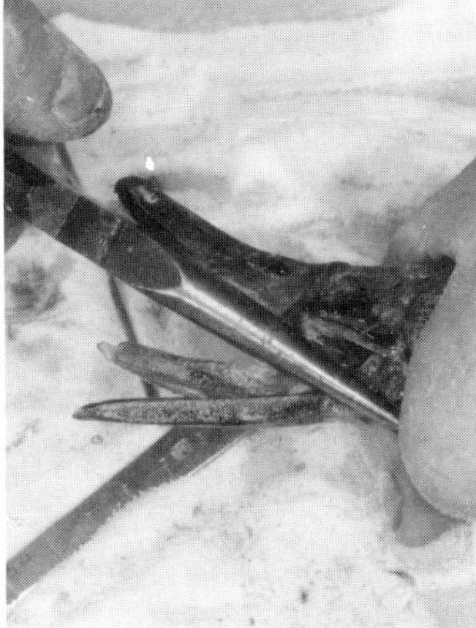

Do this *without* ripping the skin; hold the skin in place with your fingers as you pull with the pliers and it should easily come out. Now grasp the hinge on the other side of the lower

 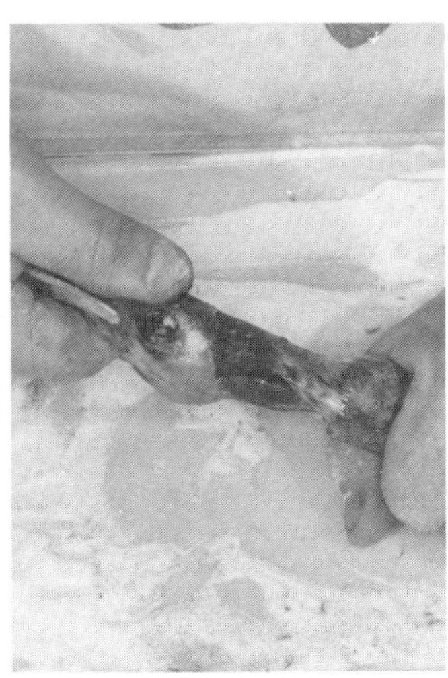

mandible and pull it out. Once this has been done, the skin can easily be slipped over the skull and removed. Another easy method is to compress and crush the skull with pliers which will also allow the skin to easily be pulled over the head and removed from the neck. That is all there is to it! The skinning is now complete and attention can be given to mechanically cleaning the skin and preparing it for chemical cleaning.

Mechanically Removing Flesh and Fat

The most common tools used by taxidermists to implement the mechanical removal of flesh and fat from the skin are:

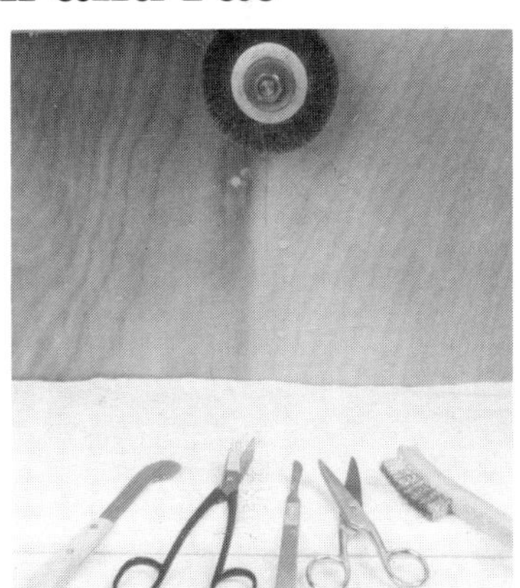

1. Ultimate Scissors
2. Knife with serrated edge
3. Scalpel
4. Curved Scissors
5. Wire brush
6. Wire wheel

The most thorough and efficient method is to utilize a "combination" of all of these tools.

Skull Cleaning

If you choose the natural skull method, the skull and head area must be cleaned first so that the head can be inverted back to the feather side out as soon as possible. Inverting the head back to feather side out will help to avoid drying out the delicate tissues surrounding the eye. It will be advantageous for you to apply *True-Tan*™ Skin Conditioner to the flesh side around the eyelids and head area before reinverting. This will keep the eyelids pliable and ready for mounting.

To clean the skull, cut a small hole in the base of the skull and remove the brain. The brain matter may be removed by scraping the brain cavity with a hook scraper or wing cleaner. Generously apply borax into the brain cavity to absorb as much fluid as possible. Continue to scrape inside the skull until *all* of the brain matter and fluid are removed.

Carefully cut away the meat (*only* the meat) from the mandibles. By removing just the meat and leaving the ligaments attached to the mandibles, you will keep the basic configuration of the skull intact. Once the skull is cleaned and excess membrane and fat are removed from the skin, coat the head and eye area with *True-Tan* Skin Conditioner and invert the head back to the feather side out.

Cleaning the Skin

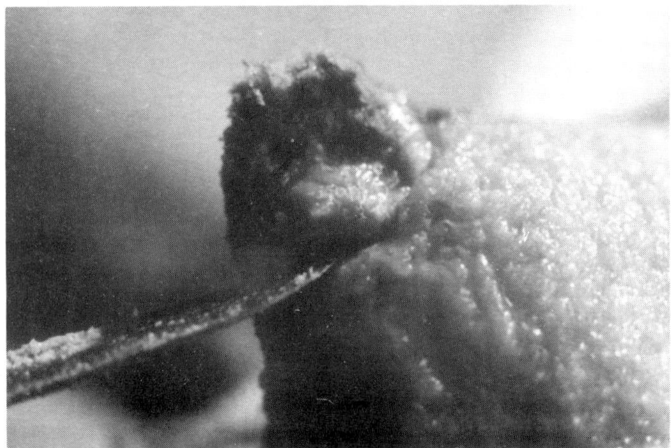

To begin removing the meat and fat from the bird skin first remove the meat and oil glands from the tail area. The oil glands are located on top of the tail quills (pygostyle). Carefully cut the glands close to the skin with the scalpel. The fat and meat must then be scraped and removed from the tail quills.

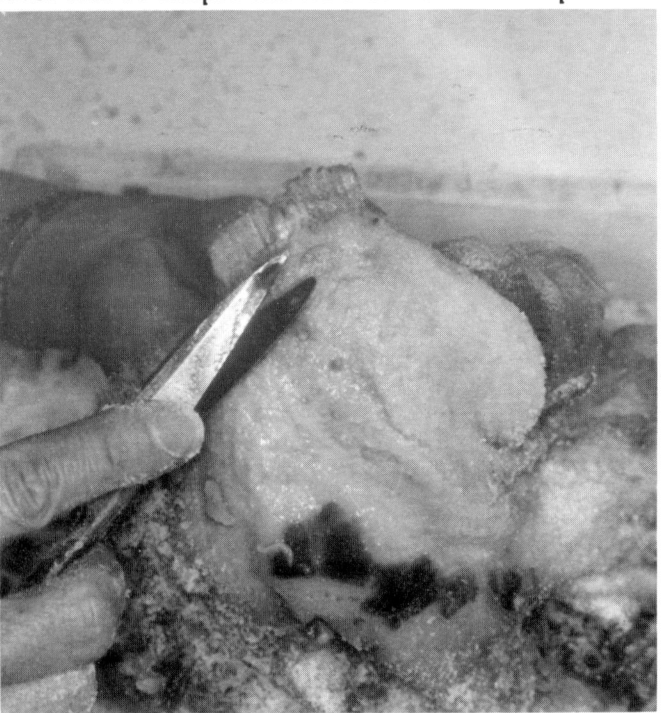

Cut and scrape between each "quill" to remove all of the material.

Now proceed to the leg area. Scissor, scrape, and remove all meat from the tibia.

Clean the bone marrow by inserting a wire down the center of the bone. Thoroughly preserve it with borax or dry preservative and invert it back to feather side out.

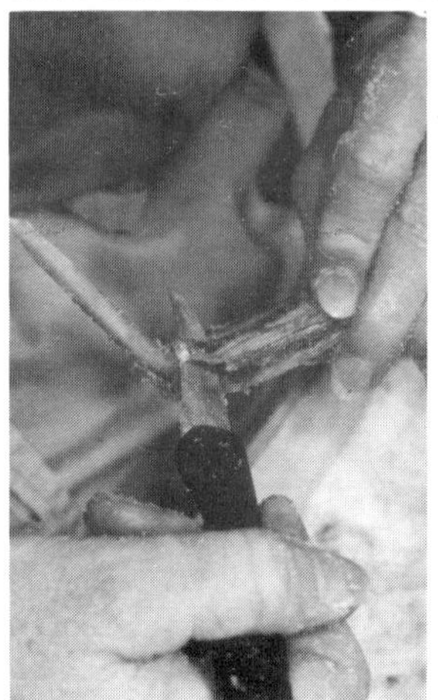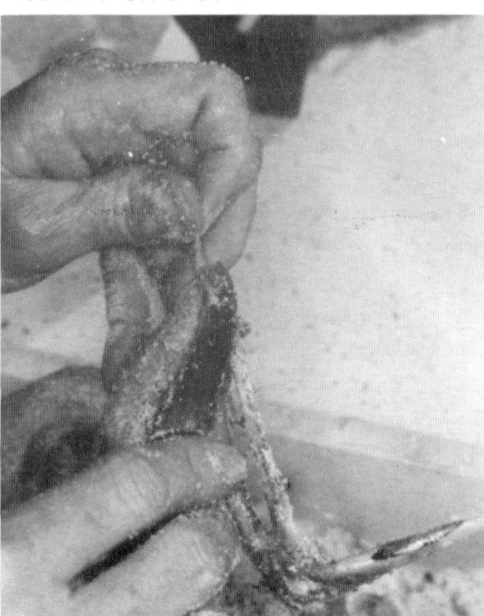

Next, finish skinning the wings down to the wrist (carpal) joint. Remove all of the meat from the humerus, radius, and ulna. Generously apply borax or preservative to the wings to facilitate removing the meat and tissue. *Care* should be taken not to cut the tendons attaching the humerus to the radius and ulna.

It must be emphasized again that it is very important to remove *all* meat and fat from a bird skin. Any meat or fat remaining on the skin will cause problems of *major* proportions. It will inhibit the penetrating action of the degreasers, preservatives, and tanning agents, and drastically affect the longevity of the mount. An improperly prepared skin invites insects, unpleasant odors, "grease bleed" and eventually skin rot. It is important to remember that there are no short cuts in this operation. *All fat and meat must be removed from the skin before proceeding to the next steps of the mounting procedure.*

Always begin the defatting process of the skin at the base of the tail and work towards the head of the bird. The feather roots are embedded in the skin and positioned with the base of the quills pointing towards the head (anteriorly). Cleaning against the quills will cause the Ultimate Scissors or the wire wheel to snag and possibly tear the skin. Even when hand scraping the fatty residue from the skin, for the most effective action, always scrape from the *tail towards the head.*

With one hand placed under the bird skin, and with fingers elevating that portion of the skin that is being defatted, use the Ultimate Scissors to remove large deposits of fat and any remaining meat. Proceed to delicate areas such as the base of the tail or areas that are difficult to clean with the wire wheel.

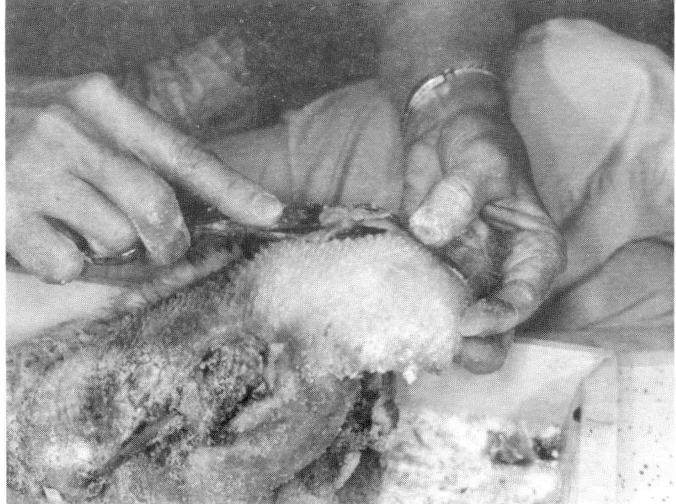

With practice, the ends of your fingers will tell you if you are applying too much pressure to the skin with the Ultimate Scissors. By perfecting this technique you will eliminate cutting holes in the skin with the scissors.

Once the large fat deposits are removed and the delicate areas are cleaned, the skin will be ready for general defatting with the wire wheel.

We advise that you use a spare bird skin for a practice run before attempting to clean a customer's bird with a wire wheel. The wire wheel is easy to use, *but* it does require practice to develop a "feel" for the amount of pressure that needs to be applied in order to avoid tearing a hole in the skin.

Always hold the skin firmly in one hand and place the other hand under the skin. It is *important* to remember to hold the skin *firmly* with one hand. This is because the rapidly turning wire wheel can actually "grab" the skin and take it out of your hand and before you realize what has taken place, the entire skin will be wrapped around the wire wheel and be spinning wildly. This will cause a heart "murmur" for sure! (Take my word for it.)

Begin defatting with the wire wheel at the side of the tail quills.

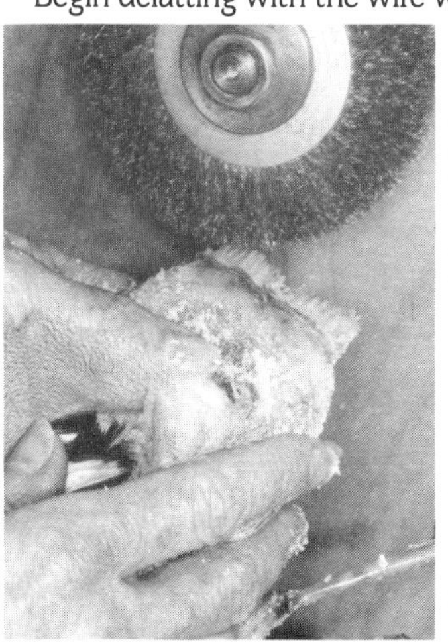
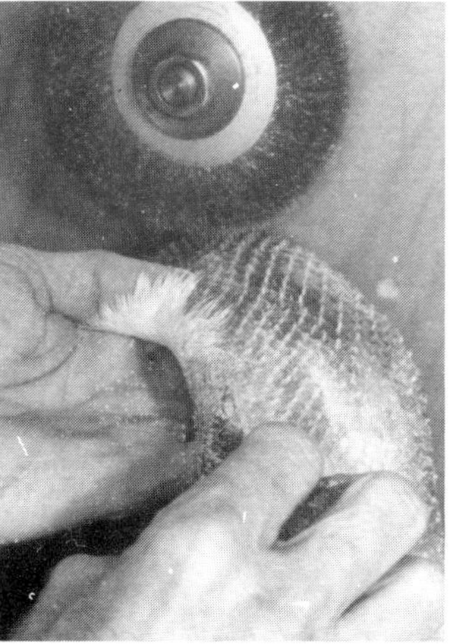

Gently place the skin against the wire wheel and slowly proceed towards the head of the bird. Pay particular attention to the edges of the breast incision. Very gently place the wheel along the edge of the incision; care should be taken not to apply too much pressure as the incision will tear quite easily. Also, you need to understand that the delicate and already thin bird skin can easily be "overthinned" with a wire wheel and/or scissors. It is a very common mistake for beginning taxidermists to use a wire wheel excessively and thin the bird skin so much that it is unmountable. When one side of the skin is cleaned, proceed to the other side, again starting at the tail section.

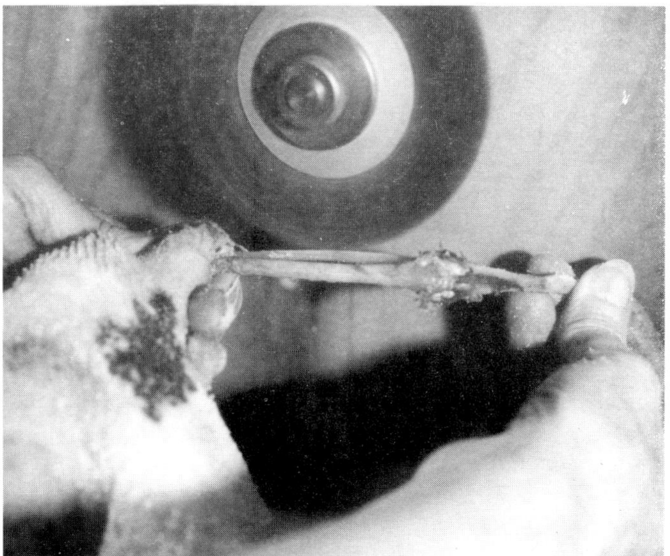

Be sure to clean the wing and leg bones.

Wire Wheel Heat

Friction caused by the wire wheel rubbing against the skin produces heat. *Do not overheat the skin.* The wire wheel can cause heat build up which will literally *ignite* bacteria growth. (Most bacteria actually live inside the fat and oil pockets, deep inside of the skin and even a slight amount of heat will stimulate their rapid growth.) The rapidly growing bacteria stimulated by heat will produce enzymes that will very quickly dissolve the skin and cause loss of feathers and a ruined mount. To avoid this problem, have a bowl of Bacteria Stat treated ice water by your side to cool the skin with.

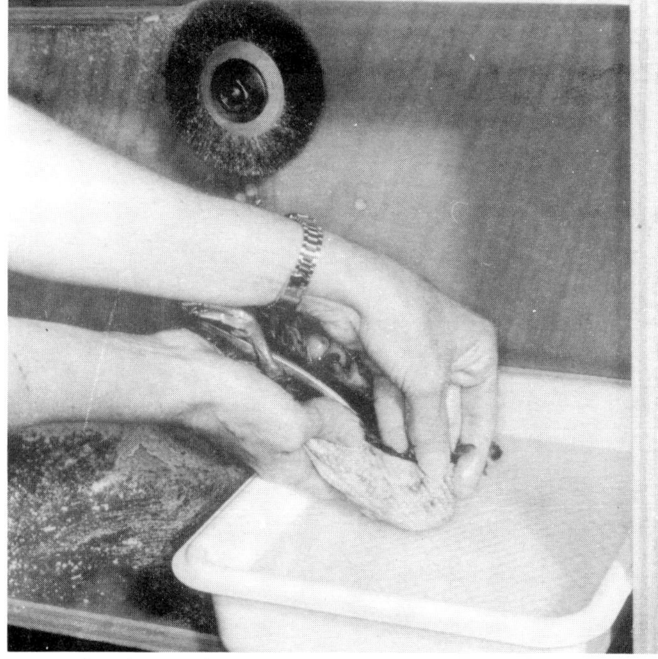

Dip the skin in the ice water *often* while wire wheeling any skin to prevent this very common problem. Dip the skin in the ice water—defat—dip again—defat and so forth.

When both side sections of the skin are cleaned, proceed to the back area. Pay particular attention to the scapular feather group area. Caution: The skin in this area is usually very, very thin and extreme caution must be taken to avoid making holes. Be gentle! Now that the bird skin is cleaned, use the wire wheel for a final "once over" on the leg bones, wing bones, skull, (if attached) and tail areas. Once this is done, the wire wheel procedures will be completed.

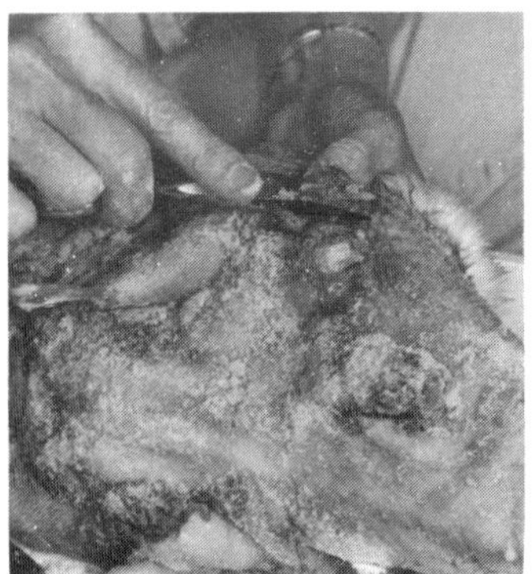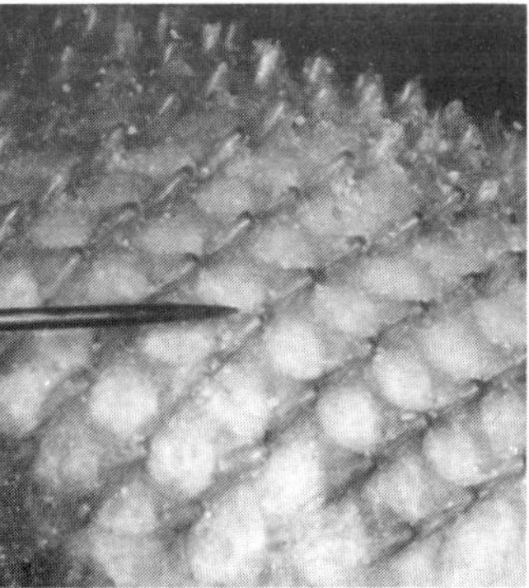

Notice this comparison of the raw skin (left), with the mechanically cleaned skin with the fat removed and quills exposed (right).

As a final step to the mechanical removal of any remaining fatty residue, use a knife with a "dull" serrated edge and gently scrape any fatty residue from the skin that might have been missed. The WASCO defatting knife works well for this operation.

With a little practice, the mechanical flesh removal and defatting process can be accomplished quite quickly and efficiently. Some important things to remember are:

1. Always start at the tail section and proceed towards the head.

2. Use Ultimate Scissors to remove large fatty deposits.

3. Use WASCO granulated borax or WASCO dry preservative during the process to help remove and absorb fluids and fatty residue.

4. Always hold the bird skin firmly when using the wire wheel.

5. Keep the skin that is being defatted by the wire wheel operation *cool*. This is done by periodically dipping that portion of the skin in ice cold Bacteria Stat treated water.

6. If you must stop the process, place the bird skin in a towel and put it in a refrigerator, or refreeze it in a plastic bag if longer than overnight.

Chemically Cleaning the Skin

The chemical treatment of bird skins removes fat, oils and soluble proteins remaining on the surface and deep within the skin. The skin can then be tanned or leatherized, which chemically changes the structural proteins within the skin into stable fiber. Simply put, the system flushes out the substances that are not needed and tans or leatherizes those that are needed. The *True-Tan*™ Skin Conditioner also "drives" and "stuffs" lubricants deep within the skin. These lubricants help to replace the substances that were removed and their presence help to prevent shrinkage.

The chemical treatment that we recommend for the professional treatment of bird skin is the *True-Tan* system. It consists of the following steps:

1. Salting
2. Degreasing with *Polytranspar*™ Degreaser (outer skin treatment)
3. Degreasing with Grease-Buster (inner skin treatment).

4. Washing with Skin Prep
5. Brine
6. Edolan-U mothproofing
7. Mild acid pickle
8. Tanning
9. Skin Conditioner

Dry Preservative vs. Tanning

Taxidermists have used dry preservative for hundreds of years to mount birds. Many of you may wonder why we recommend the above system instead of dry preservative. First, let us say that just because taxidermists have used dry preservative for hundreds of years does not necessarily mean that it is good. Remember, they used arsenic prior to Edolan-U. Keep in mind that bird skins are no different than mammal skins, in that, if they are tanned, rather than simply dry preserved, they will last longer and experience far less shrinkage.

Dry preserved skins whether mammal, reptile, fish, or bird are essentially "rawhide" and will return almost entirely to a raw state if subjected to moisture. Additionally, they will shrink excessively making feathers hard to control and will create distortion of the skin. Remember that a hide is comprised of two-thirds moisture and once the dry preservative removes it, very little replaces it. In reality, approximately one-third of the skin is all that remains after dry preserving. Common sense will tell you that the skin will be badly shriveled and shrunken if two-thirds of its original composition is removed and not replaced.

Further compounding the problem is the fact that hide paste is not used when dry preserving a bird skin. Hide paste and hide nails are used to hold the rapidly shrinking skin in place in dry preserving other species such as mammals. Bird taxidermy seems to be the *only* area of taxidermy where hide paste is not used to glue the skin to the mannikin. Why? Because it is messy, but you can be assured that it is just as important to hold a bird skin in place as it is a mammal or any other species (especially if dry preserving).

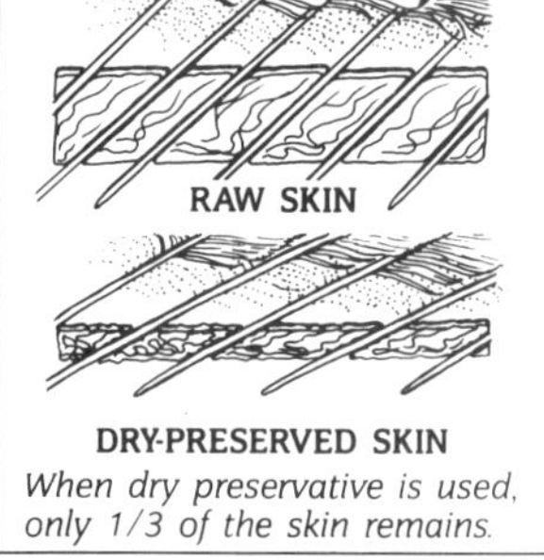

When dry preservative is used, only 1/3 of the skin remains.

There is definitely a better way! Each of the *True-Tan* steps has a definite purpose and should not be omitted if you want to obtain the best results. Later in this chapter we are going to detail how to achieve as good of results as can be obtained using dry preservative if you insist on using that method, but we strongly advise going to a tanning system.

Salting

Salt removes many of the soluble proteins from the skin, such as blood, and also drives out the moisture. It also helps in the chemical process of tanning by preparing the protein fibers for tanning.

Once the skin has been mechanically cleaned, apply an even layer of WASCO salt or non-iodized fine grade substitute to the flesh side of the skin. It is not necessary to salt bird skins for nearly as long of a period of time as mammal skins because they are so thin. Usually an hour or two will suffice. (You may salt them longer if so desired.)

Degreasing

After salting, the skin should be degreased. First, shake the salt off the skin. Two types of fat and oils are contained in a bird skin. Adipose, or "outer" fat, and interstitial or "inner" fat. *Both* must be removed.

The outer fat is the first that needs to be removed. All birds have adipose fat. Waterfowl and other water birds have large concentrations of it because it is a waterproofing agent for them.

Since it is a waterproofing agent it will also prevent chemicals from becoming absorbed into the skin, and therefore, must be removed. Adipose fat is easily removed by simply immersing the skin in *Polytranspar*™ Degreaser and allowing the solvent degreaser to dissolve it. Always add Bacteria Stat into any solution to help prevent bacteria growth. Soak the skin in the degreaser, moving it around occasionally from 20 minutes for slightly greasy birds, to an hour for very greasy birds. Take the skin out, gently squeeze it, and it is ready to have the inner fat removed.

Inner fat is comprised of pockets of fats and oils located deep within the skin. Surfactants such as Skin Prep, and solvent degreasers like *Polytranspar* Degreaser cannot effectively remove interstitial fat.

Grease-Buster will remove it! This product was specifically chemically formulated to penetrate the skin and dissolve these

pockets of fat and oils. To use it, simply make the solution by following the directions on the label and then immerse the skin into the solution. Remove the skin, rinse it thoroughly and it is ready for washing.

Washing

After degreasing, the skin should be washed in a solution of one-half of a cap full of Skin Prep and 1 Bacteria Stat tablet and clear cold water.

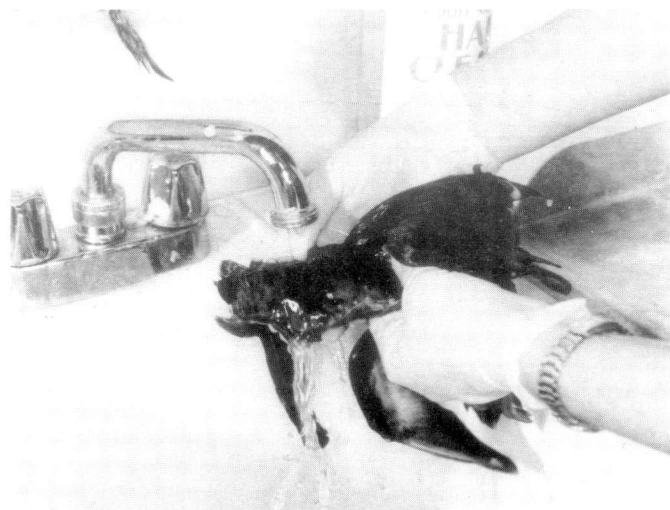

Once it has been washed and the skin is clean, then the skin should be thoroughly rinsed until no soap residue remains.

Tanning/Leatherizing

Simply put, tanning/leatherizing a skin is the process whereby you chemically change the structural proteins of a bird skin into stable fiber. After the initial cleaning of the skin, mechanical and chemical defatting, dry salting, and washing, it is a four step process to "tan" a bird skin. There is no question about it, it will be more time consuming and trouble to tan a bird skin versus dry preserving one! The same is true with mammals and fish.

"You" must make the decision as to whether or not it is worth it to tan or dry preserve. It is "our" opinion that tanning is much better because it is a more permanent process and the skin will experience far less shrinkage if it is tanned. In other words, you will produce a better, longer lasting product if you "tan" versus dry preserving the specimen. Be aware though, that it will cost more and take more time to achieve the better results of tanning.

Brine

Tanning is a four step process. First a brine is prepared to prevent the skin from acid swelling. Prepare the brine by adding one Bacteria Stat tablet and 1½ cups of salt to one gallon of "cold" water. Allow the skin to soak approximately 30 minutes. The skin should become completely pliable while in the brine.

Mothproofing

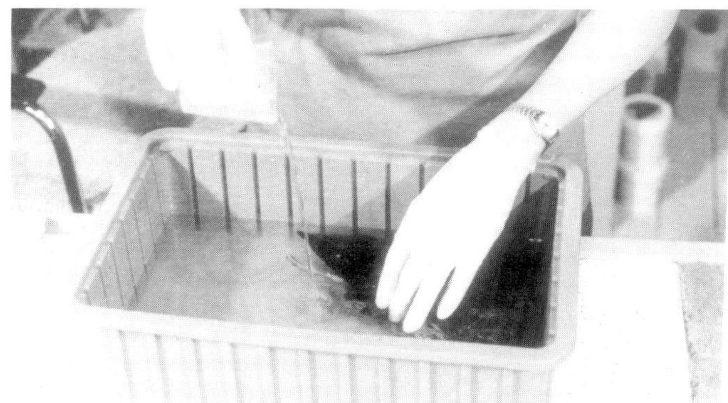

At this time, the bird should be mothproofed (treated with Edolan-U). Calculate 1% of the dry weight of the bird and that is how much Edolan-U is required to be put into the brine solution to mothproof the skin. Although mothproofing is not related to tanning in any way, it is a "necessary" process that can be done simultaneously with the tanning process and is best performed when the skin is in the brine (before the acid is added).

After the Edolan-U, which is a colorless odorless dye, is put into the brine, the acid is added a tablespoon at a time until a pH of 4.5 to 5 is reached. Remember, *it is very important that you add the Edolan-U to the brine first, and then add the acid*. At this point the dye will permanently attach itself to the keratin protein of the feathers.

Insects detest the taste of the "dyed" feathers thus it provides a permanent deterrent. (NOTE: If the skin has not been properly cleaned and prepared, such as is the case with freeze drying "flesh," insects have been known to attack the specimen in spite of treatment with Edolan-U.)

Acid Pickle

True-Tan Bird Skin Acid Crystals, which are a very mild acid, are added to the solution a tablespoon at a time. You will probably need to add about three tablespoons of the Bird Skin Acid Crystals to the solution to obtain the proper pH. Stir and then check the pH; it should read between 4.5 and 5. You may need to add more or less acid according to the condition of your water and the skin. Once the proper pH is obtained it will take about 45 minutes for the Edolan-U to chemically bond with the keratin protein in the feathers. The Edolan-U will bond faster if the temp-

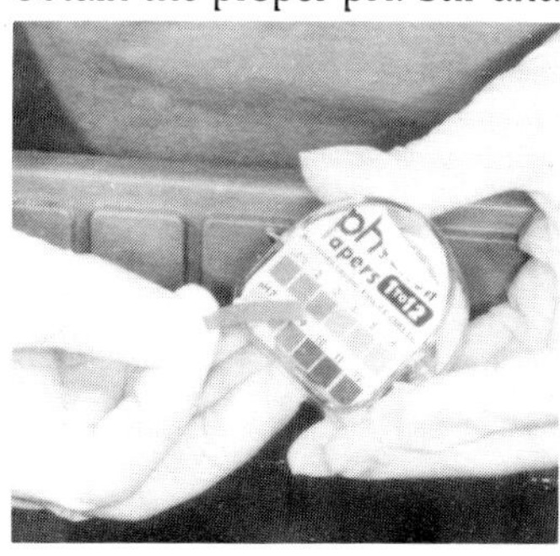

erature of the water is elevated, but the risk for generating bacterial activity is too great at warm temperatures. *Always use cold water* and wait the 45 minutes. (It will take at least an hour for the skin to become pickled anyway.)

The acid is used to further dissolve and remove the soluble proteins located within the skin. The acid will also break up the natural chemical bonds of the proteins and prepare the fibers chemically to develop an affinity for bonding with the tannins.

Acid dissolves soluble proteins and leaves the structural proteins.

Tanning

After allowing the skin to soak in the acid pickle for an hour, the third step is the actual chemical "bonding" of the tannins with the structural fibers located within the skin. After pickling

and mothproofing, add one ounce per bird of the *True-Tan*™ Bird Tan to the solution per the instructions on the container,

stir it thoroughly, and then allow the skin to soak in it for at least one hour. The points where the bonds of the natural proteins are broken down by the acid provide excellent bonding points for the tannins that are contained within the bird tanning system to attach themselves to.

After one hour, slowly add one tablespoon of baking soda to the solution to raise the pH back up to six. Allow the skin to soak for about 30 minutes more in this solution.

Lubrication

The fourth and final step after the tannins have bonded, is to "stuff" the skin with lubricants (*True-Tan* Skin Conditioner). This product penetrates into the skin and lubricates the fibers. This prevents them from becoming glued to each other and also "stuffs" and fills the skin with their presence, which prevents the fibers from collapsing and experiencing shrinkage of the skin.

Lubricants "stuff" the voids left between the fibers of the skin.

Remove the skin from the solution and gently rinse the feathers for a couple of minutes to remove any residue. (Note: If too much Edolan-U was used, it will leave a sticky residue on the feathers. This should be removed with methanol and/or soapy water.)

Wrap the skin in a clean, soft, fluffy towel to absorb excess moisture. Now tumble it in hardwood sawdust with glaze additive. Remove it from the tumbler and gently blow the feathers dry with low pressure compressed air ("dry" air—use a moisture trap) or a "reversed" vacuum cleaner.

It is then ready for the *True-Tan* Skin Conditioner. The conditioner is applied to the flesh side (not the feathers) with an artists' brush and is allowed to penetrate for at least 30 minutes (or it can sit for as long as overnight). If it is a longer period before you plan to mount it, then the skin should be refrigerated or refrozen. It can also be allowed to completely dry out. If so, it can be soaked back up at a later date and more conditioner added prior to mounting.

At this point the skin is leatherized, mothproofed, and ready for mounting. A skin treated in such a manner, "with proper care," should last "several" lifetimes!

Summary

The *True-Tan* Bird Tanning "system" is actually very easy to use. It "appears" as if it is more trouble than it really is. The summary looks like this:

Step 1. Mechanically defat the skin with Ultimate Scissors and a wire wheel.

Step 2. Thoroughly cover the flesh side with WASCO salt for at least one hour.

Step 3. Chemically degrease the bird by first soaking the skin in *Polytranspar*™ Degreaser for 20 minutes and then soak it in *Grease-Buster* and cold water for 20 minutes.

Step 4. Wash in Skin Prep and rinse in cold clean water until "squeaky" clean.

Step 5. Prepare a brine by adding 1½ cups of salt to one gallon of cold water and add 1 Bacteria Stat tablet. Soak the skin for 30 minutes in the brine.

Step 6. Add 1% of Edolan-U of the dry weight of the bird skin to the brine.

Step 7. Add approximately three tablespoons of Bird Skin Acid Crystals until the pH is lowered to 4.5 (if more acid is needed to reach 4.5, then slowly add more acid until you reach it) allow the skin to soak for one hour in the pickle. (It will mothproof while in the pickle.)

Step 8. Add one ounce of *True-Tan* Bird Tan to the solution (per bird), stir well, and allow the skin to soak for one hour, occasionally moving it about in the solution. After one hour add one tablespoon of baking soda to the solution to raise the pH back up to 6 and soak for 30 minutes.

Step 9. Rinse in cold water and tumble dry in hardwood sawdust with glaze additives.

Step 10. Use low pressure compressed air or a "reversed" vacuum cleaner to blow the feathers "completely" dry and paint a thin coat of *True-Tan* Skin Conditioner to the flesh side with an artists' brush. Allow the conditioner enough time to penetrate the skin.

Step 11. Now proceed to mount a "world class" bird, or freeze until you are ready to mount. The skin may also be allowed to completely dry out. If so, rehydrate and reapply a small amount of conditioner when ready to mount.

Dry Preservatives: Are They Effective? You be the Judge!

"Are dry preservatives effective?" The answer to this question cannot be determined until a couple of things are cleared up. First and foremost, the term "dry preservative" must be "defined," and secondly, it must be determined just exactly what is meant by "effective." In other words: What is dry preservative? And what is the product expected to do?

Dry preservative can be defined for the purposes of this chapter as a powdered chemical substance, that if properly applied to a skin, will temporarily preserve and protect the skin from bacteria growth. Some dry preservatives also contain chemicals that will, to some extent, discourage moths from attacking the specimen. Tanning, or changing the chemical composition of the structural proteins of the skin to a stable substance, or fiber, is not accomplished (to any significant degree) with dry preservatives. A skin treated with dry preservative is essentially "rawhide," and will return almost entirely to its raw state if subjected to enough moisture; a tanned hide will not!

To help keep the moisture at an acceptable level, dry preservatives are loaded down with desiccants. Desiccants or "drying agents," are chemicals that absorb, remove, and help to prevent the recurrence of moisture in the skin. It is easy to understand how this system got its name, "dry" preservative, as it works primarily because of its drying action. This is good, as bacteria must have moisture in its environment in order to exist. Without a sufficient amount of moisture present, bacteria cannot survive and decomposition or "rotting" will not occur. Thus the desiccant-laden dry preservative effectively "preserves" the skin and prevents it from rotting by keeping the skin dry.

There are many desiccants on the market, most notable of which is a chemical with the common name of borax. There are "*numerous*" different chemical formulations and variations of the borate family. Some are anhydrous, i.e., without water, while others are not. Please understand that borax varies and that the standard 20 Mule Team borax is not necessarily the same formulation as others. Borax is chemically comprised of many different ratios of ash, boron, sodium sulfate, calcium, magnesium, iron, and phosphorous. If the proper formulation and particle size is used, borax can be an excellent desiccant. Various formulations of borax have been effectively used for taxidermy "preservation" purposes for years. Boric acid is another common desiccant and is considered to be a mild antiseptic as well. It is often used by chemists as an additive and is used with other desiccants and chemicals to make a more effective dry preservative. Many taxidermists have used boric acid for years. There are various silicates and many other chemicals that are also used as desiccants. "All," however, share the common property of being "drying agents" and are effective preservatives because of that property.

So the first fact established concerning dry preservatives is that one of the main or "primary" ingredients are desiccants. These desiccants effectively remove moisture and will main-

tain a fairly moisture-free environment if properly formulated and applied.

Dry preservative is prevented from entering the skin by a natural oil barrier on the skin's surface.

Surfactants break up this natural barrier allowing the dry preservative to penetrate more easily.

The second fact that our analysis reveals is that dry preservatives contain surfactants. These are used to aid in the "application" of desiccants. Surfactants help the preservative to better penetrate the skin and allow the dry preservative to be absorbed.

Technically what happens is that the surfactant lowers the surface tension of the liquid or moisture in the skin; this counteracts the natural "repellent" characteristics of the oil in the skin and allows it to be easily absorbed. Simply put, a surfactant "wets out" the preservative and allows it to easily penetrate and be absorbed into the skin. WASCO dry preservative contains an ingredient that is designed to combat and repel insects (primarily moths). Treating skins with Edolan-U is far superior to these chemicals for this purpose, but the chemicals in dry preservative certainly will not hurt anything and we consider their being there a definite plus!

WASCO dry preservative also contains a few tannins. The tannins are derived from the salts of various metals. It is doubtful according to our tests whether there is any "great" degree of permanent bonding of the tannins to the structural proteins. Several factors are needed to ensure bonding, one being salting, and the other being treatment with acid. The salt and acid processes "flush out" the skin and break up the bonds of the structural fibers. This action opens up attachment points for the tannins. As a rule, neither salting nor acid pickling is used with dry preservatives; if you did you might as well tan them. Without using these steps and combined with the fact that they are applied in a dry, powdered state makes it highly doubtful that any acceptable degree of permanent bonding is likely to occur. This is further reinforced by the fact that as soon as any of the dry preservative treated skins in our tests were soaked up, the skins returned to their raw state. The *True-Tan*™ treated skins did not.

Shrinkage

To our way of thinking it is a definite disadvantage that a dry preserved skin is not stable and that it will return to its raw state if subjected to moisture. However, most mounts will be protected from moisture, and stability is not as an important of a consideration as shrinkage in dry preserved skins. The **biggest disadvantage** to dry preserved skins is probably the **shrinkage factor.**

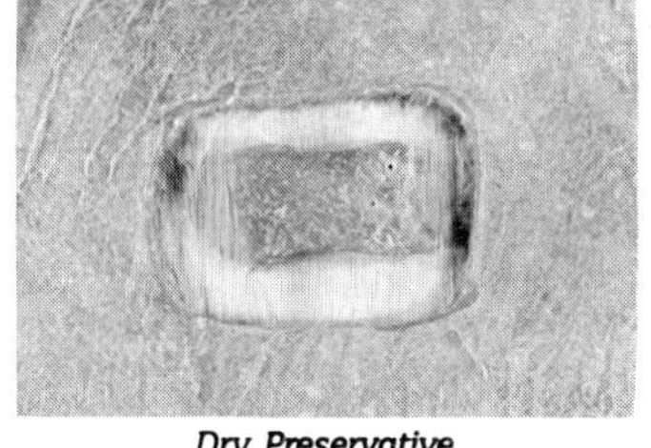

Dry Preservative

Thin shaved hide True-Tanned

Look at the comparison photos of these dry preserved and tanned mammal skins. Awesome isn't it? Leatherizing a skin will effectively prevent the massive shrinkage experienced with dry-preserved skins. Tanning accomplishes this by "stuffing" the skin with tannins and lubricants.

A good analogy of this process would be to picture a very leafy tree as a fresh skin. *True-Tan* calls for salt treating and then acid pickling a skin. These processes can be pictured as being similar to defoliating, or removing all of the soft green leaves, from a very leafy tree until nothing remained but the bare hard limbs. (See illustration below.)

The salt and acid effectively "defoliates" or draws out and dissolves the blood and various other soluble proteins, fats, oils, etc. (soft green leaves) that are located within the skin but leaves the structural proteins (hard, bare, limbs) intact. When the *True-*

Tan™ is then applied, tannins go in and attach themselves to the limbs (structural proteins) where the leaves were previously attached, and the lubricants also stuff themselves in the areas in-between. This prevents the limbs from collapsing and becoming glued together, thus eliminating any great degree of shrinkage of the skin. (See illustration above.)

Comparing dry preserved skins to this same analogy presents several major problems. First, the dry preserved skin process does not effectively "defoliate," or remove, as many of the soluble proteins and fats from the skin, because salt and acid are not used. Simply rubbing dry preservative on the surface of the skin cannot remove enough of the substances inside of the skin that need to be removed. Salt and acid are needed in order to do an effective job! The failure to remove these substances

can cause several problems. Through a natural process some of the soluble proteins, fats, and oils that are still left in the skin will turn to "natural" acid which will weaken the structural portion of the skin. In fact it can actually completely destroy parts of the skin in severe cases. Other substances that are left in the skin, in time, will form a natural glue, similar to the glues extracted from horses that ended up in the proverbial "glue factory," that will tightly cement the fibers together.

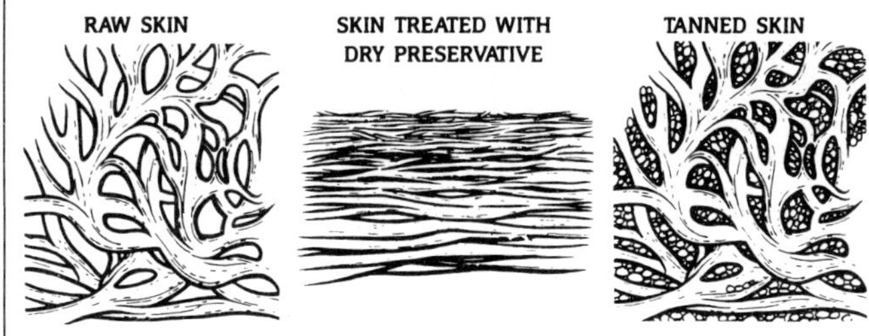

When dry preservatives are used, the voids between the corium fibers are not replaced with stable lubricants. Consequently, when the skin dries, it shrinks dramatically as the pro-collagen dries to a hard, glue-like material which cements the fibers together. The result: a shrunken, brittle hide, which could rehydrate and rot in the future. Tanning, on the other hand, floods the voids between the fibers with stable lubricants, which keeps the skin from shrinking.

Note the examples of this problem. The skin becomes harder with time and eventually completely glues itself into a solid, very hard mass.

Aside from these problems, another "primary" difficulty with dry preserved skins is caused when the moisture is removed from the skin. When the skin dries, two-thirds of the composition of the skin will have been removed. It will not be replaced with tannins or lubricants as with leatherized skins. Remember that a skin is comprised of two-thirds moisture with just one-third being comprised of structural proteins. If a tremendous amount of moisture is removed, and only a small amount of desiccant replaces it with the dry preserved method, then shrinkage most definitely will occur.

Tips to "Try" to Make it Work

If you still want to use the dry preserved method, then we suggest that you adhere to the following tips for better results:

1. Take extra care to remove all flesh, fat, and membrane, and to thin down the skin to eliminate as much shrinkage as possible. Be aware that this is somewhat dangerous, because if it is "too" thin, the skin will crack when it undergoes the tremendous shrinkage that we know will occur. Admittedly, this is a tough problem, because if the skin is not thinned enough, massive shrinkage will occur and you will have distortion of the skin, drumming, and pulling away at the eyes and detail areas; if the skin is thinned too thin, you will experience cracking.

2. Always allow the dry preservative to become thoroughly absorbed into the skin. Heavily powder the entire skin, and then wrap the skin in a towel or put it in a pillowcase and place it in a refrigerator overnight. This will help the preservative to become absorbed into the skin. Take it out the next day and shake it out and reapply before mounting.

3. Advise your customer not to get the mount wet. Remember it is still a "rawhide" and can return to its raw state. If the skin is subjected to enough moisture, bacteria will very quickly begin to multiply, and the skin can easily begin to putrefy or rot. Even "small" amounts of moisture can cause this.

4. Advise your customer to pay particular attention to avoiding putting the finished mount in places where it will be exposed to extreme temperature changes and/or humidity changes. It will cause cracking and splitting of the brittle dry preserved hide.

5. Always use a "good" hide paste and plenty of pins to *"try"* and hold the "shrinking" dry-preserved hide in place!

CHAPTER 7

The Sportsman Series Bird Mounting System

Mounting a Standing Bird with Closed Wings Using an Artificial Head and Accu-Flex Body

The development of the Sportsman Series artificial heads and Accu-Flex bird mannikins with pre-attached flexible necks has revolutionized the taxidermy industry. The advantages of using "the system" are numerous. First and foremost, the artificial heads are anatomically accurate with absolutely no shrinkage.

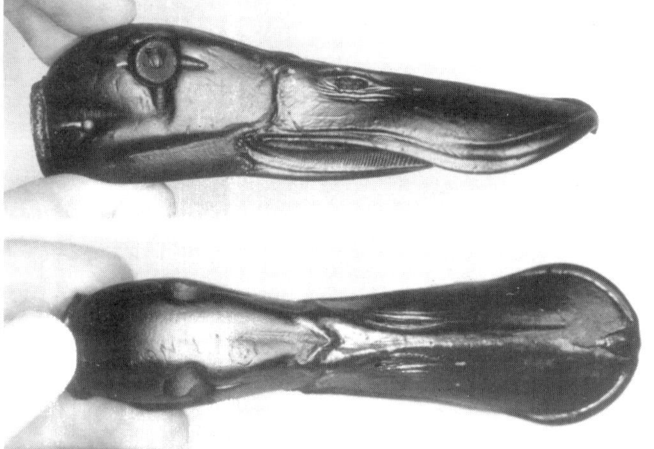

It is all but impossible and *very* time consuming to completely rebuild, a natural bill to its true dimensions with modeling compound.

In addition to anatomical accuracy, the artificial heads dramatically increase the speed of completing a mount. There is no time consuming neck incision to sew and—best of all—there is no messy skull to clean when using artificial heads. Also when the skull and bill are removed, the feathers will stay cleaner during the tumbling and mounting process. The artificial head can be attached to the Accu-Flex body and pre-attached neck in seconds. The accuracy and the time saved by not having to wrap a body and neck is truly unbelievable. Another advantage is that the head can be prepainted separate from the mount and thereby avoid getting any paint on the feathers.

In a nutshell, the incorporation of this "system" will save time and money—plus produce award winning results. One important fact to remember is that success with this system of bird mounting is "guaranteed," *if* the steps and instructions are followed exactly.

The key to assembling a "first class" mount without a great amount of difficulty and hassle is organization. All the necessary materials and tools should be assembled *before* the mounting procedure begins. Words cannot begin to describe how frus-trating it is to have to stop working on a bird and attempt to find a specific tool or supply. Not only is it time consuming, but the wasted time adversely affects the condition of the bird skin.

Preliminary Procedures

First, assemble all of the necessary supplies. This would include the Accu-Flex body, artificial head, eyes, clay, cotton, thread, preservatives, wires and references.

Artificial Head

Carefully inspect the head for any imperfections. Make certain the eye orbit will accept the proper eye size. If any alterations must be made, they should be done at this time.

Setting the Eyes

In order to capture the expression of a live bird and to avoid problems with seating depth and with recreating the delicate contours of the upper and lower eyelids, it is important to have reference pictures close at hand.

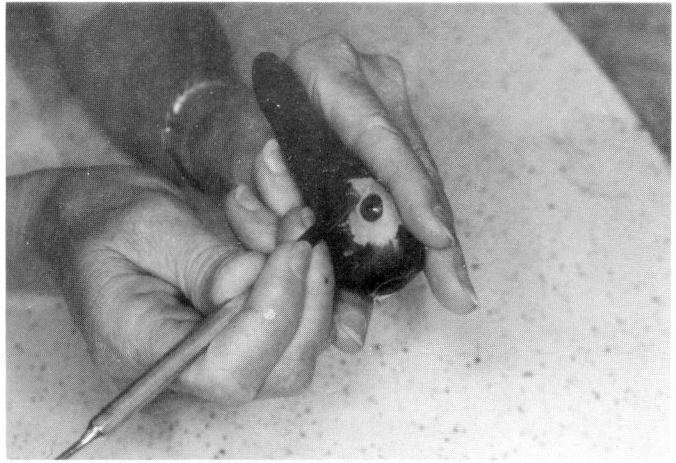

First, the eye sockets should be filled with WASCO clay or Sculpall to hold the glass eye in position. Once the eye socket is filled, carefully place each glass eye into the socket.

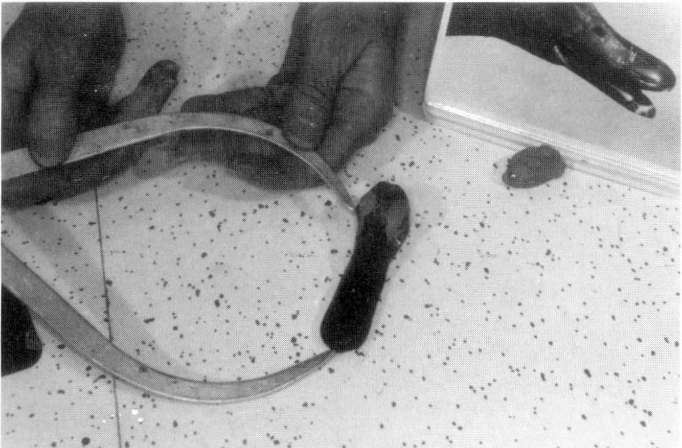

Use calipers to check the symmetry once the eye is seated in the socket.

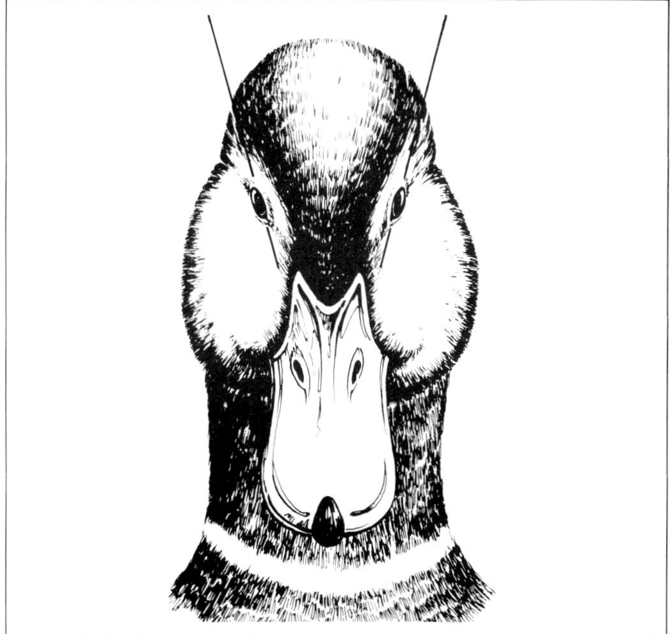

To duplicate the "look" of a live bird, the eyes should be set with a slightly forward angle and a slightly downward cant. Seating depth is also critical to give the bird a natural life-like appearance. Setting the eye too deep gives the illusion of the bird not being able to see, or worse, the appearance of being emaciated and about to expire. Setting the eyes too shallow will give the bird a "bug-eyed" or "popeyed" appearance.

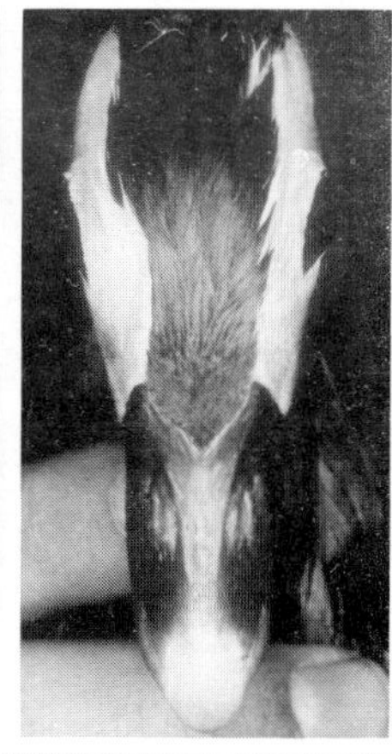

Remember, CHECK your references to be certain of the proper seating depth.

Once the eyes are set, paint the bill using a *Breakthrough* waterfowl finishing schedule for the particular specie being mounted. Painting at this point prevents overspraying airbrush paint onto the feathers. The exception to this procedure would be if Winsor & Newton artist's oils are used to paint the bill. Since overspray is not a factor when using oil paints, the bill may be painted after the bird is mounted and has dried if so desired.

Accu-Flex Mannikin

The Accu-Flex mannikin should be inspected and compared to the actual carcass to ensure a proper fit. All Sportsman Series mannikins have a unique "lock-in" system of wire insertion points to accurately locate each section of the bird. Sharpened wires should be inserted through the marked "lock-in" insertion points on the mannikin to form pilot holes for the wires to be run through later in the mounting process.

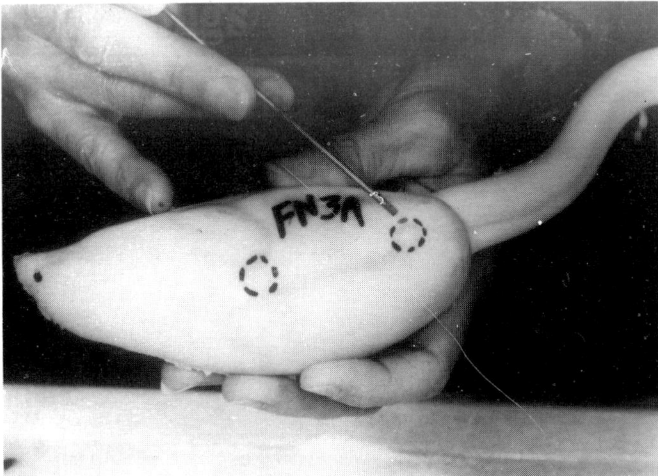

Inserting a wire through these points "before" the bird is assembled will greatly facilitate installing the wires that are attached to the humerus and tibia later in the mounting process.

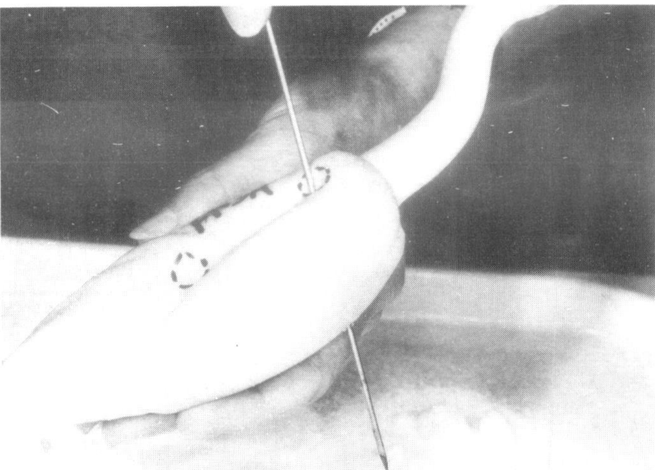

To prepare pilot holes for the wing wires, insert and angle each wire so that it exits the body on the opposite side of the lower quadrant of the breast and then remove the wire.

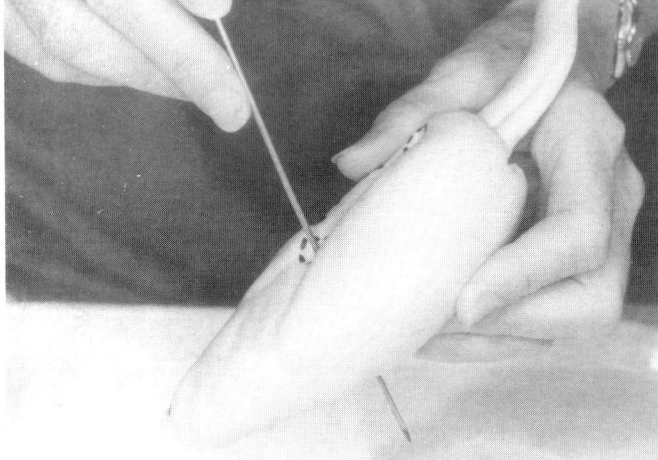

To prepare pilot holes for the leg wires, simply push the wire completely through the mannikin at the designated insertion point and then remove it.

Leg, Wing, Tail and Neck Support Wires

Wires must be cut and prepared for the legs, wings, tail and neck. The leg wire should normally be cut approximately twice the length of the leg bone (digits, tarsus, and tibia).

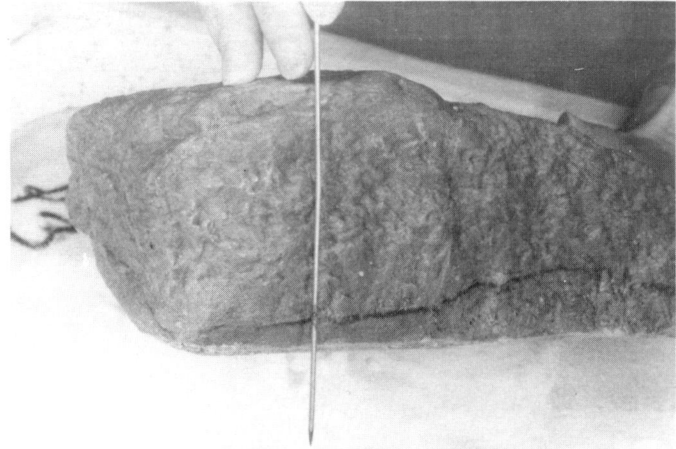

At this time, consideration should be given to the particular base that the bird will be attached to. If the base is unusually thick, additional leg wire length should be allotted to compensate for this. The wire should be of a heavy enough gauge to completely support the bird without any sign of wobbling.

WIRE SIZE CHART
8 ga.● 10 ga.● 12 ga.● 14 ga.● 16 ga.● 18 ga.● 20 ga.●

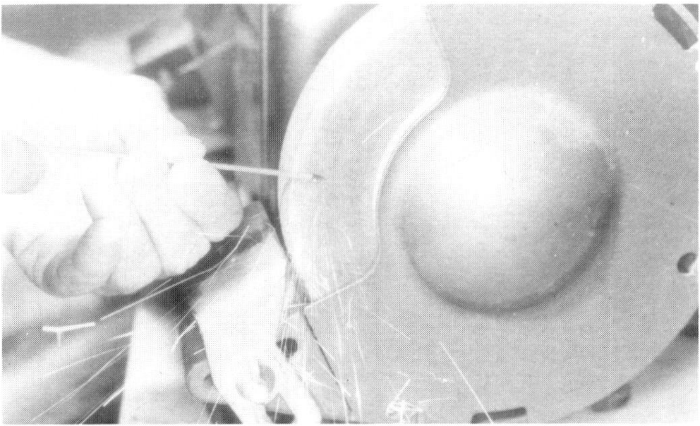

The leg wires should be sharpened at both ends with a bench grinder or file. Do not make the points too sharp, because when the wire is being run down the tarsus it will have a tendency to snag along the delicate tissue of the interior of the leg.

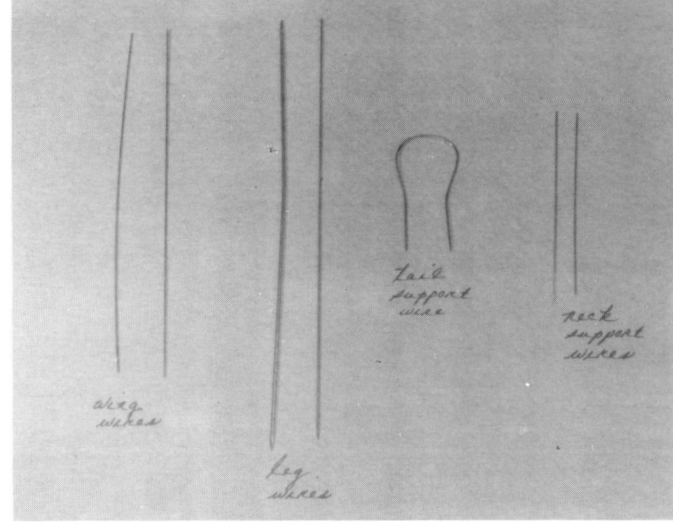

For the wing wires, use a very light gauge wire (20 gauge). Cut the wire three times the length of the humerus. This wire is strictly used to stabilize the humerus and should only be sharpened at one end.

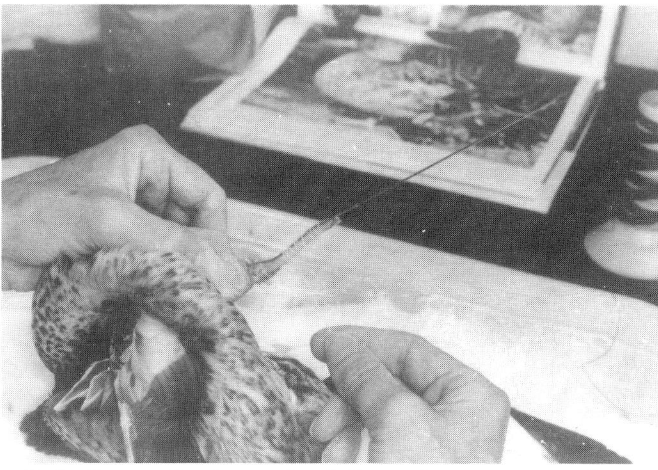

Wiring the Bird Wings

The tail support wire is cut from a medium gauge wire; it is sharpened at both ends and bent into the shape of a horseshoe.

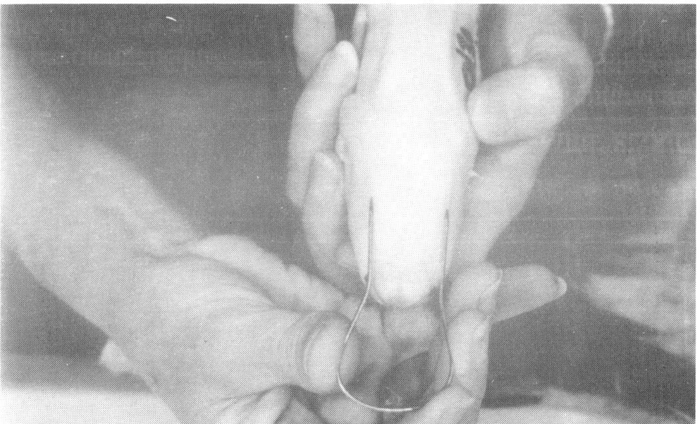

The tail support wire should be long enough to insert a sufficient amount of the wire into the mannikin to support the tail feathers.

The neck support wires are cut from light gauge wires. They should be approximately seven inches long and sharpened on one end.

The base should be completed and prepared for the attachment of the mounted bird prior to mounting the bird. In the case of a water scene, where the bird would not be put into the scene until it is completed, a suitable drying board should be prepared to accommodate the mounted bird. It is imperative that the feet dry in the same position as the final pose and this should be taken into consideration when placing the bird on a drying board. After these preparation procedures, the bird should be ready for assembly.

Assembling the Bird

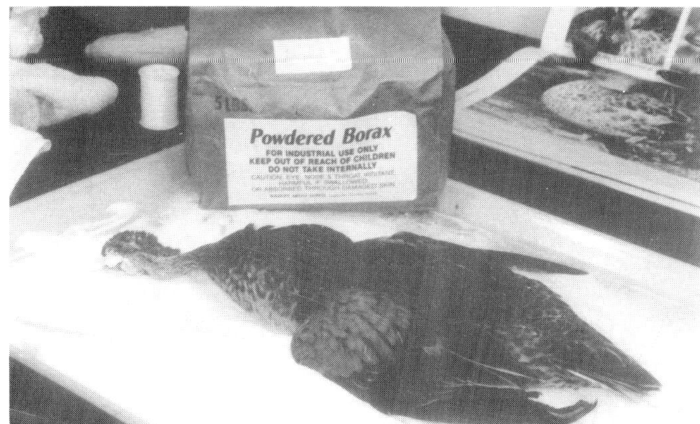

For assembly purposes, the clean, tumbled bird should be placed in a tray of new granular borax. The borax will help keep the feathers clean during the mounting process.

To facilitate adjusting and stabilizing the wings, it is advantageous to wire the humerus with light 20 gauge wire. Place the unsharpened end of the wire against the humerus extending the wire down to the elbow, then secure the wire to the humerus with cotton thread.

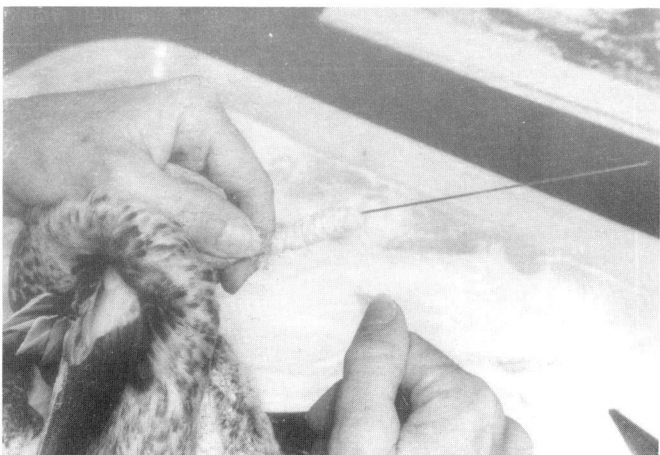

Rebuild the muscles of the humerus to their original dimensions by wrapping cotton around the wire and bone. The cotton is held in place by wrapping thread around it as it is placed on the humerus. Once the muscles are rebuilt to their original dimensions they may be wrapped with soft tissue paper or WASCO clay.

Wiring the Bird Legs

When installing the leg wire down the tarsus and digit, be careful not to pierce the outside skin of the leg and expose the wire. To avoid having difficulty when inserting the wire down the tarsus and digit, be sure to round the sharp point of the wire with a file. (Do not make it too dull or it will not pierce the cartilage at the knee or the base of the foot.)

Hold the tarsus in one hand and expose the tibia. Place the wire on the back (posterior) side of the ankle joint and gently push the wire down the back of the tarsus. When the joint between the tarsus and digits is reached, straighten the joint to allow the wire to continue past the joint.

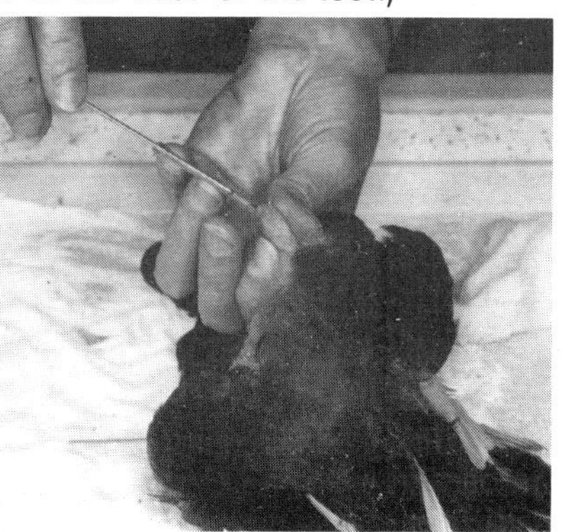

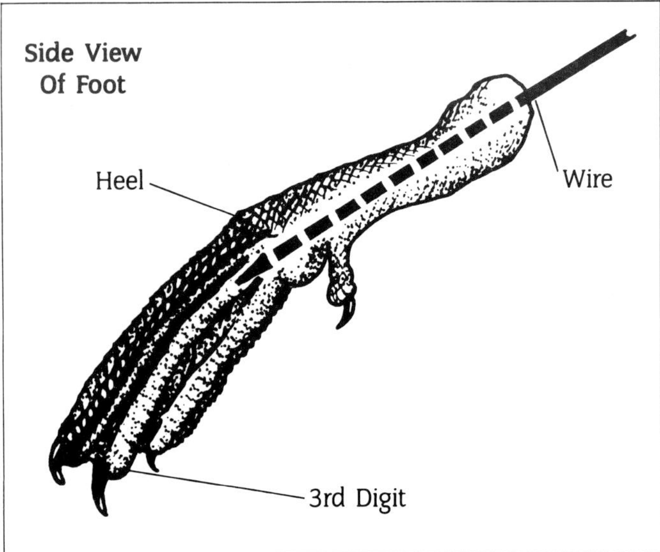

Once the wire is past the joint, continue pushing the wire down the third digit.

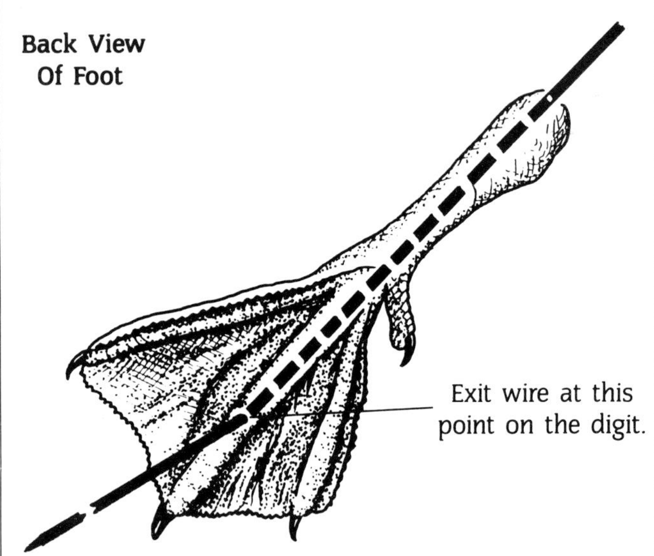

When the wire is approximately midway down the third digit, push the wire through the skin at the base of the foot so that it will exit at this point.

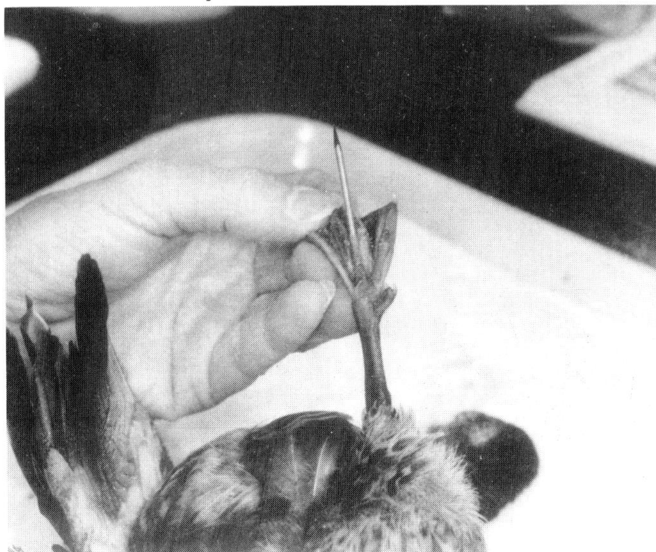

Expose only enough wire to permit the bird to be attached to a suitable base.

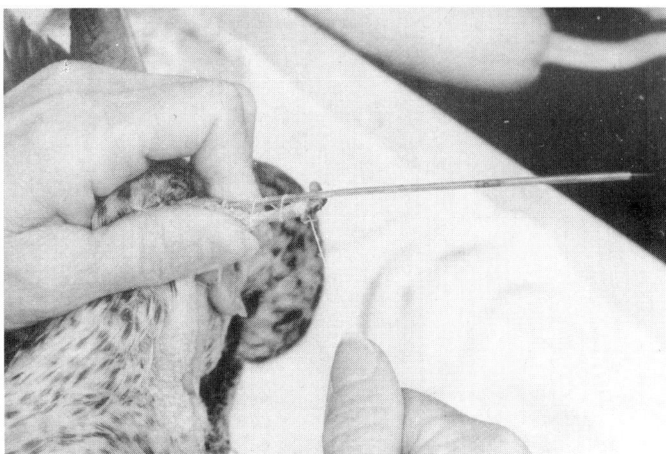

The wire is attached to the tibia by wrapping it with thread.

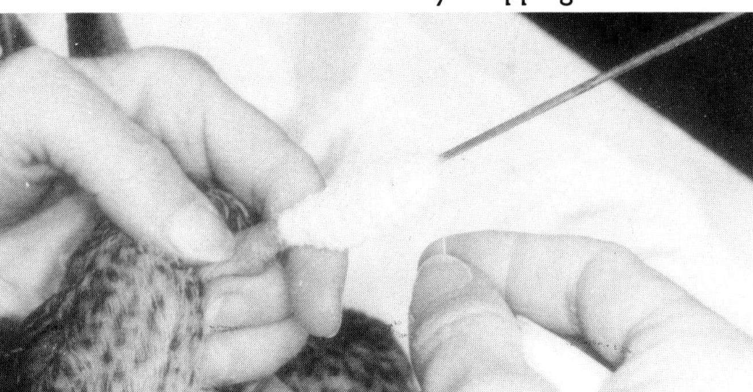

The muscles are then rebuilt to their original dimensions with cotton which is held in place with thread. Again, tissue paper or WASCO clay can be put on it to give it a softer touch. With both wings and legs wired, the skin should now be ready to accept the Accu-Flex mannikin.

Installing the Accu-Flex Body

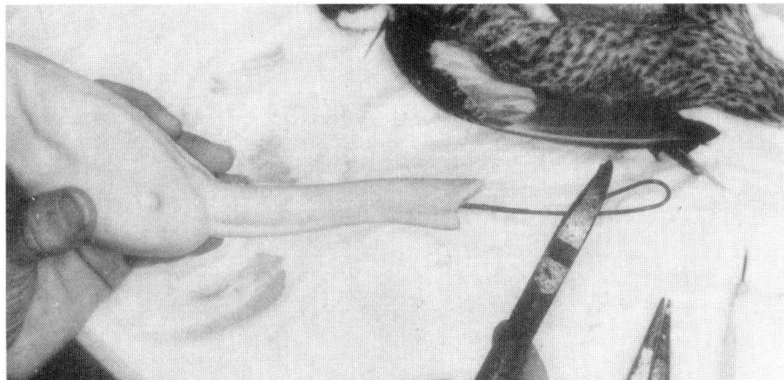

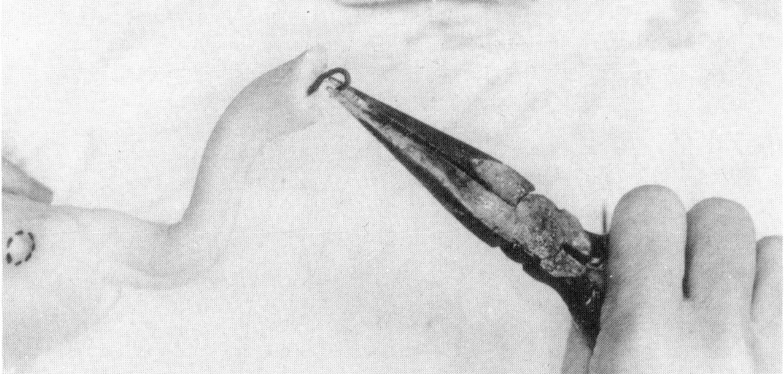

To install the Accu-Flex body, first cut the neck wire off about one inch from the end of the neck and curl it into a loop. Looping the end of the neck wire in this manner will keep it from snagging on the neck skin as the neck wire is being pushed through the skin. Later in the mounting process, the loop will be inserted into the hole in the back of the skull and hot-melt glued into place.

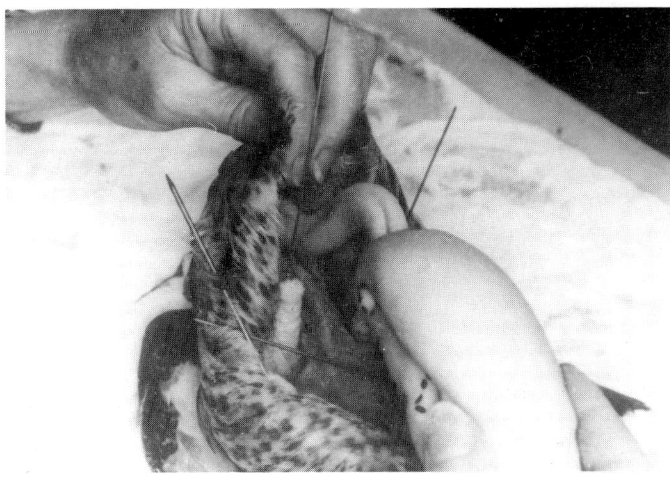

Next, insert the Accu-Flex mannikin through the breast incision and work the skin down over the artificial neck. Do not make adjustments to the neck until the wing wires are attached to the mannikin.

Now insert the wing wire through the pilot hole at the insertion point (lock point) on the mannikin. The wire should be angled so that it exits the body on the opposite side of the lower section of the breast. Pull the humerus snugly against the body.

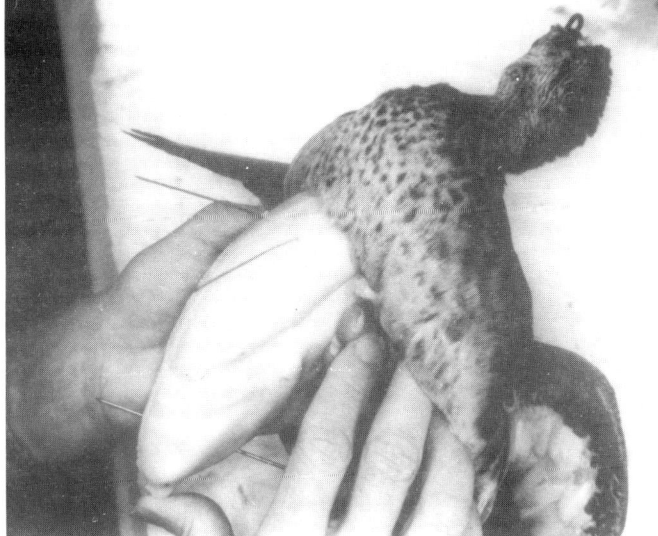

Insert the opposite wing wire in the same manner, exiting in the lower quadrant of the opposite breast. Make certain both wings are securely in place and fit snug against the body.

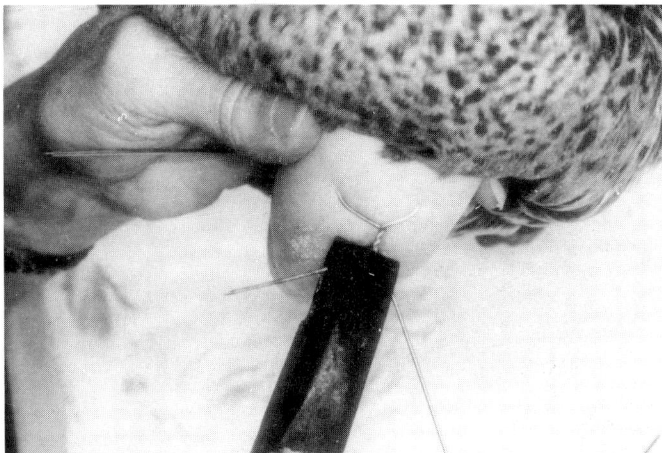

Cross the exposed wires, and with a pair of lineman's pliers, make several twists on the wires, locking them together.

Bend the remaining wire into the mannikin.

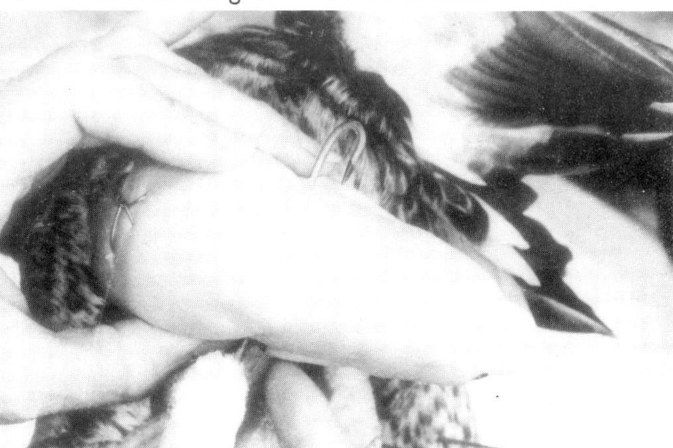

Insert each leg wire into the pilot holes at the insertion points. Run the wire through the body, making a horseshoe bend on the end of each wire with lineman's pliers.

Push the wire with the bend on it back into the body to anchor the wires. Once each leg wire is anchored, hold the exposed end of the wire below the digits and push the tibia *snugly* against the body. This will securely attach each leg and prevent the bird from wobbling.

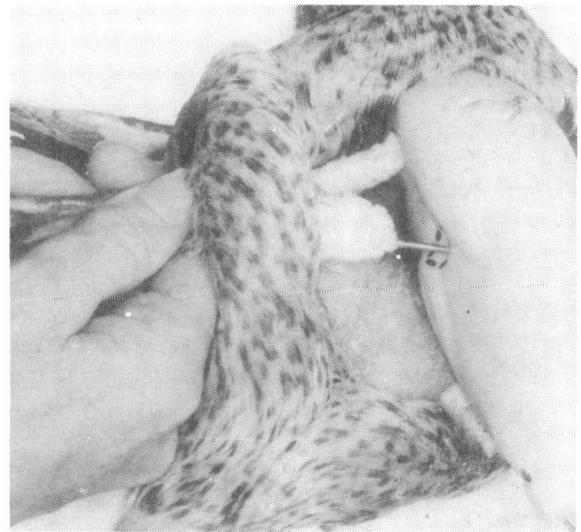

NOTE: If a bird is to be standing on one leg and a very heavy gauge wire is used, it may be necessary to hot glue the wire at the point where the horseshoe was made and pushed back into the mannikin.

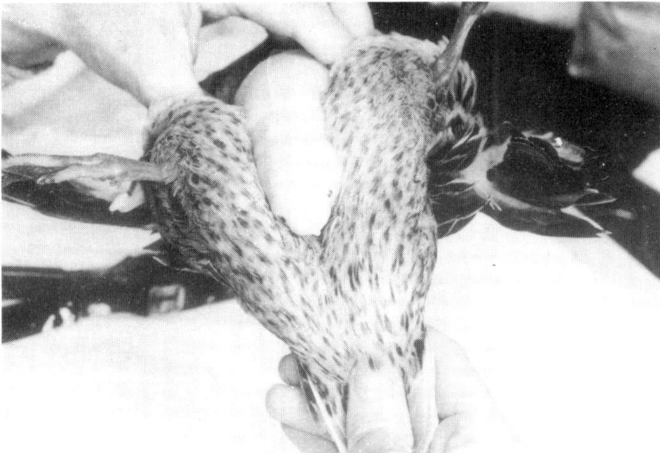

With the wings and legs secured, gently pull the skin around the back section of the body and carefully adjust the tail feathers.

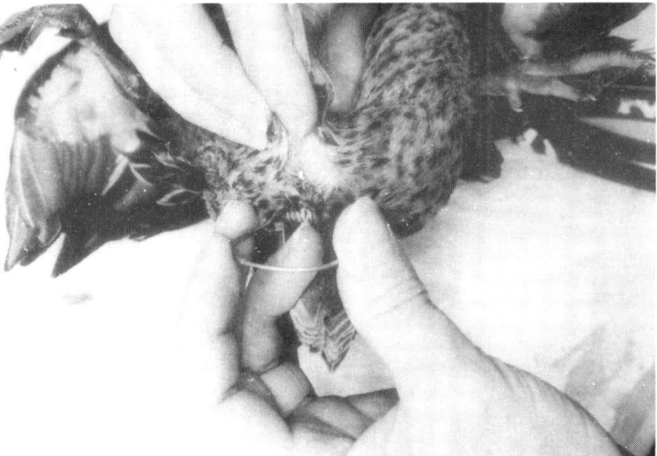

Separate the under tail coverts from the tail feathers. Insert the horseshoe-shaped tail support between the under tail coverts and the tail feathers. The tail support should not be pushed snugly against the body, but positioned with enough distance between the body and the base of the tail feathers to align and adjust the feathers naturally in the rear (posterior) section of the bird.

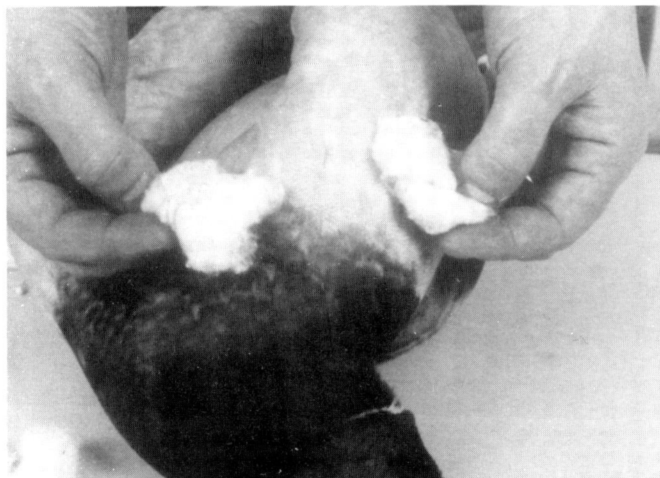

If needed, add cotton to the side pockets and crop at this time.

Now, starting at the neck, adjust the skin and position the feathers. When the skin and feathers are in their approximate positions, sprinkle dry preservative into the open incision. Close the incision using a double strand of nylon thread and a three-cornered sewing needle. Now close the incision by starting at the top of the incision and sewing toward the tail section using a baseball stitch.

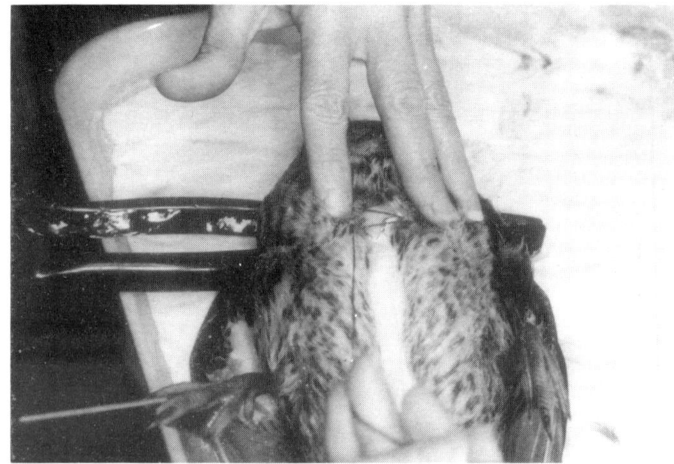

Make small neat stitches, being careful not to get the feathers caught between the stitches. After the incision is closed, adjust the feathers along the breast and flank areas.

Attaching the Artificial Head

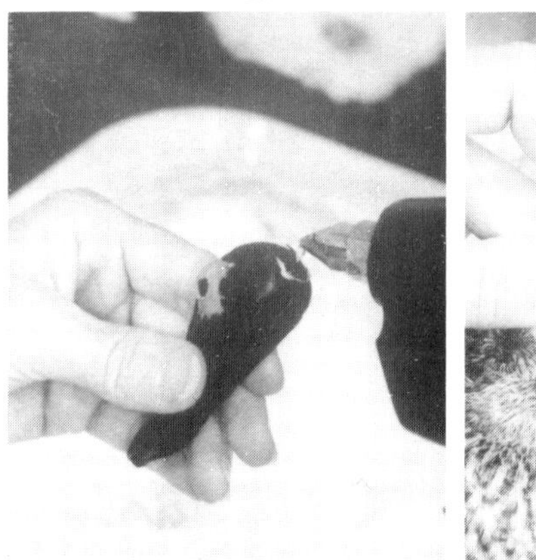
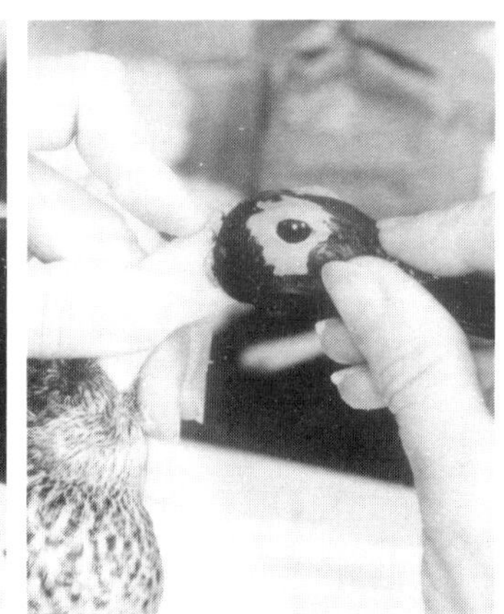

Fill the hole in the back of the artificial head with hot-melt glue and insert the looped end of the wire directly into the glue filled hole. Hold the artificial head and neck in the proper position until the glue cools. Once the glue has cooled, it will securely lock the two together.

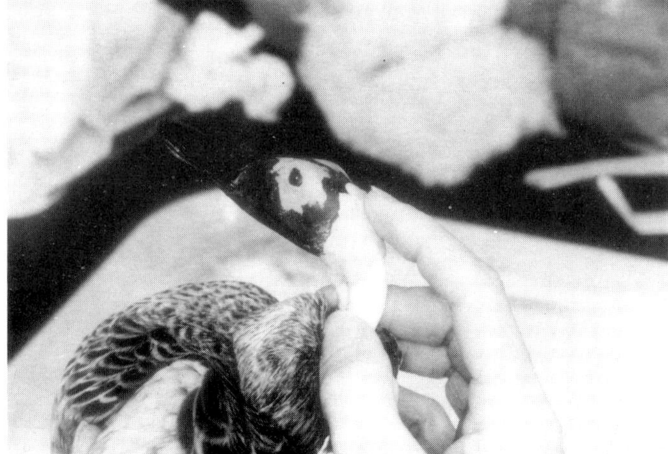

Any additional sculpting around the base of the artificial head can be done at this point with WASCO Clay. Now, carefully pull the neck skin over the artificial head. (NOTE: Keeping the skin of the head and eyes damp with wet cotton during the mounting process will make this step easier.)

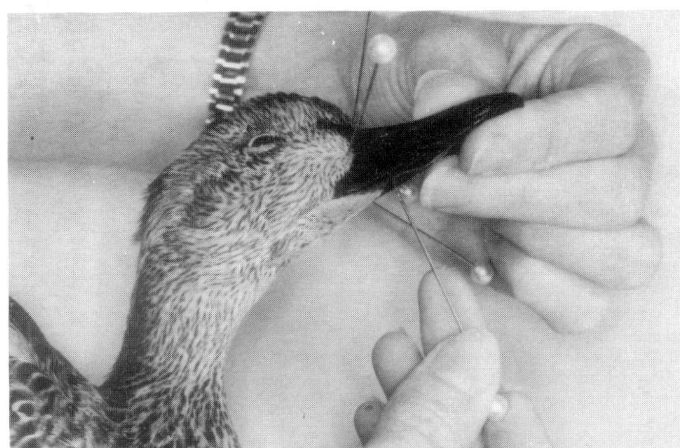

Temporarily pin the skin in place around the head and eyes.

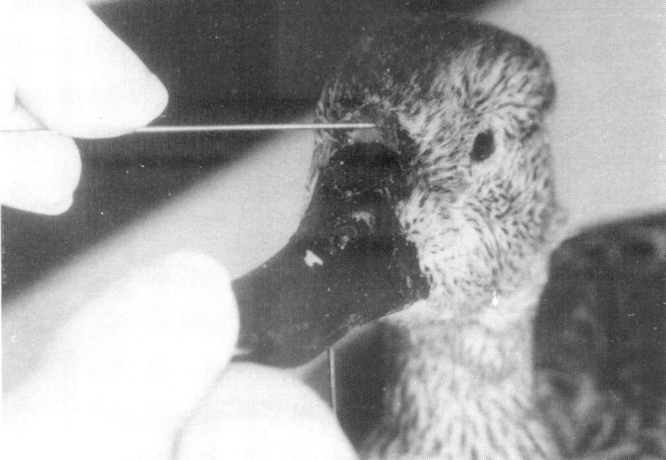

Then, find the center of the forehead and pull the skin flap down into the notch at the top of the bill. Using magnifying type glasses for this delicate operation will be quite helpful.

Apply Nu-Glue (*Polytranspar* Epoxy Adhesive) with an artists' brush (do not use an expensive one), carefully lifting the skin and painting glue around the edge of the bill.

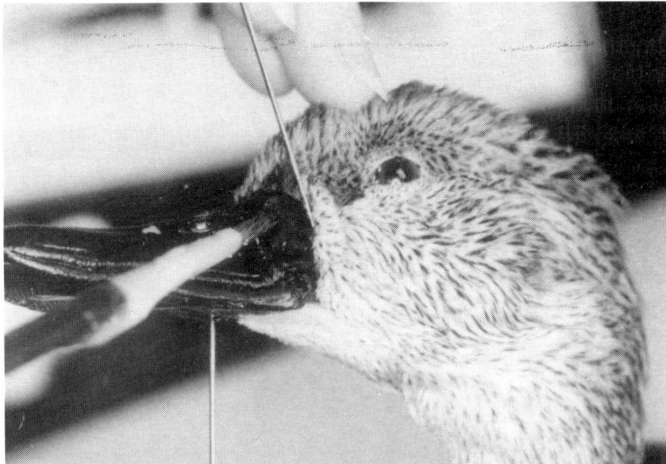

Now press the skin into place along the edge of the bill and pin it securely until dry. Then work down each side from that point, being careful to put the glue on the head rather than on the skin and also to keep the glue out of the feathers. Remember, it is easier to keep the skin clean and free from excess glue than to try to remove Nu-Glue from the feathers.

If the glue does get on the feathers, it may be removed with *Polytranspar* Thinner. The primary advantage of using Nu-Glue is that it has a longer "setup" time than Instant Bonding Glue which will allow additional "working time" for positioning the skin. Because Nu-Glue takes approximately 45 minutes to

one hour to set up, there is no need to frantically adjust and position the skin. If an adjustment does need to be made, it will be less of a hassle to make because of the longer set up time. Also, because Nu-Glue was originally developed and formulated for use with mammal skins, the adhesive capabilities will be far superior to other types of glues.

Once the Nu-Glue has set, adjust the head to the desired position. Additional support wires may now be inserted into the neck to hold it solidly in position. Use small sharpened wires and push them through the neck and into the mannikin.

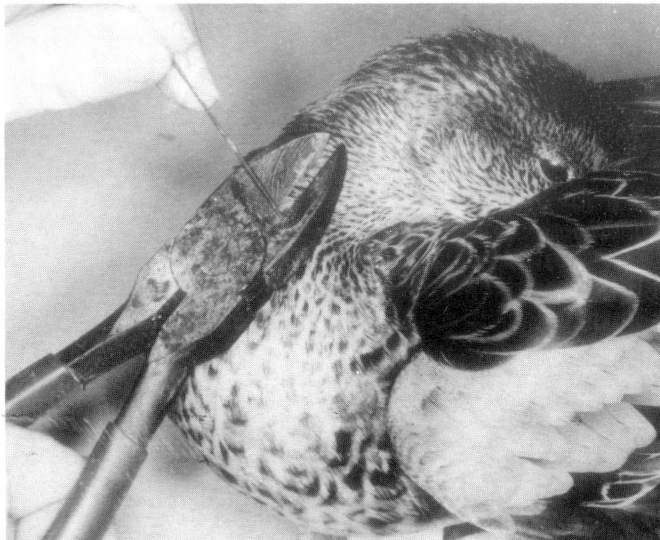

The wires should be pushed all of the way down flush with the surface of the neck where they will not be visible.

Now with the neck and head in position, make final adjustments to the eyelids. Adjust the eyelids with a small feather adjuster. Press the excess tissue of the lower lid between the glass eye and the filler used in the eye socket.

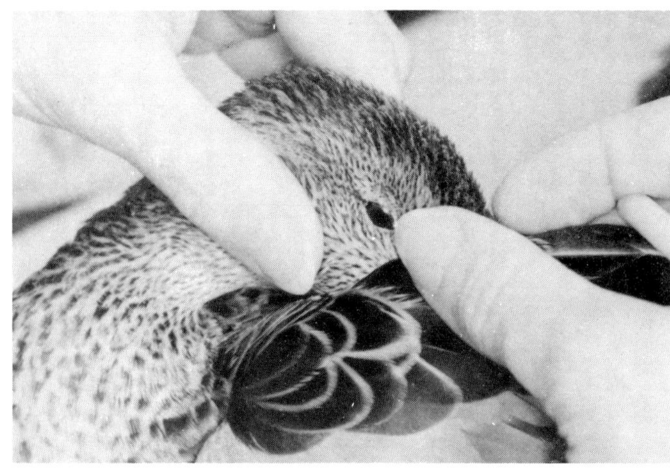

If clay was used as a filler, once the proper contours of the upper and lower lids are achieved, place a small insect pin in the front and rear corners of the eye. These pins will keep the lids in place and will be removed when the mount is dry.

Attaching the Bird to the Base and Posing

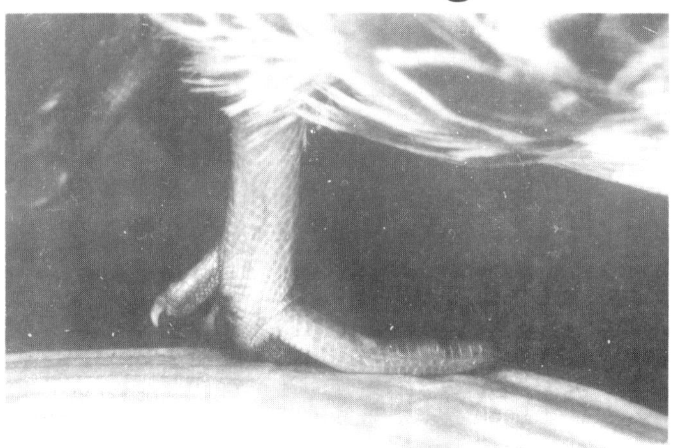

Attach the bird to the base making sure that the feet are securely fitted against the base. Absolutely no support wires should be visible between the bottom of the foot and the base. If the bird is posed with its heel slightly elevated, then the leg wire will be "hidden" because the wire exits midway down the middle toe.

Complete attention should now be given to posing the bird and securing the wings and neck. Check the bird from all angles for balance and symmetry. Once the bird is in the desired position, begin aligning the feathers and placing the feathers in their proper feather tracts. "Feather tracking" the bird is best accomplished with a feather adjuster. Simply place the tip of the feather adjuster next to the skin and lift up.

This will cause the misaligned feathers to fall back into their proper tracts.

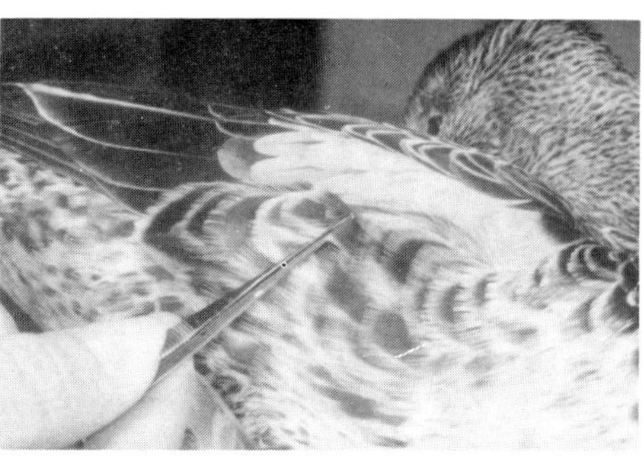

Tweezers also work well to align the feathers. Pinch the end of the misaligned feather and gently pull on the feather. One gentle pull will usually get a stubborn feather back into place.

Setting the Feet

In order to keep the feet full and plump looking, it is necessary to inject the feet. Many taxidermists use formalin (formaldehyde) for this step. When using this chemical, it is imperative to wear protective goggles, surgical gloves, and to have a good ventilation system. This chemical is extremely dangerous. It is a good idea to use the formalin just before you are ready to close your studio for the day; the fumes should have enough time to dissipate overnight. Remember, *formalin has a devastating effect on the eyes, skin and respiratory system. Always use the recommended safety precautions!*

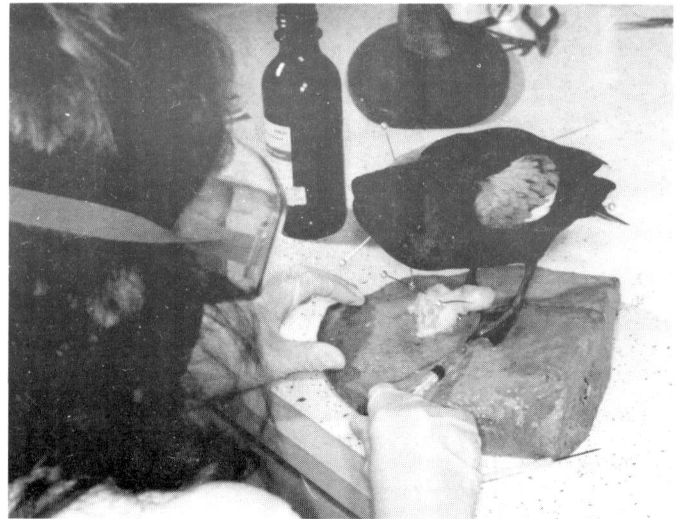

When injecting the feet with formalin, use a small, insulin type needle. This small needle will prevent the formalin from leaking out of the puncture site. To ensure that the foot will remain full and plump, apply another injection of formalin on the second day if necessary.

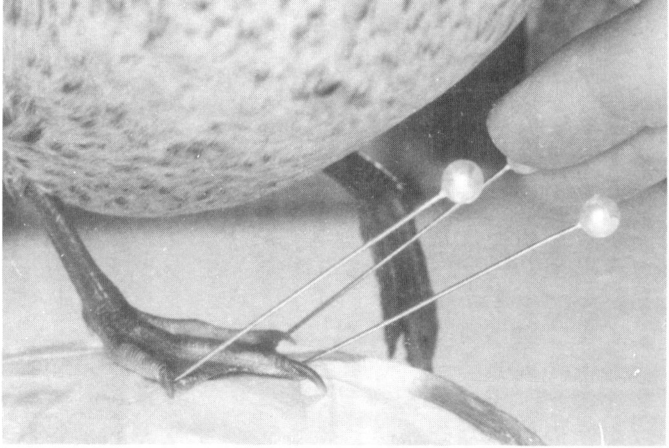

Whenever possible, it is best to pose and position the feet on the permenant base that the bird is to eventually remain on. If the feet can be glued and pinned to their permanent base, it will ensure that the feet will remain in their desired position.

Five minute epoxy works well to glue the feet. Apply the epoxy to the bottom of the digits, and in the case of waterfowl, to the bottom of the webs.

Insert the leg wire through the pre-drilled hole in the base, bend the exposed wire, and clinch the wire into the bottom of the base. A Foredom tool may be used to rout out a groove for the wire to fit flush in. If it is not feasible to extend the wire completely through the base, cut the wires short enough so that they will not go completely through the base.

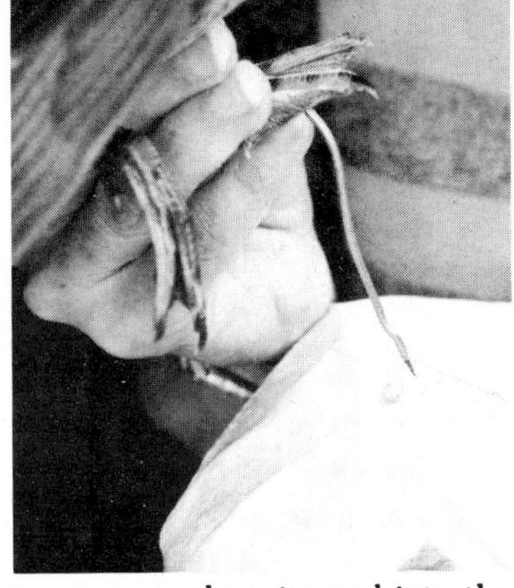

Next, apply five minute epoxy to the wire and into the hole; then insert the wire into the pre-drilled hole in the base.

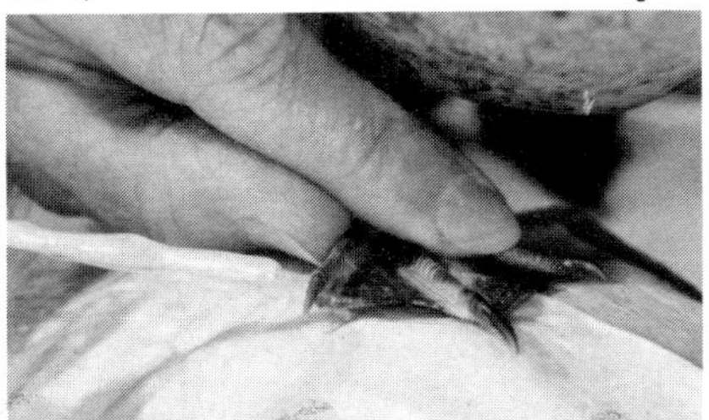

The epoxy will hold the bird securely in place on the base. The epoxy should then be applied to the foot which will permanently hold the foot in place on the base.

With the wire secure, press the foot down on the base. Immediately, pin it with a small diameter pin. Try to place the pin on the very edge of the digit, or if possible, angle the pin across the toes thus avoiding making any pin holes in the feet. Some taxidermists opt to simply glue the feet down with Instant Bonding Glue which avoids the use of pins altogether. After the feet are pinned and/or glued, the bird is ready for final grooming and detailing.

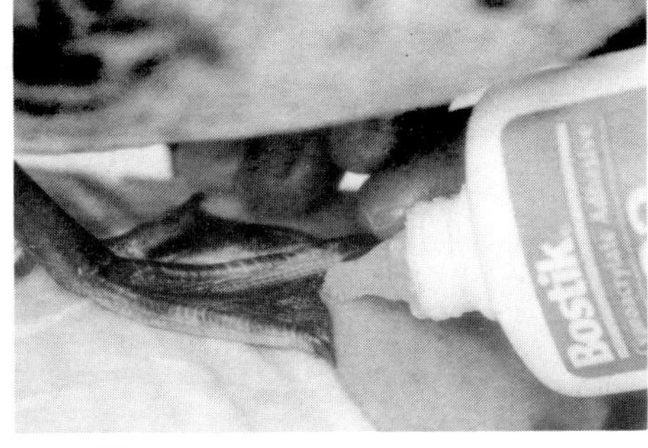

Grooming and Detailing Standing Birds

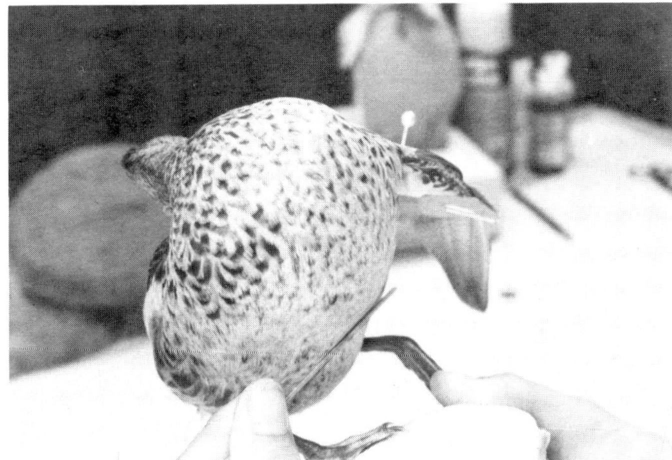

No other phase of the mounting process is as pleasurable or relaxing as grooming and "preening" a mounted bird. All of the hard work put into the mount during the defatting, skin preparation, and assembly stages will finally all come together during the grooming process. It is in this stage that the bird finally returns to its "lifelike" resemblance of the reference photos that were used throughout the process as a comparison.

Subtle changes in expression or attitude may be achieved by as little of a change as a slight fluff of the feathers. Also, in this stage the anatomical features of the mannikin are clearly defined, giving the bird's musculature lifelike shapes and contours.

When grooming and detailing, strict attention must be given to what the bird is being portrayed as doing. Are the bird's feathers supposed to be tight, as in a tense bird, or should they be fluffed as in a more relaxed pose? Study reference as it pertains to the particular pose that you have chosen.

Detailing A Standing Mount

The bird should be checked for anatomical balance and symmetry. Once this is established, the feathers should correspond in the same fashion. For example, if a bird is standing squarely, looking back over it's left shoulder, the feathers in the chest and breast should reflect this stance by flowing to the left.

Checking to make sure that distinct color patterns in different areas on the left and right side of the bird are properly aligned, is a foolproof way of checking symmetry.

For example, if a wood duck is standing, looking straight ahead, the white and black bands of the side pockets should be positioned exactly the same on both sides of the bird.

Once all of the feathers are in their appropriate positions, the skin should be pinned in place to define and lock-in the underlying anatomical detail. When studying any of the pictures in the reference book, "Prairie Wings," it becomes evident just how much detail can be placed into a mounted bird.

The incredible amount of anatomical detail sculpted into the Sportsman Series bird mannikin makes it easy to recreate the live bird with a minimum amount of effort.

The detail is put into the mounted bird by simply working the skin into the proper position and then placing small pins in the areas that are to be defined.

Using hide paste requires extra effort but pays big dividends in this process. Pinning and tucking loose skin in the abdominal area defines the subtle contours of the legs and caudal areas.

Continue this procedure until the bird has been completely defined.

Head

A bird's complete expression can be changed simply by shifting a little skin around the eye, partially closing or opening the eyelid, or fluffing the feathers in the crown or cheek. Remember though, the feathers should correspond to what the bird is doing. If the bird is standing in a relaxed pose, you would not slick the feathers down and give the bird a taut, "greasy kid look." Fluff the feathers up for such a pose.

Use the feather adjuster to stroke the crown feathers from behind the skull toward the bill. This fluffing action will help align the feathers and also help to prevent the feathers from looking flat. If the areas on the cheek or throat look flat, or if they need more fill, take a long, sharp pin and pierce the skin and push the pin underneath the skin. Then, with the point of the pin, lift the skin and work the feathers until they align themselves properly.

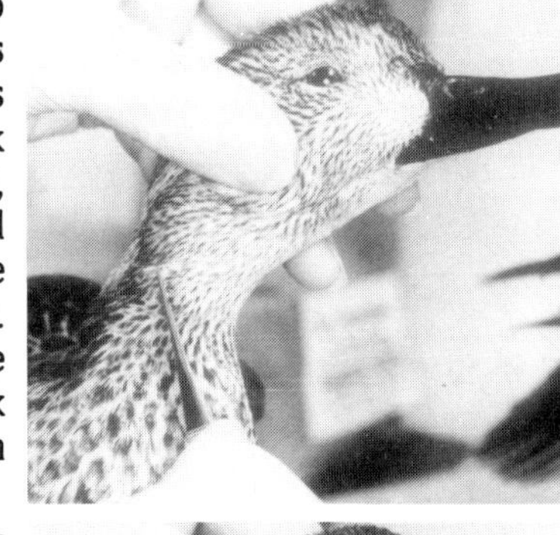

If necessary, the ear opening may be used to insert small amounts of fill. Birds such as wood ducks and hooded mergansers may even require a very, very light touch or "whisper" of hair spray to keep the long plumage in place.

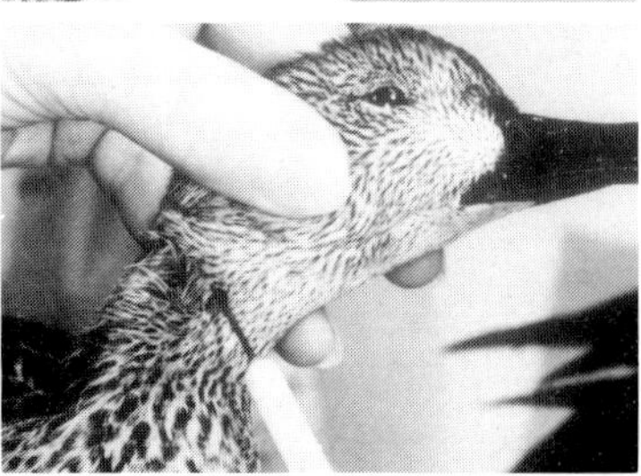

The *Breakthrough* Bird Taxidermy Manual

Neck

The neck feathers require special attention if the neck is to look proportionate. The tendency is to pull or work the feathers down the neck to the chest. Working the skin in this manner will cause the neck to look long, thin, and stretched, and will make the chest look puffed and square. However, if the skin is worked up to the back of the skull and then allowed to fall naturally, the transition from the back of the skull to the body will be much smoother and more natural looking.

When the skin is properly positioned on the neck, the feathers in the neck skin should be arranged using a feather adjuster and tweezers.

Chest and Breast

The position of the chest and breast feathers will be determined by the pose. If the bird is turned and preening its wing, the neck and chest feathers should reflect this pose. The important aspect of positioning the chest and breast feathers is that they must follow and depict the action and movement of the underlying muscle structure. If the muscles are turning the bird's head and neck to the left and extending the left wing, you would not want the chest and breast feathers shifting off to the right side of the bird.

To avoid having birds which have distinct markings on their chest and breast from appearing off balance, it is important that the feathers align or "track" properly.

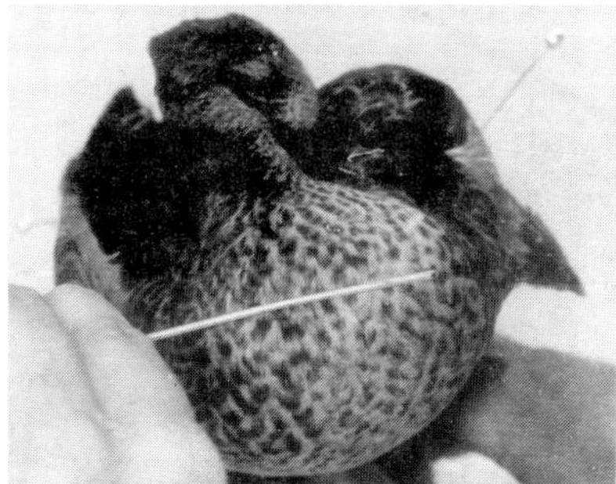

Wings

Notice how the feathers overlap in each feather group.

The feather groups of the wings (primaries, secondaries, tertials, wing coverts, and scapulars) are often difficult to properly adjust. One thing is for certain; any feathers out of place on the wing will show up like a flashing neon light. The "easiest" way to groom the wings is to approach them in a very methodical fashion and to know the topography and anatomy of the species.

A basic understanding of how a wing folds, and how the individual feathers interact so that they will close, perfectly conforming to the curvature of the bird's back, is mandatory. This is why carcass sketches, accurate wing placement points, and accurate attachment of the wings to the mannikin were so important during the mounting process. The wings are anatomically designed by God to fit and to conform to the natural curvatures of the back. If the bird is not assembled and wired properly, no amount of masking, pinning or carding will make the feathers fall into place.

Remember that the primary feathers slide underneath the secondaries, which slide underneath the tertials, etc. "Individual" feathers in

each of these areas also slide one underneath the other. The first primary feather slides underneath the second; the first secondary feather slides underneath the second, and so forth. Understanding the way the wing closes and the way that the feathers lie, will facilitate the adjustment and proper placing of the feathers into their appropriate positions.

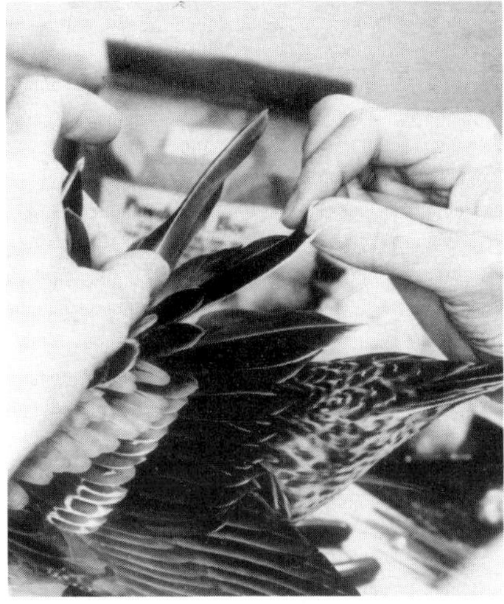

Take each wing separately; start with the primaries and place them into their proper position, and then progress to the secondaries. Place them in their proper sequence, and then adjust the tertials and wing coverts, in that order. Positioning the feathers in the wings will be less confusing if this sequence is followed.

Side Pockets, Scapulars, and Back

The side pockets, scapulars and back feathers are generally quite cooperative. Usually a combination of a gentle lifting motion with the feather adjuster and a slight pull with the tweezers will align these feathers. Again, reference is the key to success!

Tail and Tail Coverts

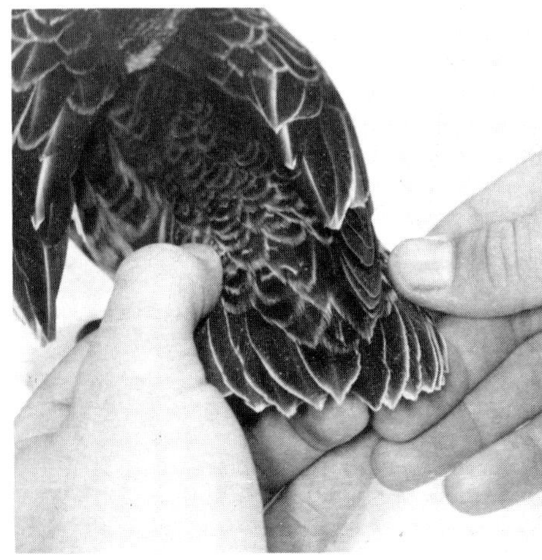

The tail and tail coverts fall into place quite easily as long as the primary tail feathers are in their proper order. As with the feathers in the wing, the tail feathers close in the same manner, with the outermost feathers sliding underneath the next feather, and so forth.

Final Touches

When all of the feathers are in position, use a feather duster or large make-up brush to smooth the overall appearance of the feathers and the mount.

If there are any unruly feathers on the bird that simply will not be coaxed into place, then cut a piece of carding material and place it over the stubborn feather. The carding material is secured to the problem area by sticking a T-pin through the carding material into the mannikin. Be careful about applying too much pressure, as this will flatten the feathers.

Drying

The completed mount should be placed on a shelf or wall, away from the main flow of traffic to dry. The bird is vulnerable at this point, and accidentally bumping or dropping the bird will mean spending time redoing the procedures.

It is best *not* to attempt to force dry a bird by applying or placing the bird too close to a heat source. This will usually cause some excessive shrinkage in the delicate tissue around the eyes and bill.

During the drying process, periodic checks should be made to make certain all feathers are drying in their proper positions. In most cases, the bird should be thoroughly dried and ready to complete in ten days. Drying time, however, will vary from region to region depending upon the humidity in a particular locality.

Using Artificial Bird Feet

This method allows the taxidermist to replace the bird's natural feet with a realistic-looking substitute that will not shrink. The first steps of this procedure should be carried out immediately after the wiring of the wings.

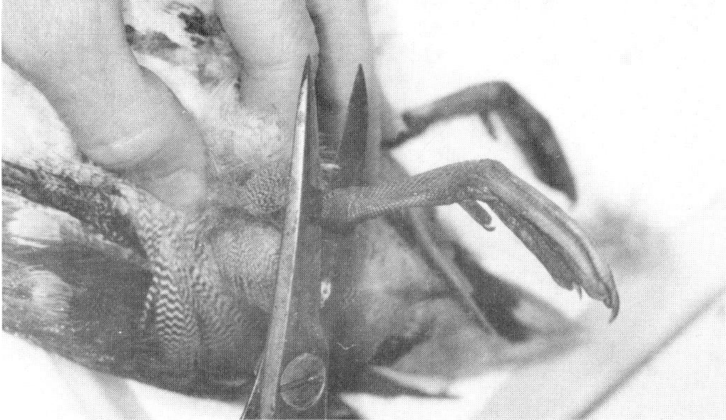

The natural feet should first be removed. This is done by cutting the foot off just above the ankle, where the scales of the foot meet the skin. The tibia can be left in place (attached to the skin) or completely removed, but it should be retained for reference when wrapping. If the tibia is removed, insert a leg wire through the area where the tibia was. If the tibia is still in place, insert the wire directly behind it, pushing enough wire out to go completely through the pre-drilled artificial foot, and also to secure the foot to the base.

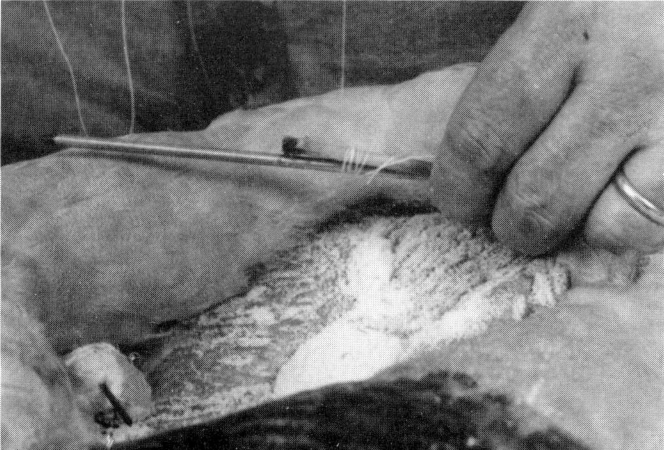

If the tibia was not removed, secure it to the leg wire at this point with cotton thread.

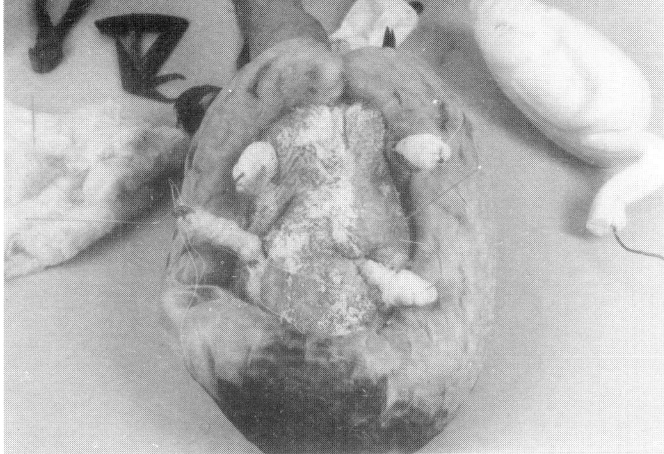

The leg muscles and tissue should now be duplicated with cotton held in place by thread. When wrapping, use the tibia (removed or in place) as a reference for length.

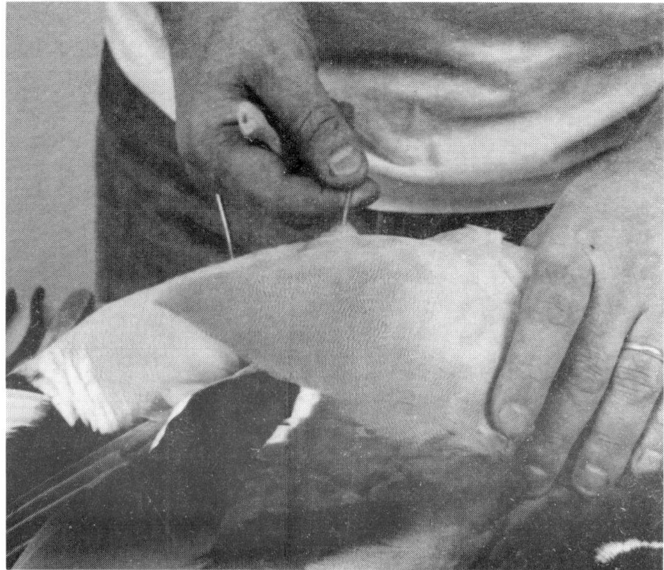

Once satisfied with the size and shape of the leg muscles, insert the wing and leg wires into the lock points on the bird mannikin and close the ventral incision.

Before pre-positioning the feet on the base, test fit the feet on the leg support wires to ensure easy installation later in the mounting procedure. If the wires will not fit into the pre-drilled holes in the artificial feet, take an appropriate size drill bit and enlarge the hole for an easy fit. Once the artificial feet have been "test-fitted," pre-position the feet onto the base by heating the toes and webbing with a heat gun or hairdryer until pliable. Place them into their approximate position, forcing the toes and webbing into place until the feet cool and become rigid.

With the feet correctly conformed to the base, inspect the feet for any flaws such as air pockets and missing toenails. Due to the nature of the foam, tiny parts such as the toenails can easily be broken during the release of the mold or during shipping. These areas are easily filled and sculpted back into place with Sculpall or All-Game. Allow plenty of time for the sculpting compound to dry before proceeding.

With the feet now shaped to fit on the base and all flaws repaired, it would be best to completely paint them at this time following the appropriate *Breakthrough* Waterfowl Finishing Schedule. This eliminates masking off feathers on the finished mount to prevent overspray.

Allow the paint to completely dry and slip each foot onto its perspective leg wire. Attach the bird to the base by inserting the leg wire through the pre-drilled hole in the base, bend the exposed wire and clinch the wire in to the bottom of the base. Re-check the contour of the foot on the base.

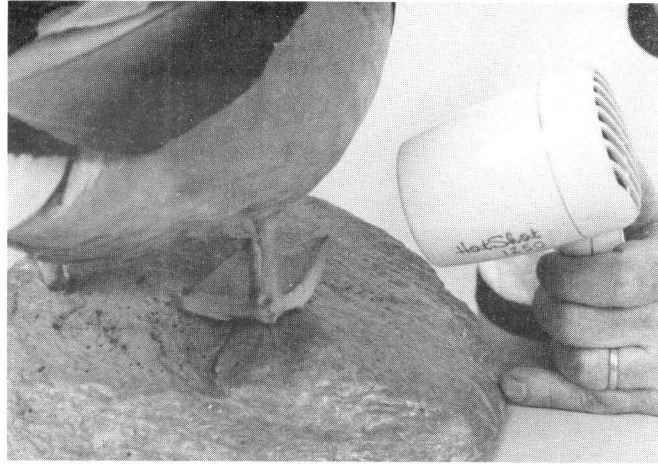

Re-heat and re-position if necessary at this time.

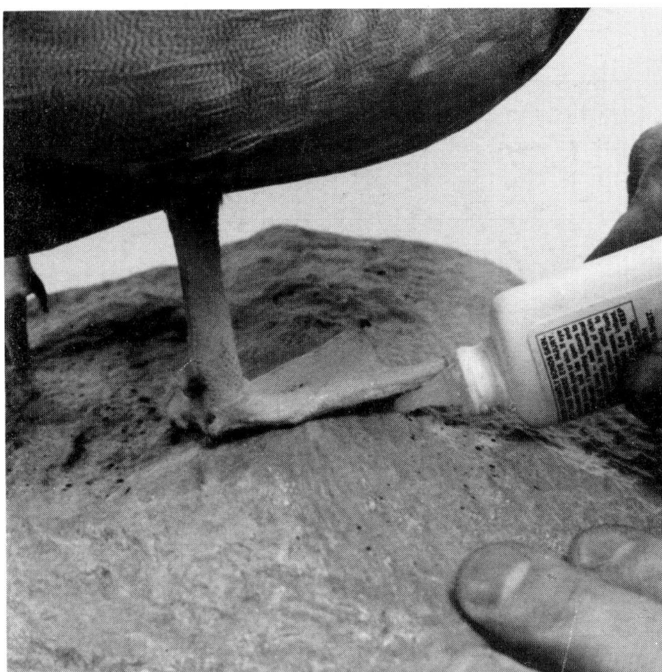

With the foot now resting flat against the base, lift the toes and webbing slightly and place a small amount of Instant Bonding Glue to the base and bottom of the foot.

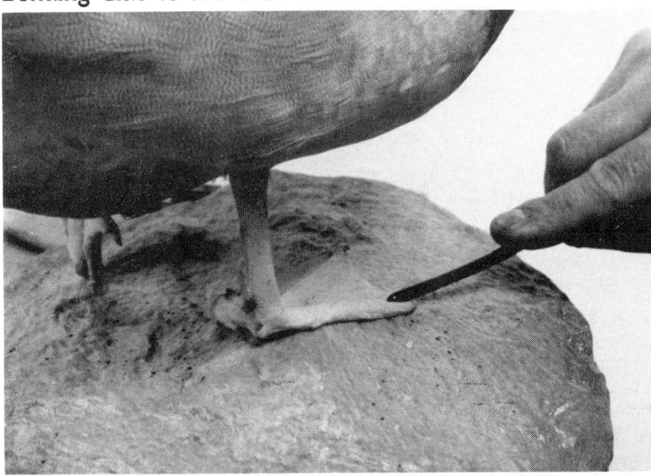

Press into position and hold until the glue is completely dry. Place a small amount of Instant Bonding Glue at the top of the leg (where the leg meets the body) and taxi the skin and feathers in this area into position, allowing the glue to bond with these areas.

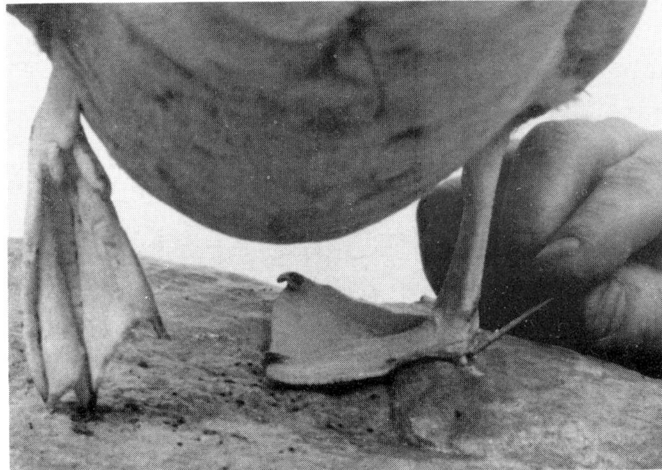

If the wire shows between the heel and the base, a small amount of Sculpall can be used to fill in this area.

If using a "resting" or closed foot, the leg support wire will not exit the artificial foot. To determine how much of the exposed leg support wire is needed to secure the artificial foot to the bird, simply hold the artificial foot next to the exposed leg wire so that there will be a sufficient length of wire to insert midway down the tarsus of the artificial foot. Once the wire is cut, apply Instant Bonding Glue to the leg wire and insert into the "resting" foot.

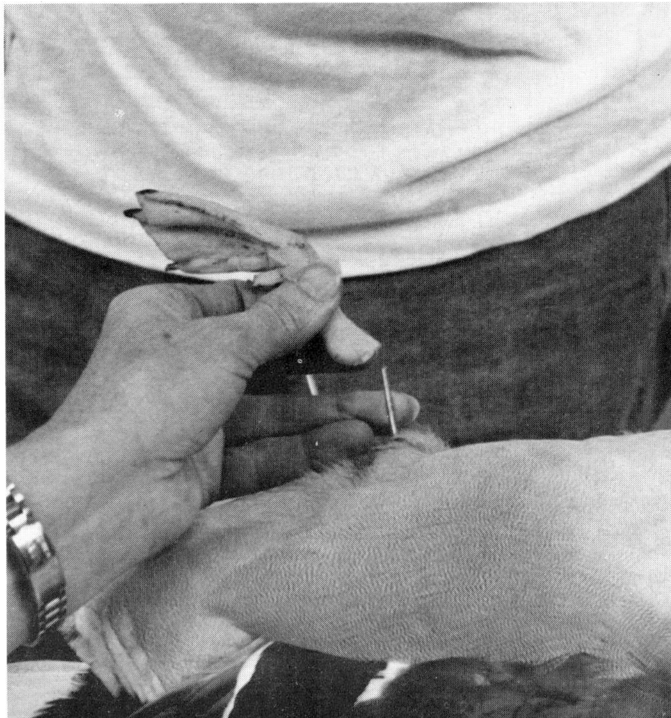

Don't delay when inserting the glue covered wire into the artificial foot, as the Instant Bonding Glue sets extremely fast.

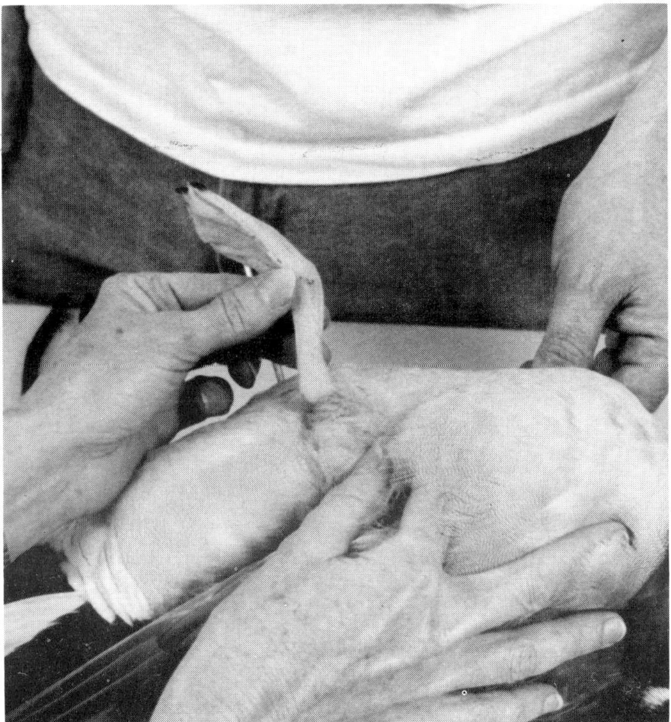

At this point, check the appearance and positioning of the feet. Make any alterations including paint touch-up.

The completed mount now has a pair of pre-painted feet that do not require dangerous formaldehyde injections, and will not shrink or deteriorate as time passes. What more could a taxidermist and their clients ask for?

CHAPTER 8

Mounting a Flying Bird Using an Artificial Head and an Accu-Flex Body

Birds in flight have captivated man since the beginning of time. The effortless flight and intricate maneuvers that are performed each season by millions of birds on their annual "southerly" migrations serve as a constant reminder to all sportsmen and taxidermists of Mother Nature's "gift of flight."

Every muscle of a birds body is utilized during flight; capturing the power and complexities of these muscles with your mount is mandatory if the taxidermist is to successfully "pull-off" a convincing flying mount.

The Sportsman Series "system" of Accu-Flex bodies, necks and artificial heads will enable the taxidermist to incorporate expression, attitude and anatomical detail into the flying bird with a minimum amount of effort. Because the Sportsman Series mannikins are so anatomically accurate, the skin and feathers simply need to be pinned in place in order to show the underlying detail.

Neck placement and position is critical in capturing the proper attitude of a flying pose. With the pre-attached flexible necks, all the guesswork has been removed because the necks are shaped and fitted to the proper dimensions for each individual specie. All that is needed is an accurate reference photo so that the neck can be placed in the proper position.

needed to produce an award winning flying mount is to use the recommended tools, equipment and supplies, carefully follow the instructions, and *"always"* use your flying bird references when posing the mount.

Preliminary Procedures

Everything needed to ensure success when mounting a flying bird (Accu-Flex mannikin, pre-attached flexible neck and artificial head) has been incorporated into "the system." All that is

To avoid complications and loss of time, assemble all the necessary tools and materials *before* the mounting procedure begins. Bird taxidermy should be fun and "relaxing." Don't turn

bird work into a nightmare because tools and supplies were not assembled and prepared in advance.

A typical list would include the following: Accu-Flex body, artificial head, eyes, clay, cotton, thread, preservatives, wires, tools, references and a bird skin that has had the skin prepared as instructed in the skin preparation chapter.

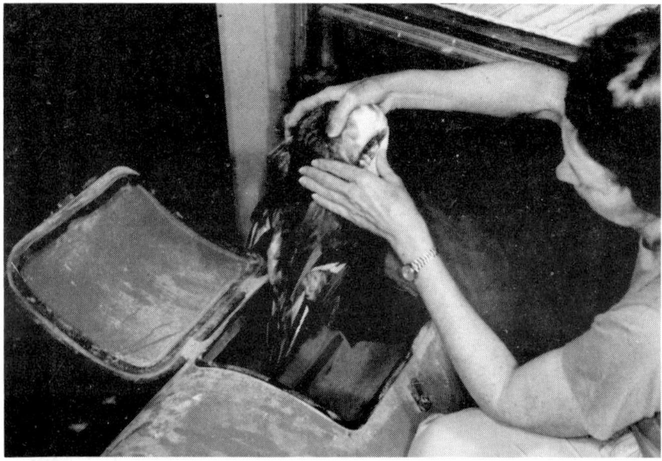

The skin should be tumbled (above) and then blown "completely" dry with compressed air (below).

MOUNTING PROCEDURES:
Artificial Head

The mounting procedure begins with the focal point of any mount, the "head." Carefully inspect the head for any imperfections. Make certain the eye orbit will accept the proper eye size. If any alterations must be made, they should be done at this time.

Paint the Bill

Paint the artificial bill using a *Breakthrough* Waterfowl Finishing Schedule for the particular specie being mounted. Painting at this point prevents overspraying airbrush paint onto the feathers. The exception to this procedure would be if Winsor & Newton artist's oils are used to paint the bill. Since overspray is not a factor, the bill may be painted after the bird is dry if so desired.

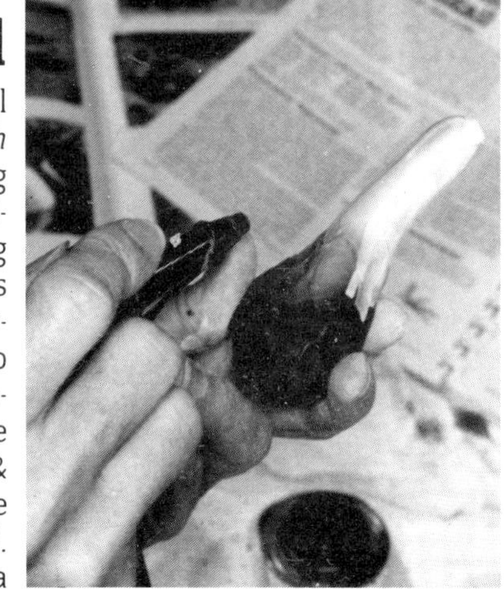

Setting the Eyes

In order to capture the expression of a live bird and to avoid problems with seating depth and with recreating the delicate contours of the upper and lower eyelids, it is important to have good reference pictures close at hand.

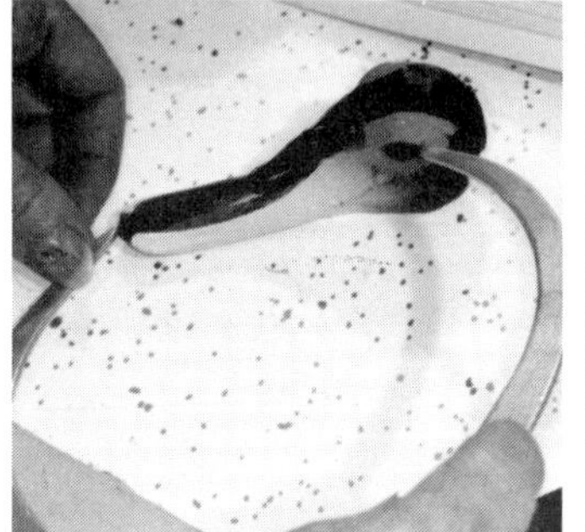

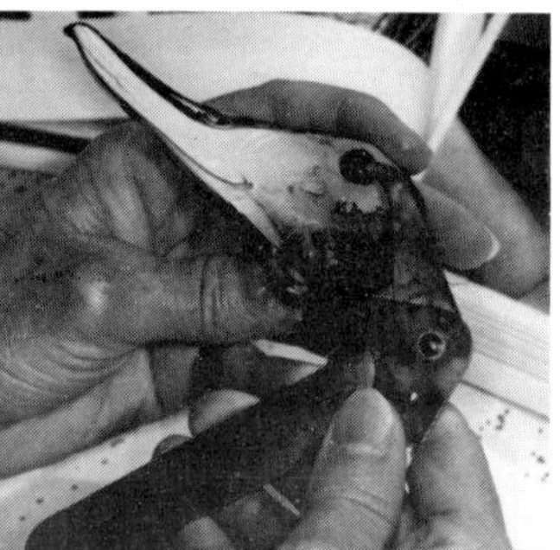

First, the eye sockets should be filled with either WASCO clay or Sculpall to hold the glass eye in position. Once the eye socket is filled, carefully place each glass eye in the socket. Use calipers to check the symmetry once the eye is seated in the socket. To duplicate the "look" of a live bird, the eyes should be set with a slightly forward angle and a slightly downward cant.

Seating depth is also critical to give the bird a natural life-like appearance. Setting the eye too deep gives the illusion of the bird not being able to see, or worse, the appearance of being emaciated and about to expire. Setting the eyes too shallow will give the bird a "popeyed" appearance.

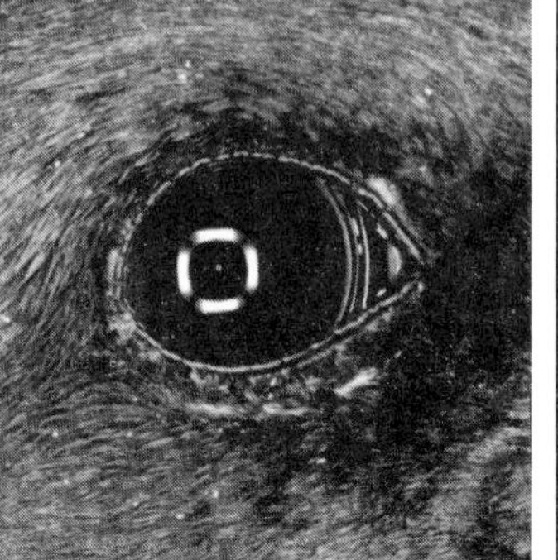

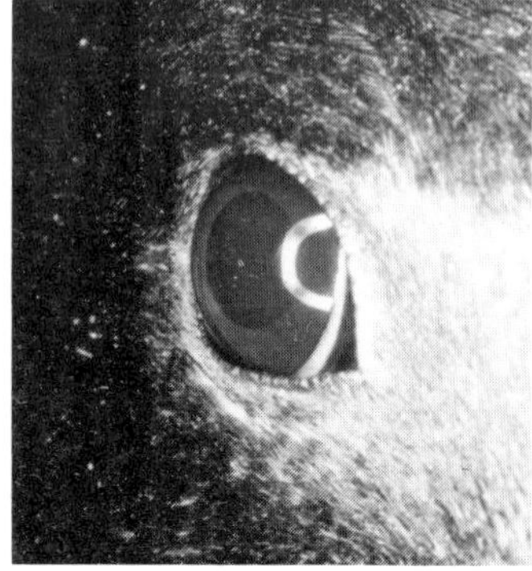

Remember, *check* your references to be certain of the proper seating depth. Fortunately, the Sportsman Series artificial head eye sockets are already molded to the correct seating depth and angle for Tohickon 123 series eyes. The back of the glass eyes are filled with WASCO clay or Sculpall and then simply installed to sit flat in the eye socket.

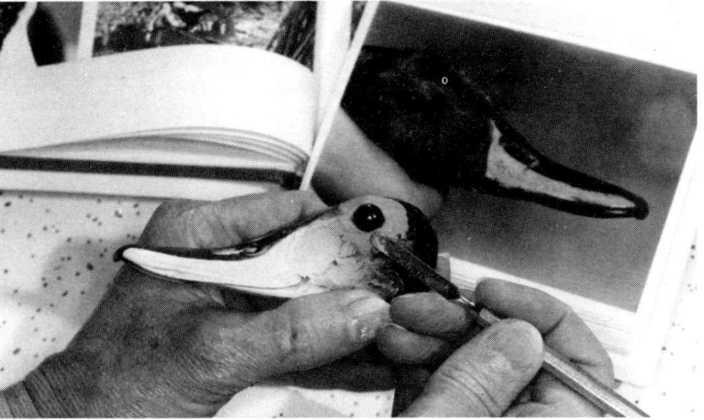

The eye lids and eye structure can be molded in once the corner seating depth has been reached. Use reference for this vitally important phase.

Accu-Flex Mannikin

Flying birds generally exhibit a flattened breast if they are stretched out in flight. A Sportsman Series mannikin can be modified to fit those poses quite easily.

Simply mark the area of the breast and grind 3/16 of an inch off of the breast area with a bench grinder, as shown above.

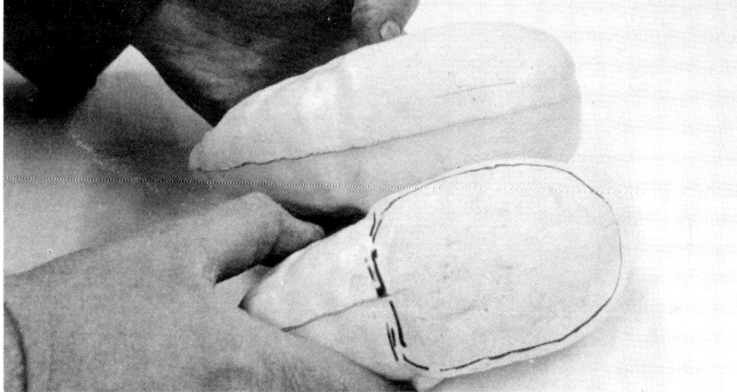

Note the comparison of a standard mannikin and one altered for the flying pose. Remember, "easy" does it. No more than 3/16" on all duck mannikins is sufficient.

Flattening the breast is necessary as the breast takes this position when a duck swims or flies as opposed to a standing mount. The breast drops forward or flattens as it pushes forward as in swimming and/or flying similar to a person sitting in a chair. The body actually widens when we sit. When a duck assumes a swimming or flying pose, the mass of the body meets the mass of water or air and assumes a flatter, more buoyant look.

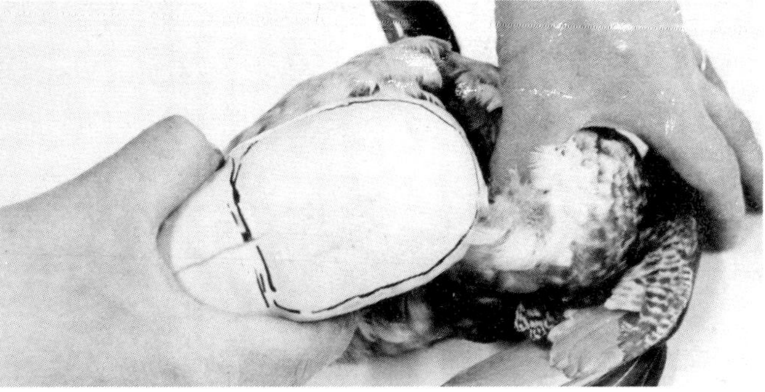

Once the breast has been altered, the mannikin is ready for insertion of the wires through the pilot holes.

The pose for the pintail used in this chapter did not require a flattened breast and was used as is. The Accu-Flex mannikin should be inspected and compared to the carcass to ensure a proper fit. A wire should be inserted through the marked insertion points on the mannikin to form pilot holes for the wires to be run through later in the mounting process. Inserting a wire through these points "before" the bird is assembled will greatly facilitate installing the wires that are attached to the humerus and tibia later in the mounting process.

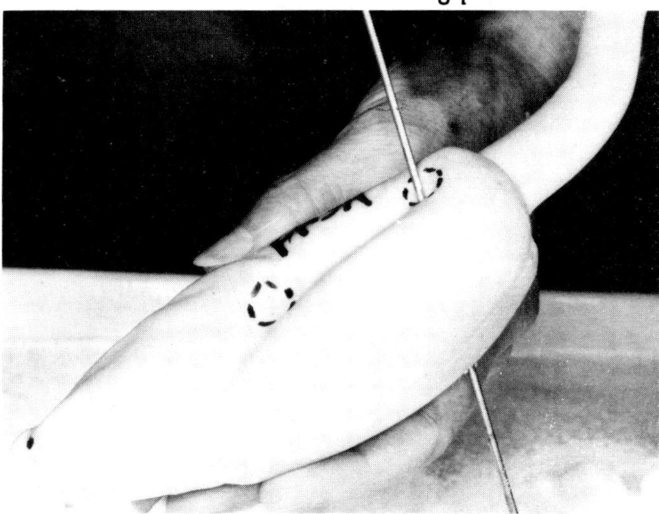

To prepare pilot holes for the wing wires, insert and angle each wire so that it exits the body on the opposite side of the lower quadrant of the breast and then remove the wire.

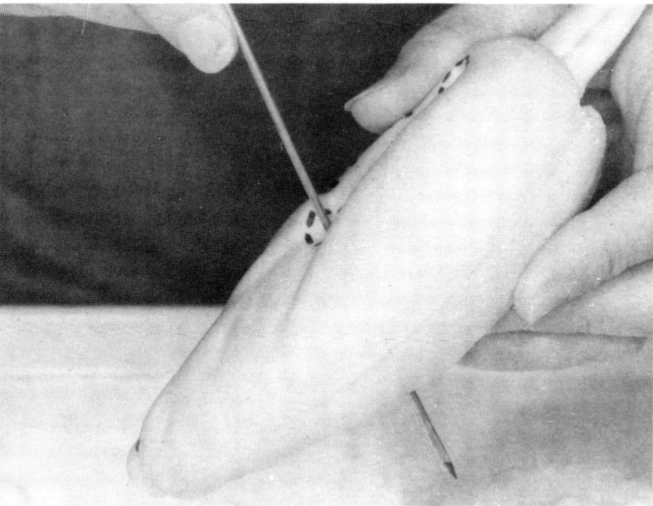

To prepare pilot holes for the leg wires, simply push the wire completely through the mannikin at the designated insertion point and then remove it.

Leg, Wing, Tail & Body Support Wires

The next step is to cut and prepare the leg, wing, tail, and body support wires. The leg wire should normally be cut approximately twice the length of the leg bone (digits, tarsus, and tibia).

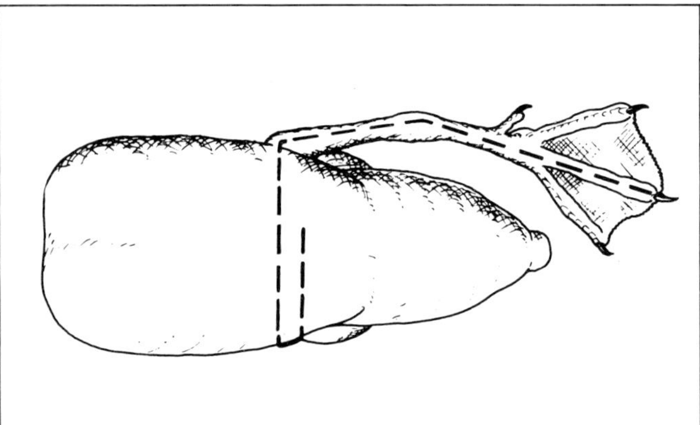

The wire should be of a heavy enough gauge to support the leg when it is placed in the desired position. The leg wires should be sharpened at both ends with a bench grinder or file. Do not make the points too sharp, because when the wire is being run down the tarsus it will have a tendency to snag along the delicate tissue of the interior of the leg.

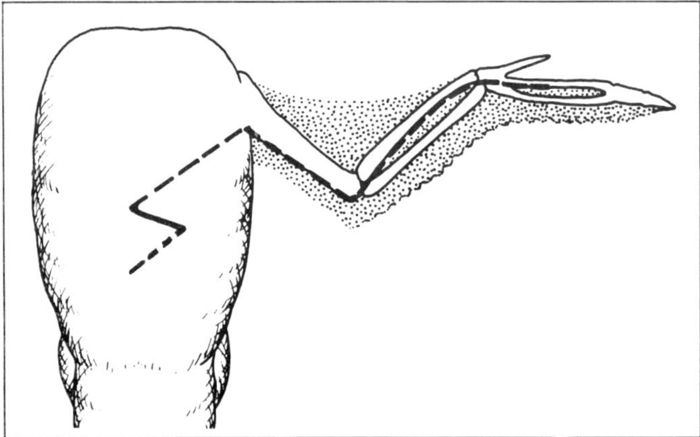

For the wing wires, use a heavy enough gauge wire to support the open wings without allowing them to droop. The wire should be cut twice the length of the humerus and radius-ulna. Sharpen the wing wires the same way the leg wires were sharpened.

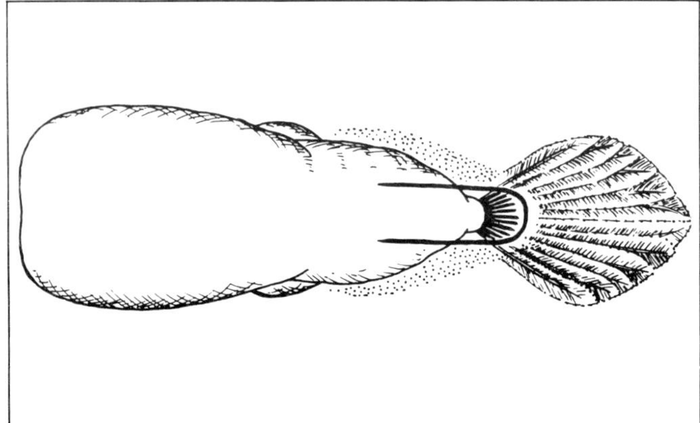

The tail support wire is cut from a medium gauge wire; it is sharpened at both ends and bent in the shape of a horseshoe. The tail support wire should be long enough to insert a suffi-cient amount of wire into the mannikin to support the tail feathers.

The body support wire should be of a heavy enough gauge to support the bird without wobbling. The length of the wire will be determined by the pose of the individual bird. The wire is sharpened at both ends and should be long enough to allow ample clearance of the bird from the wall.

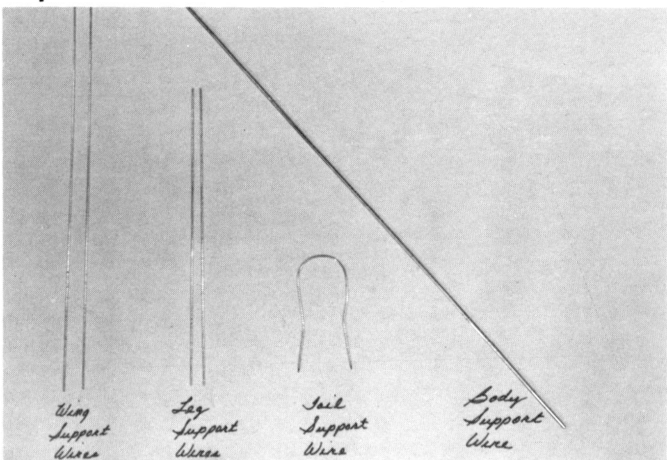

After these preparation procedures, the bird should be ready for assembly.

Assembling the Bird

For assembly purposes, the clean and tumbled bird skin should be placed in a tray of "clean" WASCO granular borax. The borax will help keep the feathers clean during the mounting process.

Wiring the Bird Wings

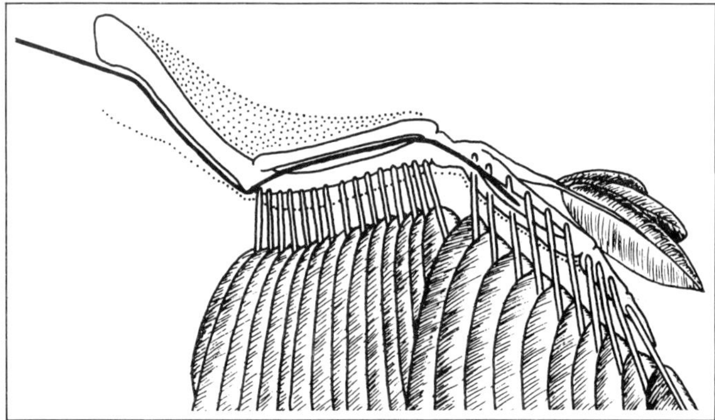

To prevent the wings from wobbling and to keep the wings open, it is important to securely fasten the wire to the wing bones.

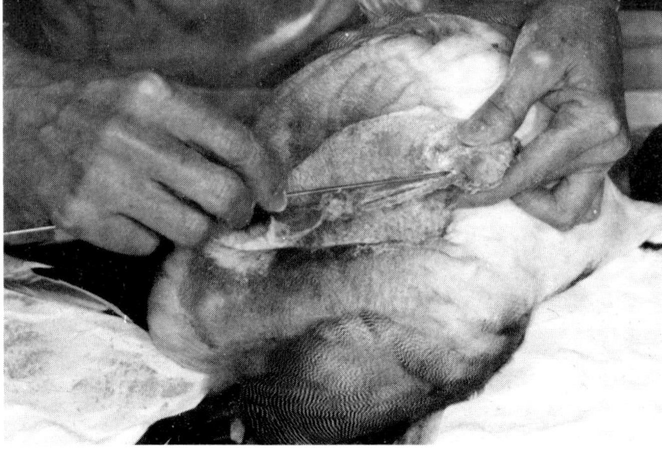

With one hand, straighten the carpal (wrist) joint. Then with the other hand, gently push the wire between the joint into the metacarpal (palm of the hand).

Be careful not to run the wire too far down the metacarpals as the pressure exerted by the wire will distort the primaries. Once the wire is in place, slightly bend the carpal (wrist) joint to lock the wire in place. Position the wire between the radius-ulna, and then bend the wire to conform to the natural bend of the elbow. Continue by placing the wire next to the humerus.

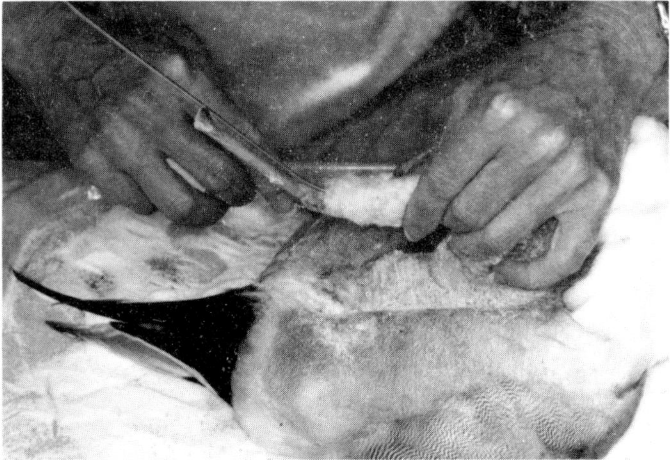

With the wire in place, and properly positioned, secure the wire to the radius-ulna and humerus with cotton thread. It is important to build up the muscles that were removed from the radius-ulna and humerus to their original dimensions.

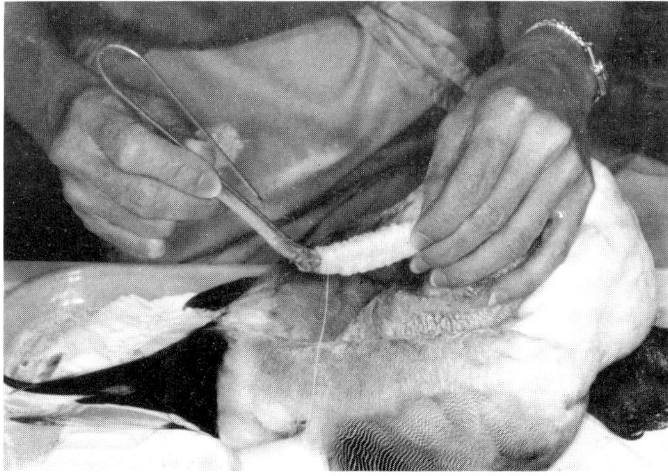

The feathers will not dry in their proper positions if there is no support material under the skin. Rebuild these muscles by wrapping cotton around the wing bones. The cotton is held

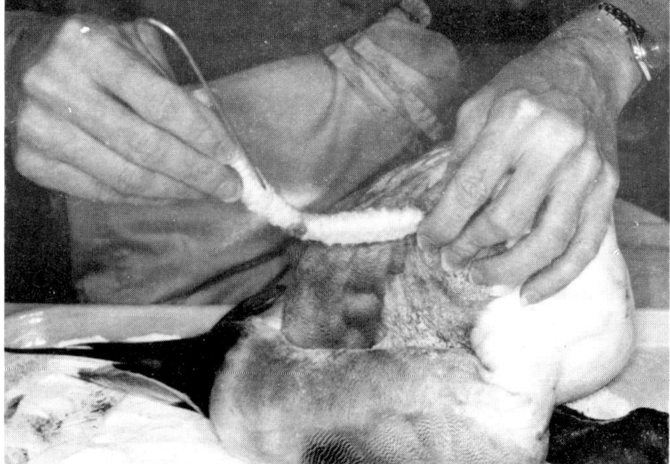

in place by wrapping thread around it as it is placed on the wing bones. Once the muscles are rebuilt to their original dimensions, they may be wrapped with soft tissue paper or WASCO clay.

Wiring the Bird Legs

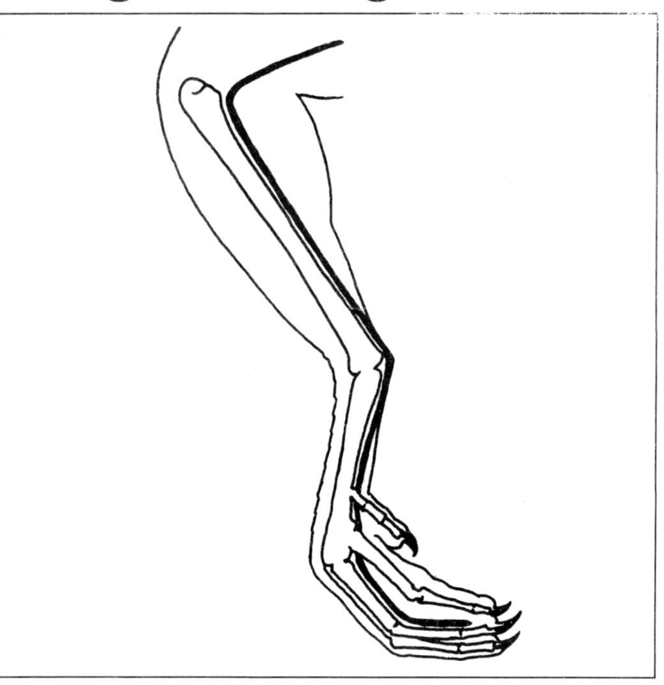

A wire is run through the legs in order to support them and hold them securely in position. Flying birds do not require that the leg wire exit the foot.

When running the leg wire down the tarsus and digit, be careful not to pierce the outside skin of the leg and expose the wire.

Hold the tarsus in one hand and expose the tibia. Place the wire on the back side of the ankle joint and gently push the wire down the back of the tarsus. When the joint between the tarsus and digits is reached, straighten the joint to allow the wire to continue past the joint.

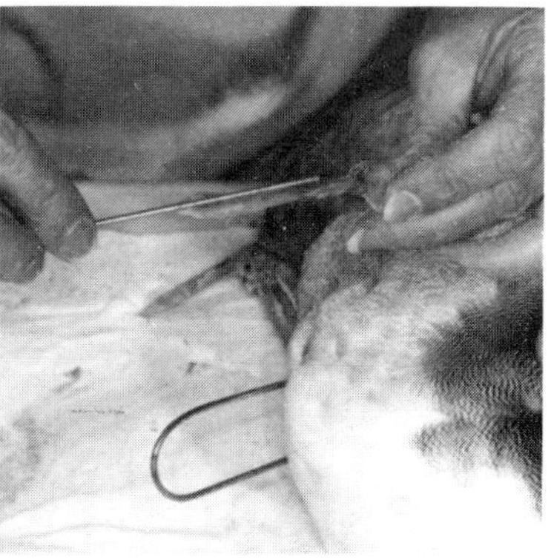

Once the wire is past the joint, continue pushing the wire down to the tip end of the third digit.

Running the wire to the very end of the digit will support the foot so that it will remain in the proper position.

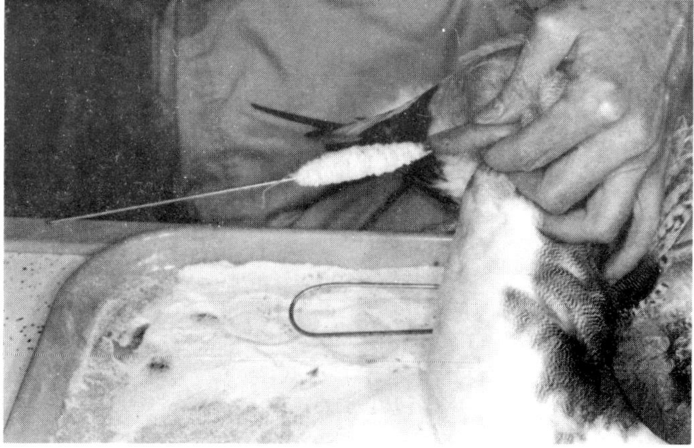

The wire is secured to the tibia by wrapping the tibia with cotton thread. The muscles are then rebuilt to their original dimensions with cotton batting which is held in place with cotton thread.

With both wings and legs wired and muscle structure rebuilt, the skin is now ready for mounting.

Attaching the Body Support Wire

The pose will determine *where* and *when* the body support wire will be placed in the flying bird. For their first mount, many sportsmen choose the standard "banking" pose. If the bird is in "good" condition with respect to the primary and secondary feathers, this pose is an excellent way to display the bird. The banking pose shows a nice profile of the bird's head and, in addition, displays the brilliant colors in the back and wings. There are many other attractive poses such as "dropping-in with the wings back" or "jumping off the water" that show the side view or profile of the bird. In this type of pose, the support wire is exited out the "off" side of the bird. If the bird is placed in a "coming-in" pose, or a pose where the breast or front of the bird is the focal point, the wire should exit out the back of the bird. The pintail in this chapter was "flushing up out of the water" displaying the bird's back; therefore, a support wire exiting the breast was used.

Many times a pose is selected where the taxidermist must use ingenuity and mechanical expertise to hide the support wire from view.

The following illustrations will show the various methods used to attach the body support wires and will also explain *when* to attach the body support wire during the mounting process. Remember, some poses will require that the support wire be attached to the mannikin *prior* to wiring the bird, in other poses the wire will be inserted *during* the mounting procedure, and some positions will call for installing the support wire *after* the bird is assembled.

Back View

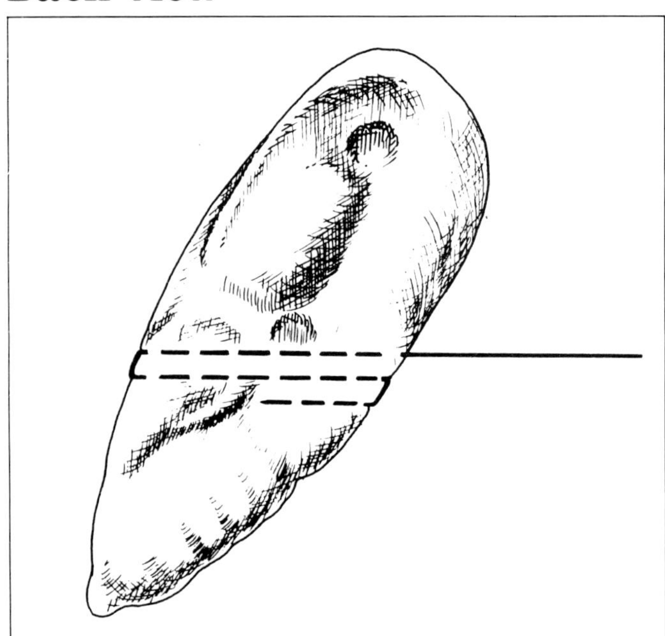

The support wire is attached *before* the bird is assembled.

Insert two-thirds of the wire completely through the mannikin from the breast side, and bend the support wire in a tight horseshoe. (This bend should be made so that when the wire is pushed back through the body there is a small portion of one end of the wire that can be clinched back into the body to lock the wire in place.) Push the horseshoe end snug against the back of the mannikin. The short portion of wire that exits the breast slightly below the entry point of the support wire, should then be clinched back into the mannikin. This will securely lock the support wire into place.

Front View

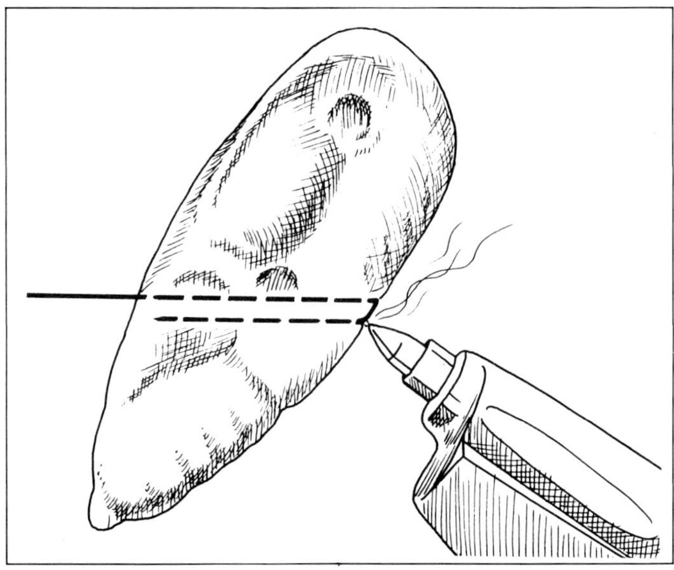

Completely assemble the bird and insert the body support wire just prior to closing the incision.

Before inserting the support wire, make a small incision in the skin where the wire will be inserted through the mannikin. Gently push the support wire through the incision in the skin and continue forcing the wire through the mannikin, exiting the wire out the center of the breast. Push approximately two-thirds of the wire through the mannikin. When making the horseshoe bend in the wire, make the bend shallow enough so that a small portion of the wire does *not* exit the back of the bird. The "short" end of the horseshoe will remain embedded in the mannikin. To secure the wire into the mannikin, place a small amount of hot-melt glue over the horseshoe bend on the breast of the mannikin.

With the wire secure, close the incision and proceed with the mounting procedure.

Profile or Side View

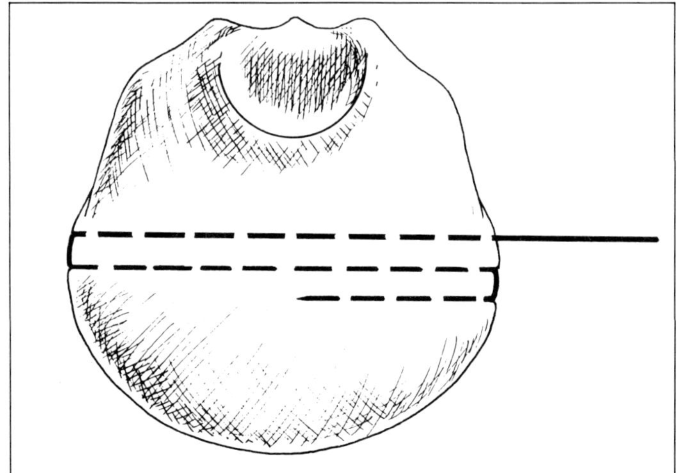

It is best to insert the body support wire *after* the bird is assembled and the incision is closed.

First, bend the support wire in a tight horseshoe. This bend should be made so that when the wire is pushed through the body there is a small portion of one end of the wire that can be clinched back into the body to lock the wire in place. Remember, it is important that the wire is cut long enough to attach the bird to a suitable panel or piece of driftwood. After the horseshoe bend is made, insert the wire under the wing (in the short, down feathers between the wing and the side pockets) on the show side. Push both ends of the horseshoe-shaped wire through the bird, exiting the wire under the wing on the opposite side of the entry point. Make certain the horseshoe bend of the wire is pushed tightly against the body. Also, be careful that the long feathers of the side pockets are *not* caught under the bend of the wire. With the wire in place, clinch the short end of the wire back into the mannikin. This will securely lock the support wire in place.

The bird is now ready to proceed with the remaining mounting procedures.

Installing the Accu-Flex Body

To install the Accu-Flex body, first cut the neck wire off about one inch from the end of the neck and curl it into a loop. Looping the end of the neck wire in this manner will keep it from snag-

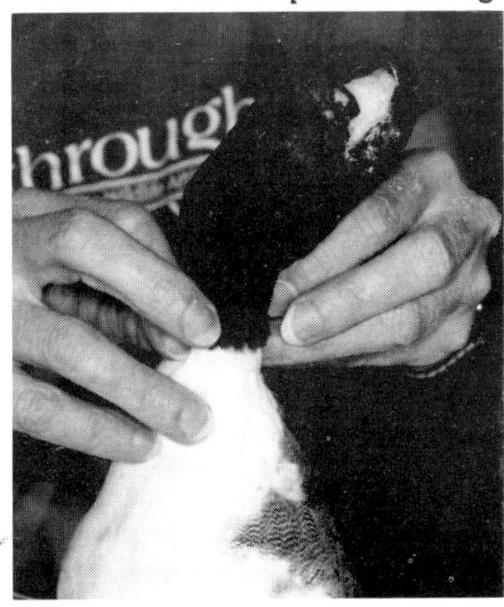

ging on the neck skin as the neck is being pushed through. Later in the mounting process, the loop will be inserted into the hole in the back of the artificial skull and hot glued into place.

Next, insert the Accu-Flex mannikin through the breast incision and work the skin down over the artificial neck. Do not make adjustments to the neck until the wing wires are attached to the mannikins.

Securing the Wings

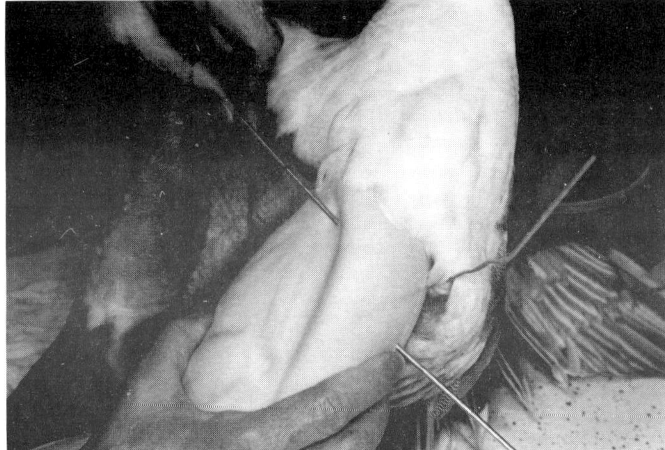

Now insert the wing wire through the pilot hole at the insertion point (lock point) on the mannikin.

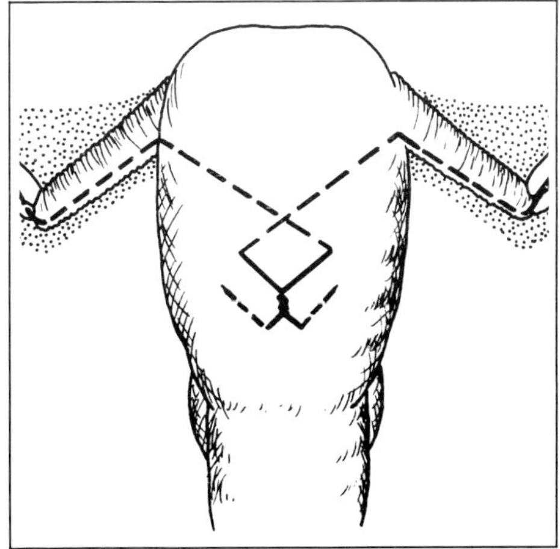

The wire should be angled so that it exits the body on the opposite side of the lower quadrant of the breast. Pull the humerus snugly against the body of the mannikin.

Insert the opposite wing wire in the same manner, exiting in the lower quadrant of the breast. Make

certain that both wings are securely in place and that the humerus is "snug" against the body. Cross the exposed wires with a pair of lineman's pliers and make several twists on the wires, locking them together.

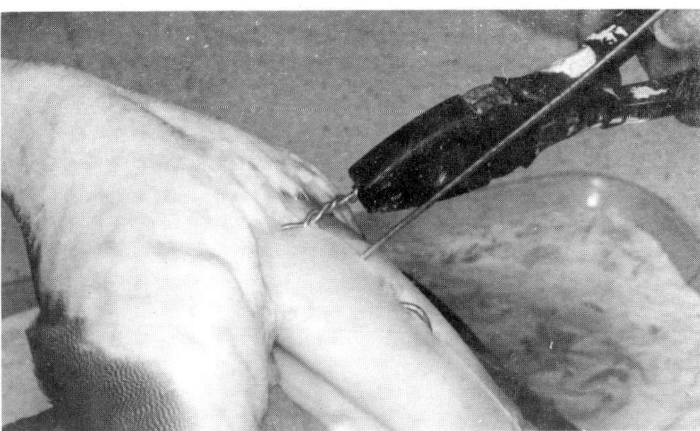

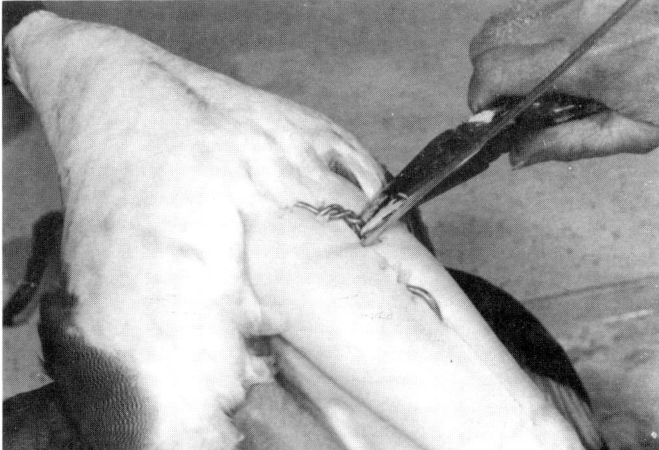

Bend the remaining wire into the mannikin.

Securing the Legs

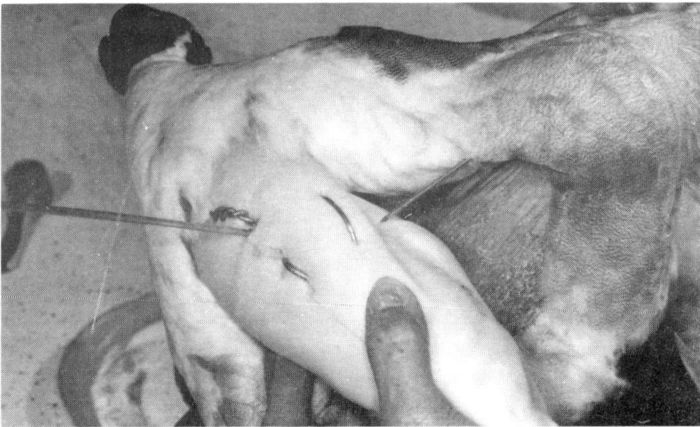

Insert each leg wire into the pilot holes at the insertion points.

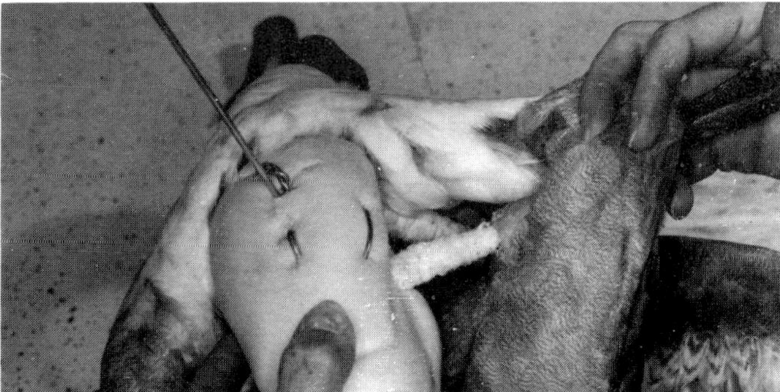

Run the wire through the body and push the tibia snugly against the body. Hold the tibia firmly against the body and

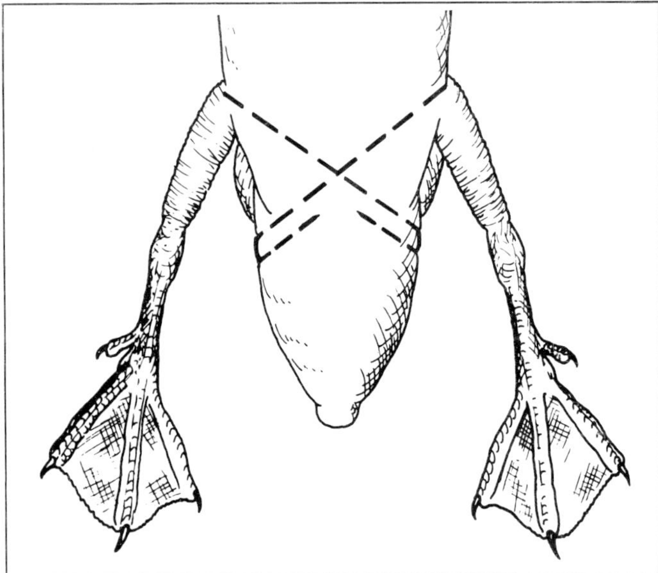

bend and clinch the end of the wire into the body. It is important that the tibia is fitted snugly against the body to prevent the legs from wobbling. The same method that was used on the wings (twisting the ends of the wires and bending them into the mannikin) also works well to secure the legs.

With the wings and legs secured, gently pull the skin around the rear section of the body and carefully adjust the tail feathers. Separate the under tail coverts from the tail feathers.

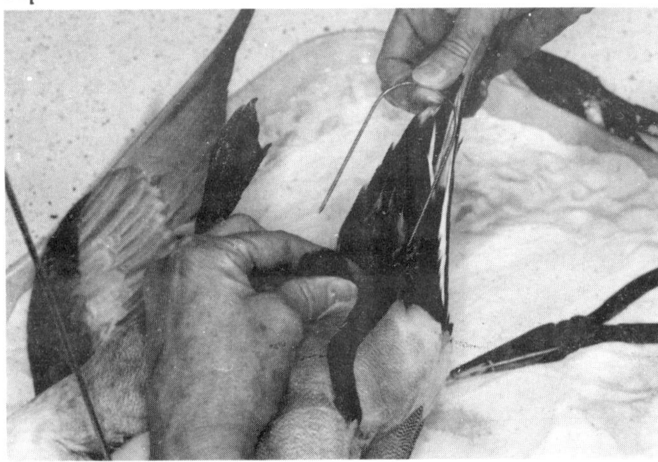

Insert the horseshoe-shaped tail support wire between the under tail coverts and the tail feathers. The tail support should not be pushed snugly against the body, but positioned with enough distance between the body and the base of the tail feathers to align and adjust the feathers naturally in the rear section of the bird.

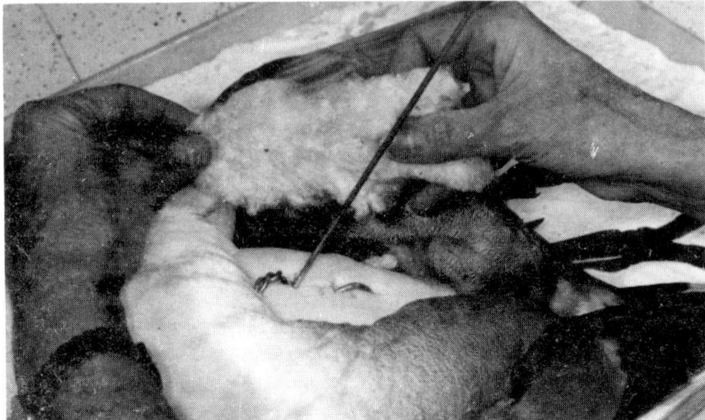

Starting at the neck, adjust the skin and position the feathers. Add cotton batting where needed to fill out the side pockets

or other areas that may need additional filling. (If using the dry preservative method, when the skin and feathers are in their approximate positions, sprinkle dry preservative in the open incision "now"!)

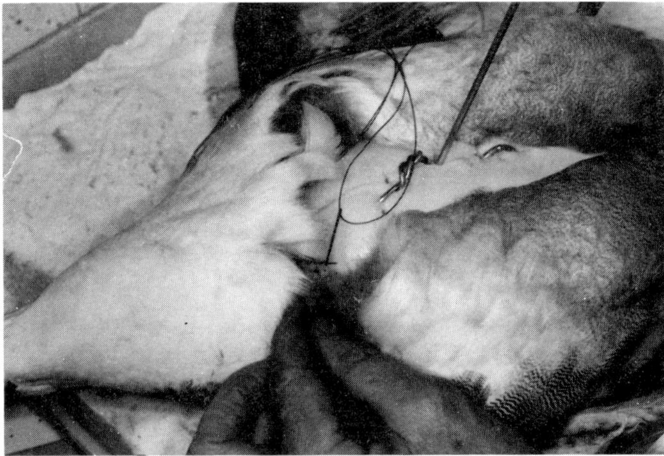

Close the incision using a double strand of nylon thread with a three-cornered sewing needle. Begin closing the incision by starting at the top of the incision and sewing toward the tail section using a baseball stitch.

Make small neat stitches, being careful not to get the feathers caught between the stitches. (See the following illustrations of the baseball and other stitches.)

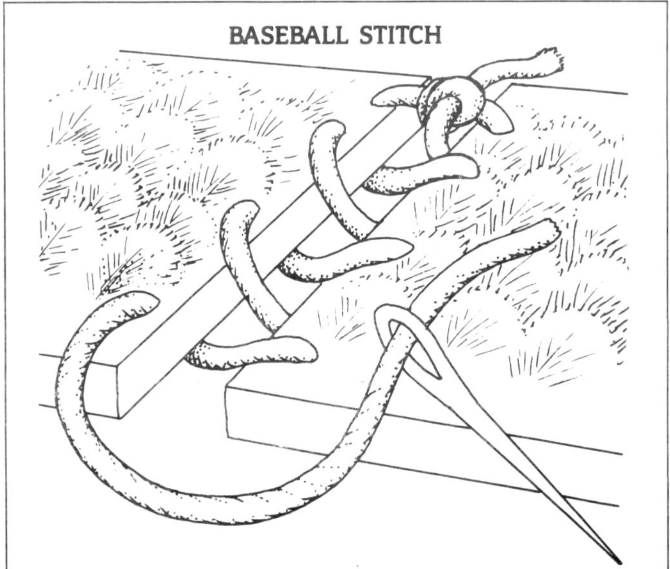

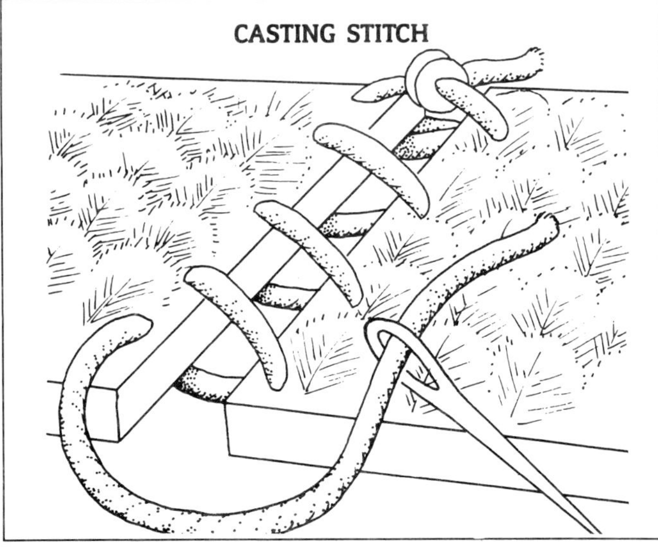

The *Breakthrough* Bird Taxidermy Manual

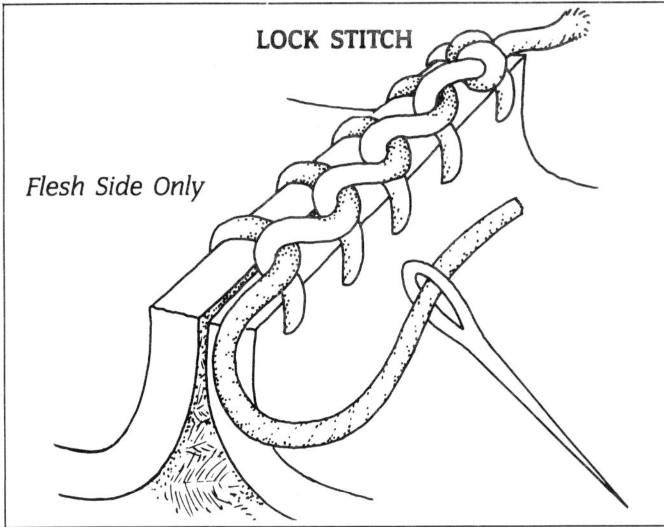

After the incision is closed, tie a knot in the thread, and then insert the needle through the body exiting the back. This will pull the knot inside of the skin. Then cut the thread close to the skin. The feathers will hide the tip end of the thread (see photos below).

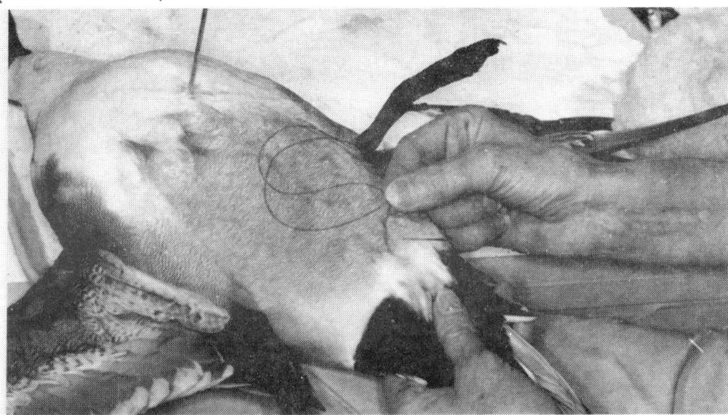

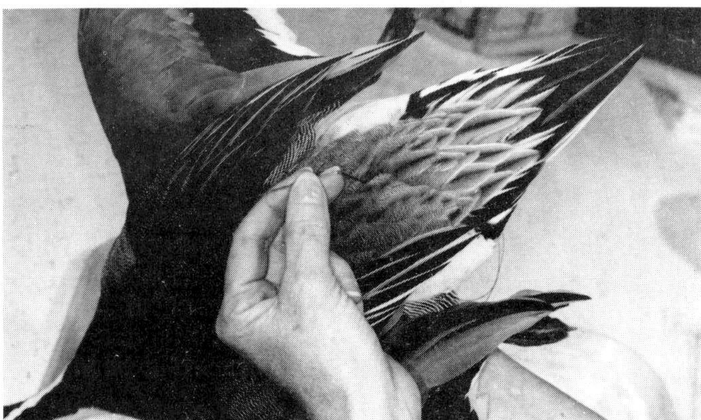

Attaching the Artificial Head

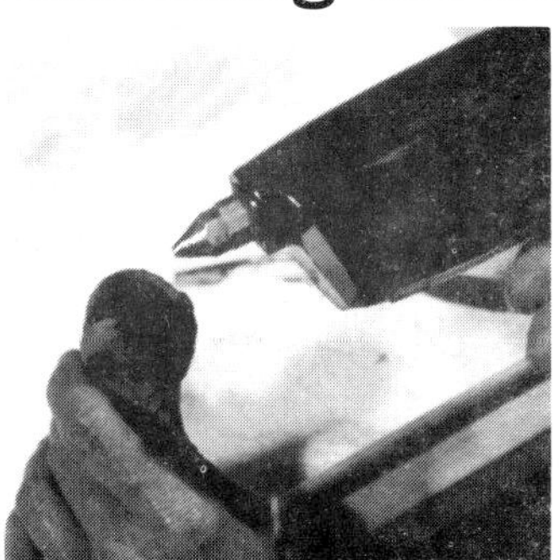

Fill the hole in the back of the artificial head with hot-melt glue and insert the looped end of the wire directly into the glue-

filled hole. Hold the artificial head and neck in the proper position until the glue cools. Once the glue has cooled, it will securely lock the two together.

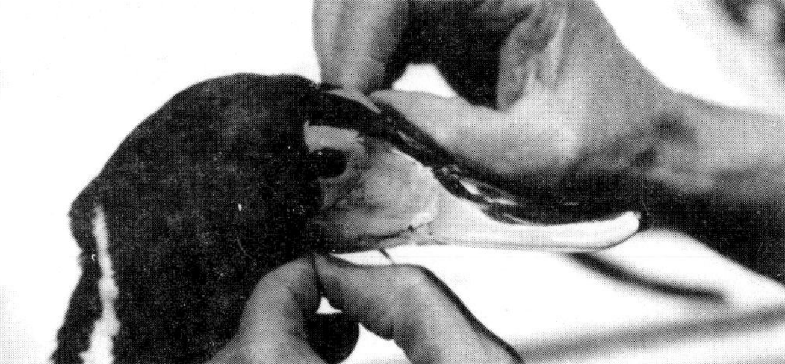

Any additional sculpting around the base of the artificial head can be done at this point with WASCO clay. Carefully, pull the skin over the artificial head. (Keeping the skin of the head and eyes damp with wet cotton during the mounting process will make this step easier.)

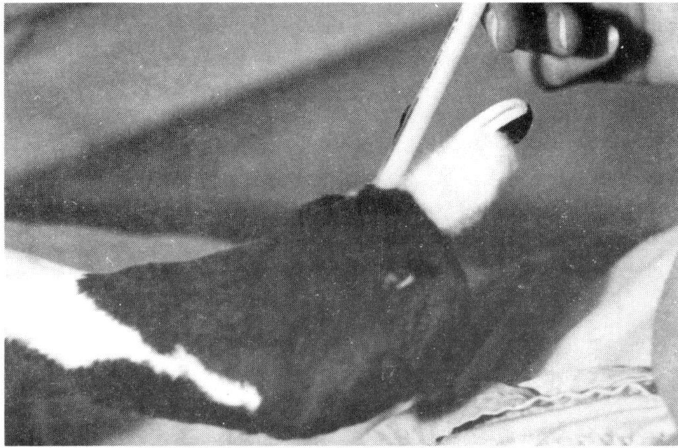

You should at this time fill the cheeks or other areas of the head with cotton batting to fill out areas in order to achieve a fuller look to enhance a pose if so desired.

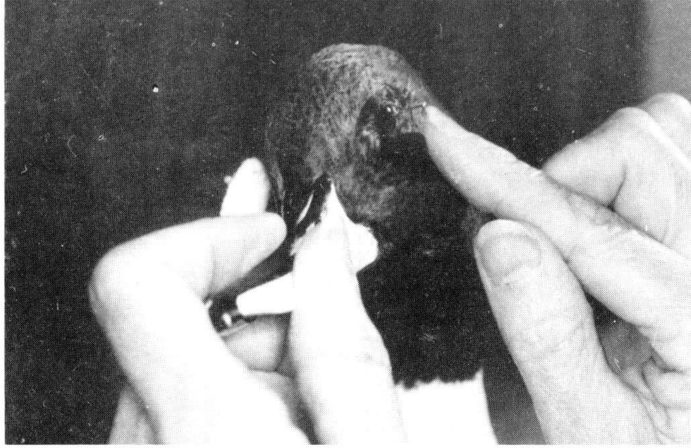

Once everything is filled to its proper proportions you may temporarily pin the skin in place around the head and eyes.

Find the center of the forehead and pull the skin flap down into the notch at the top of the bill. Using magnifying type glasses for this delicate operation will be quite helpful.

Apply Nu-Glue (*Polytranspar*™ Epoxy Adhesive) with an artists' brush (do not use an expensive one). Carefully begin lifting the skin and painting glue around the edge of the bill.

Now press the skin into place along the edge of the bill and pin it securely until dry. Then work down each side from that point, being careful to put glue on the head rather than on the skin and also to keep the glue out of the feathers. Remember, it is easier to keep the skin clean and free from excess glue than to try to remove Nu-Glue from the feathers.

If the glue does get on the feathers, it may be removed with *Polytranspar*™ Thinner. The primary advantage of using Nu-Glue is that it has a longer "set up" time than Instant Bonding Glue which will allow additional "working time" for positioning the skin. Because Nu-Glue takes approximatley 45 minutes to one hour to set up, there is no need to frantically adjust and position the skin. If an adjustment does need to be made, it will be less of a hassle to make because of the longer set up time. Also because Nu-Glue was originally developed and formulated for use with mammal skins, the adhesive capabilities will be far superior to other types of glues.

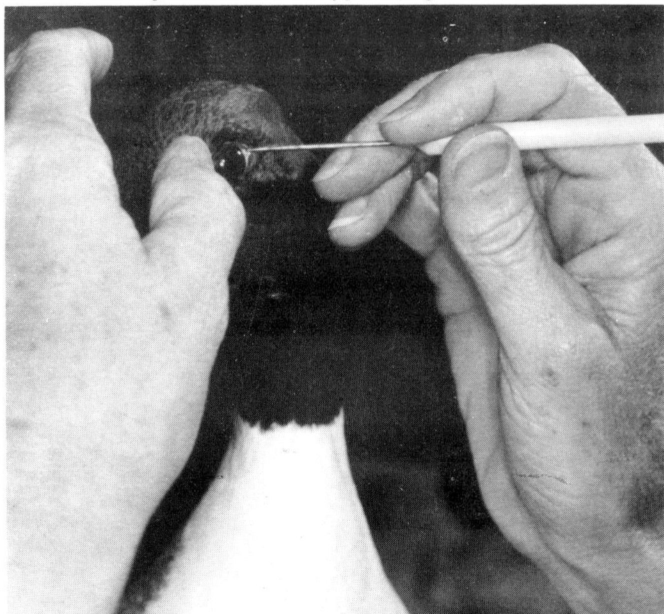

Once the Nu-Glue has set, adjust the head to the desired position. Make final adjustments to the eyelids with a small feather adjuster. Press the excess tissue of the lower lid between the

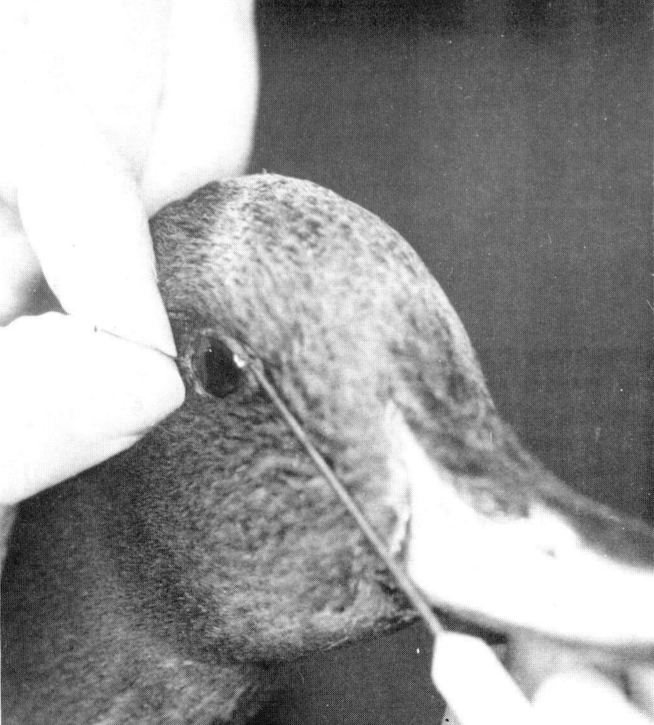

glass eye and the filler used in the eye socket. If an oil-based clay was used as a filler, once the proper contours of the upper and lower lids are achieved, place a small insect pin in the front and rear corners of the eye. These pins will keep the lids in place and will be removed when the mount is dry.

Temporary Base Attachment

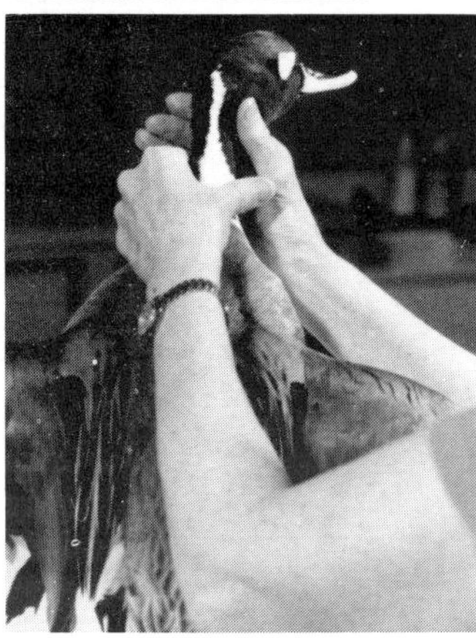

Next, the bird is attached to the permanent wall support (panel or driftwood) or to a drying rack. Utilization of a drying rack will allow total accessibility to pose and card the bird from all angles.

Posing the Bird

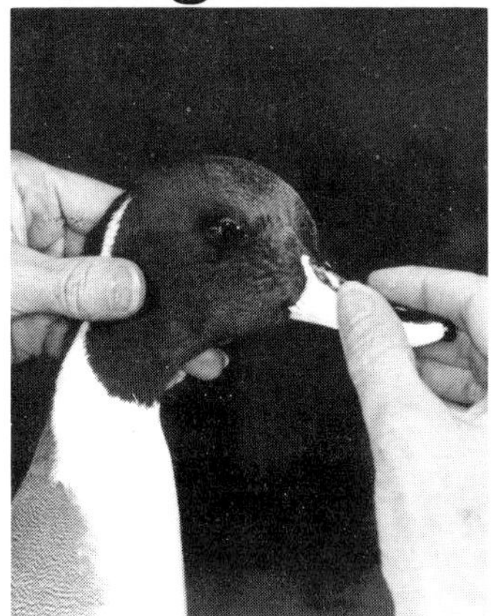
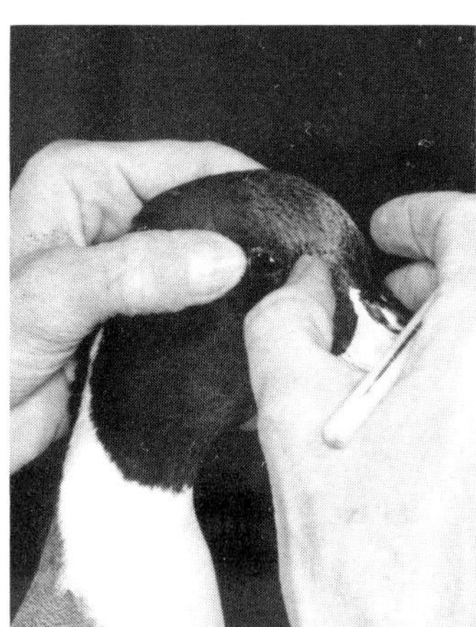

Check the bird from all angles for balance and symmetry and adjust as needed. Once the bird is in the desired position, begin aligning the feathers and placing the feathers in their proper feather tracts. "Feather tracking" the bird is best accomplished with a feather adjuster and tweezers.

Simply place the tip of the feather adjuster next to the skin and lift up. This will cause the misaligned feathers to fall back into their proper tracts. Tweezers also work well to align the

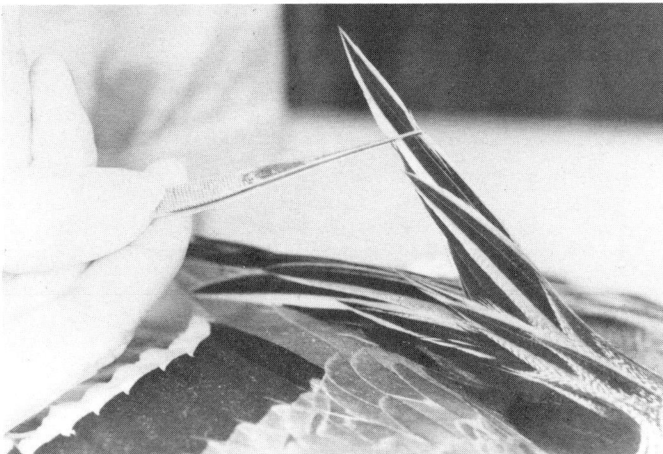

feathers. Pinch the end of the misaligned feather and gently pull on the feather. One gentle pull will usually get a stubborn feather back into place.

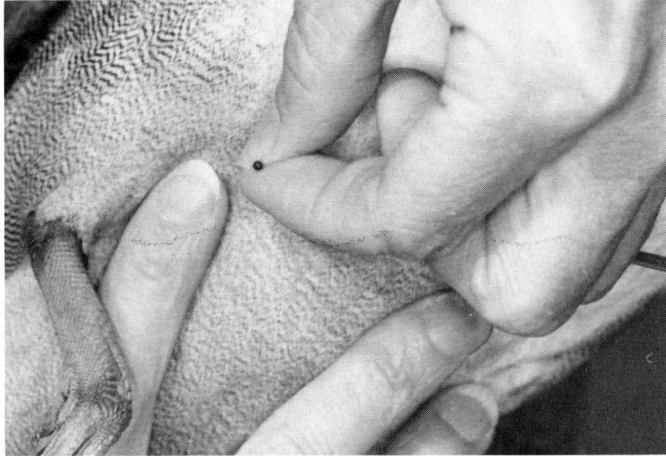

Once all the feathers are in their appropriate positions, the skin should be pinned to show the underlying anatomical detail. When studying any of the pictures in "Prairie Wings," it becomes evident how much detail can be placed in a flying mounted bird, especially with the incredible amount of anatomical detail sculpted in the Sportsman Series bird mannikins.

The detail is put into the mounted bird simply by working the skin and placing ceramic headed pins in the areas that are to be defined.

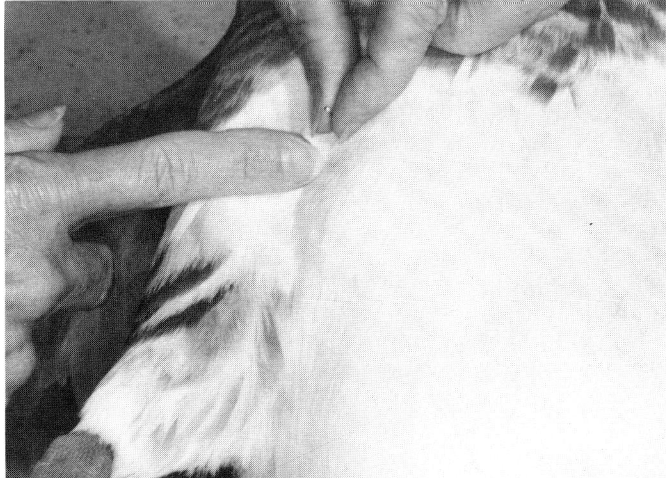

If a bird is "dropping-in with the wings back," the area under the wings (side pockets), breast, belly and tibia are exposed to the viewer. Extending the tibia and pinning the feathers between the tibia and the belly will show an incredible amount of detail and truly gives the bird a lifelike appearance.

Pinning and tucking loose skin in the abdominal area defines the subtle contours of the legs and caudal areas.

DETAILING A FLYING MOUNT: Wings

The wings (primaries, secondaries, tertials, wing coverts, scapulars) are often difficult to track. One thing for certain, any feathers out of place on the wing show up like a flashing neon light. The easiest way to groom the wings is to approach them in a very methodical fashion.

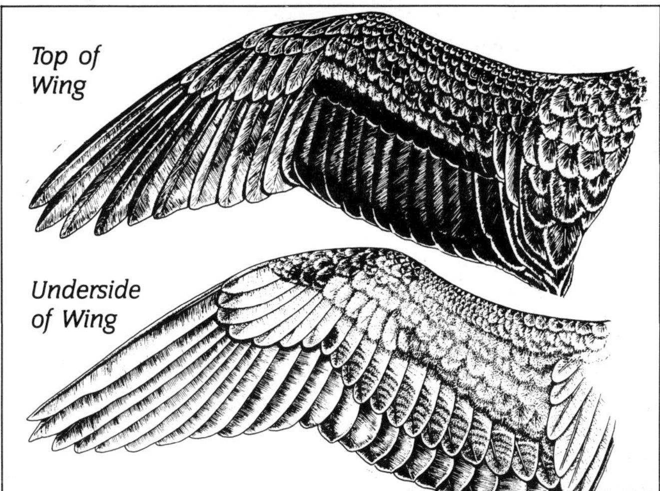

A basic understanding of how a wing opens and how the individual feathers interact during flight is mandatory. This is why carcass sketches, accurate wing placement points and accurate attachment of the wings to the mannikin were so important during the mounting process. The wings are anatomically designed by the Lord to fit and conform to the natural curvatures of the back; so if the bird is not assembled properly, no amount of masking, pinning or carding will get the feathers to fall in place.

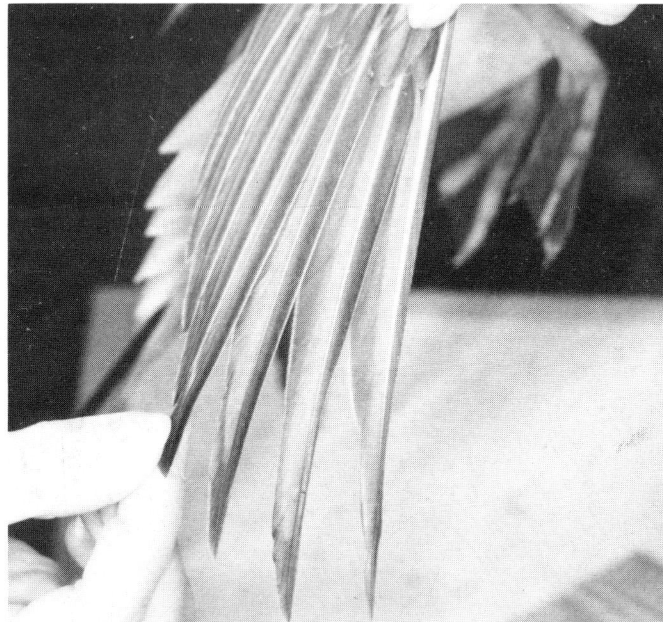

The primaries slide under the secondaries which slide under the tertials. Individual feathers in each one of these areas slide one under the other. The first primary feather slides under the second; the first secondary feather slides under the second, etc. A thorough understanding of the way the wing closes and extends (and how the feathers lie) facilitates tracking and placing the feathers into their appropriate positions.

By taking each wing separately, starting with the primaries and placing them in proper sequence, then progressing to the secondaries, tertials, and wing coverts, positioning the feathers in the wings becomes less confusing.

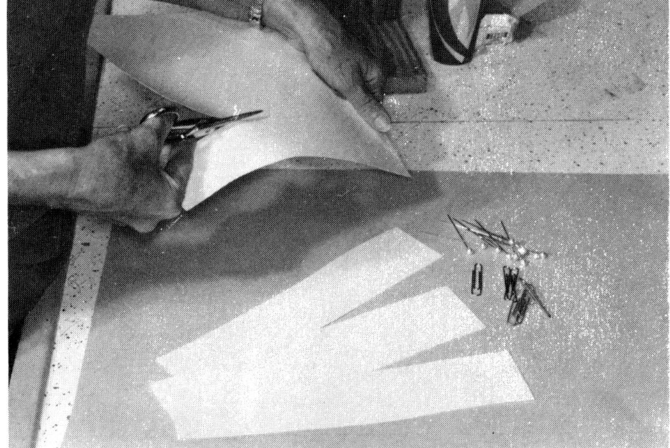

Once the feathers are aligned, the open wings on flying birds must be carded to keep the feathers flared and in their proper position. Medium weight poster paper, cut in 3 inch wide strips, and slightly longer than the open wing, works well.

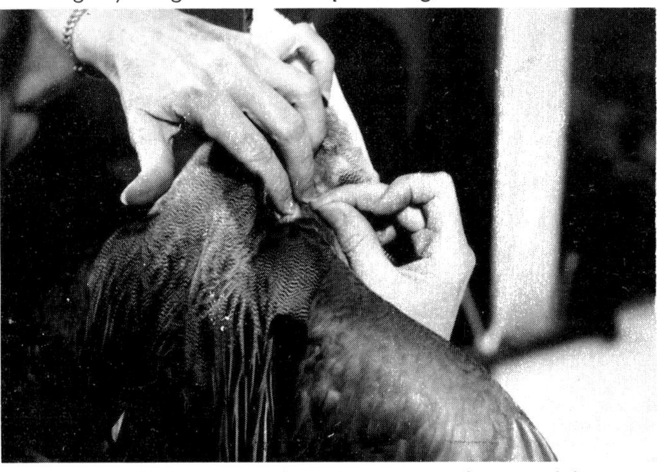

In order to keep a smooth appearance to the top of the wing and to better simulate the position of the skin under the lesser coverts, it is necessary to pin the skin located under the scapulars to the mannikin immediately in front of the humerus. Pinning the skin in this manner facilitates aligning the lesser coverts of the wing.

With the skin pinned, align the feathers in the wing starting with the primaries and progressing to the secondaries, coverts and tertials. Once the feathers are aligned, the carding should be placed on the wings.

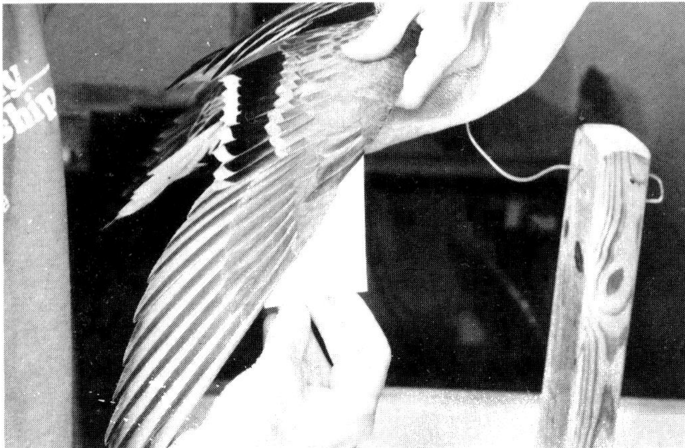

Place the first card on the underside of the wing.

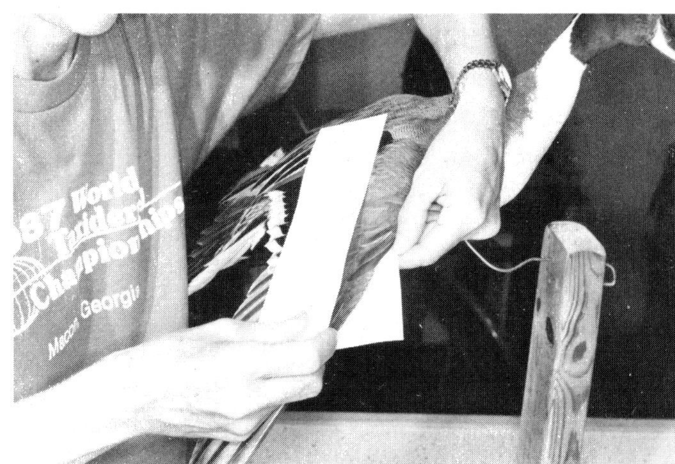

With one hand, hold the carding under the base of the first primary feather. Place the second strip of carding over the top of the wing.

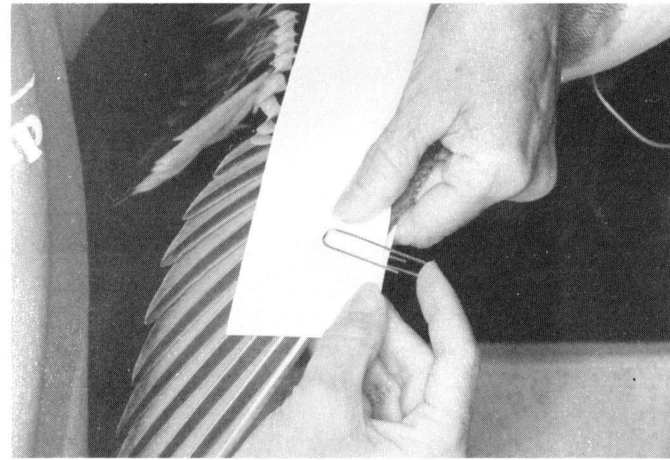

Hold both pieces of carding on the base of the first primary feather with one hand, and with the other hand, place a jumbo paper clip over the carding material and primary to hold both the carding and the primary in place.

Then, approximately every inch, place a Jumbo Head Pin through both pieces of carding material. As the Jumbo Head Pins are placed in the carding, keep checking and adjusting the feathers to make certain they are staying in position.

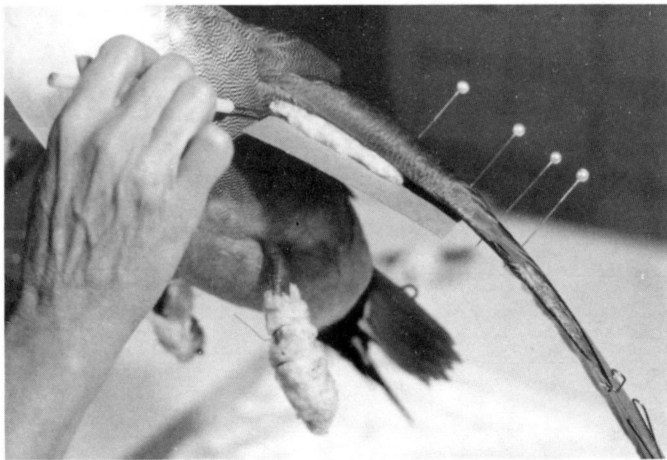

Use cotton batting to fill between the carding and wing. The soft batting will prevent flattening out any of the feathers. Depending on how elaborate the pose, additional carding or pins may be needed to maintain the desired angles of the wings.

To add curvature to the wing tips to portray "lift," the primary feathers will have to be heated and then bent to the desired position. Some taxidermists use hair curler tools to heat them, others use steam; however, both will work.

Tail

The tail plays a very important part in the bird's ability to maneuver during flight. Understanding the many functions of the tail will enable positions to be incorporated into the mount to add realism to the pose.

One of the reasons the tail support wire was *not* pushed snugly against the body earlier was to allow the artist the opportunity to adjust the tail later in order to accommodate the many different positions and poses that may be utilized. For example, slightly bending the tail support wire and flaring the tail will nicely compliment many poses.

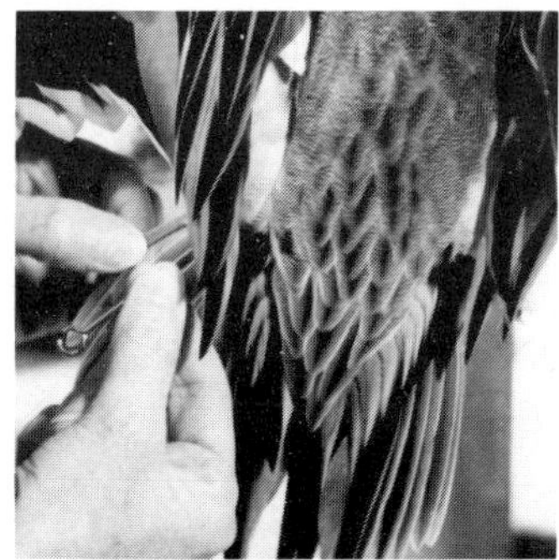

To keep the tail feathers spread open, simply place an upholstery pin next to the inside of the outermost tail feather on each side of the tail. Push the upholstery pin through the skin into the mannikin. Strips of carding and T-pins may be used to keep the feathers evenly spread.

Do not place carding over the upper tail coverts. The carding will flatten these feathers and change the overall shape of the bird's back.

Breast

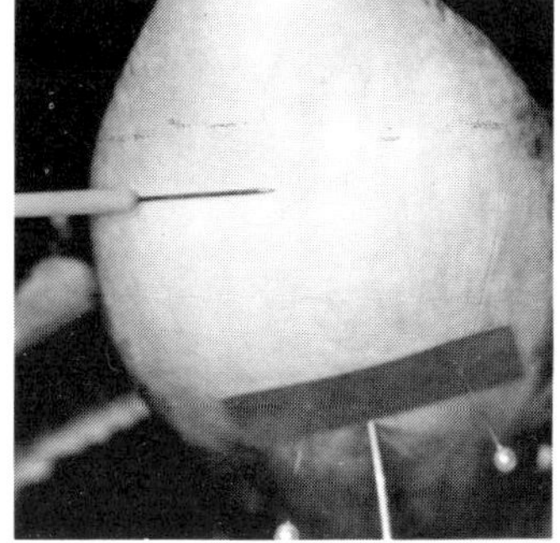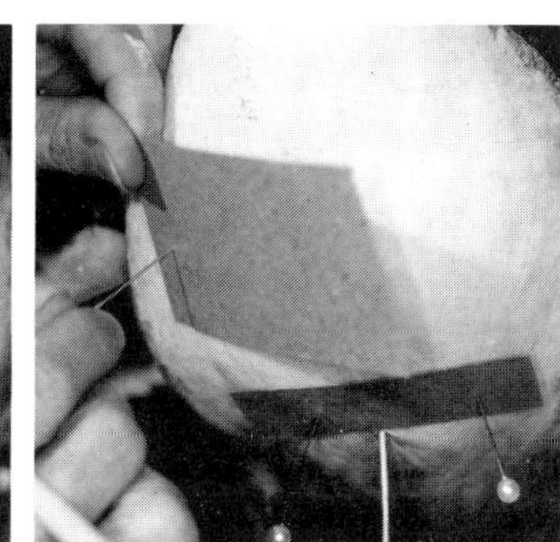

Adjust the position of the feathers of the breast to perfection and card as shown in these photos. Always install a card adjacent to the support wire to help hold the feathers closely and neatly around it.

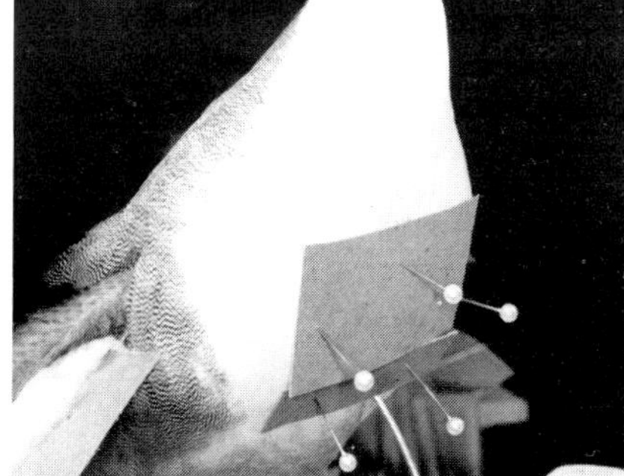

Feet

When a bird is flying, the feet are used to supplement tail action, and they act as a rudder for steering, braking and balancing. The feet are an integral part of the flying mount and an appropriate amount of time should be spent studying the positioning of the feet.

In order to keep the feet full and plump looking, it is necessary to inject them. Many taxidermists use formalin (formaldehyde) for this step. When using this chemical it is imperative to wear protective goggles, surgical gloves, and to have a good ventilation system. This chemical is extremely dangerous. It is a good idea to use the formalin just before you are ready to close your studio for the day as the fumes will have overnight to dissipate. *Remember, Formalin has a devastating effect on the eyes, skin and respiratory system. Always use the recommended safety precautions!* Use gloves, goggles, apron, and *plenty of ventilation.* Many taxidermists prefer to use a mixture of 20% alcohol and 80% glycerine to inject into the feet. Although this mixture is not as effective as formalin, it is much safer.

When injecting the feet with formalin or the alcohol/glycerine solution, use a small, insulin type needle. This small needle will prevent the formalin from leaking out of the puncture site.

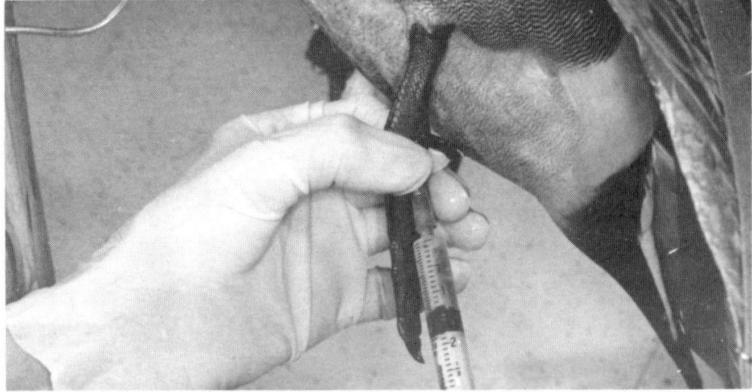

To ensure that the foot will remain full and plump, apply another injection of formalin on the second day if necessary.

After the feet have been injected and the fumes have dissipated, it is necessary to card the feet in order to keep the digits and webs from closing and losing their position.

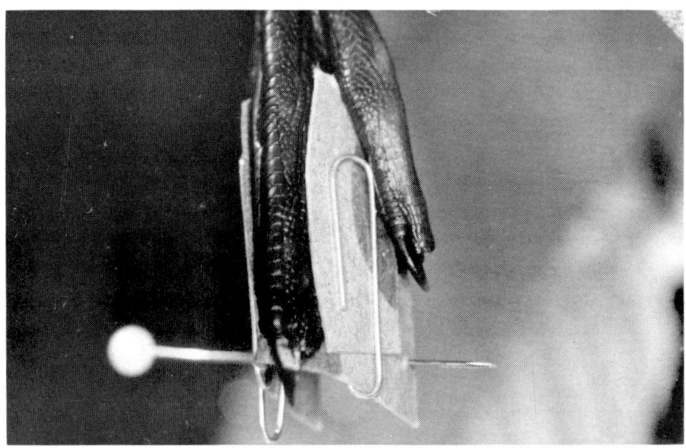

Cut the carding to fit between the digits. Place a piece of carding both under and over the web. The carding is held in place with paper clips. Make certain that the carding fits exactly between the webs as the carding will leave an indentation on the webs if it is not properly fitted.

Now that the feet have been properly positioned, and feathers aligned, wings secured, the bird is ready to be placed out of "harm's way" on a shelf or wall to dry.

Drying

The completed mount should be placed, *away from the main flow of traffic*, to dry. The bird is vulnerable at this point, and accidentally bumping and/or dropping the bird will mean having to spend valuable time redoing procedures.

It is best *not* to attempt to force dry a bird by applying or placing the bird close to a heat source. This will usually cause some excess shrinkage in the delicate tissue around the eyes and bill.

If the carding is removed from the wings on a flying bird before it has had time to adequately dry, the feathers *will* shift from their positions, and the entire carding process will have to be redone.

During the drying process, periodic checks should be made to make sure that all feathers are drying in their proper positions. In most cases, the bird should be thoroughly dried and ready to complete in ten days. Drying time will vary from region to region depending on the humidity in a particular locality.

CHAPTER 9

How To Make Your Own Mannikin

Wildlife Artist Supply Company has provided taxidermists with a vast array of anatomically accurate mannikins to fit virtually every popularly mounted specie of bird, from Bobwhite quail to wild turkeys.

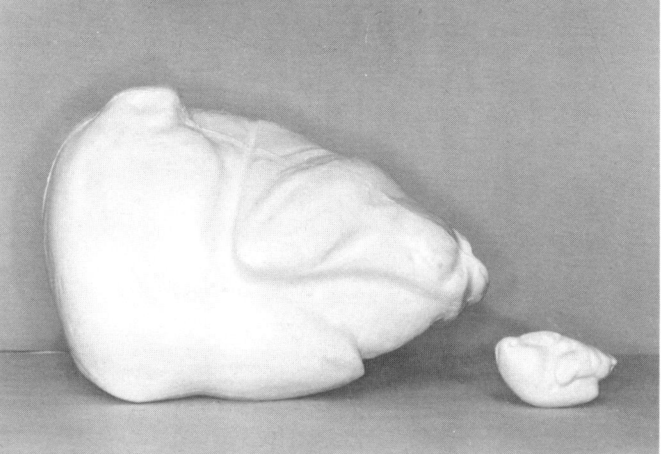

Extremely accurate commercial mannikins (such as these Sportsman Series mannikins) are available for most any specie of "commonly" mounted birds.

But, just as there are anatomical differences between people, the same principle holds true for birds. Fortunately for the taxidermist, commercial mannikins are available in adequate selections to accommodate the anatomical differences of the commonly mounted waterfowl and upland game species. However, on occasion, a taxidermist will encounter a species or unique pose that a commercial mannikin is simply not available for. An example would be when a particular pose would best be accomplished by using a slight bend or twist to the bird's body that cannot be achieved with a commercial type mannikin. When this need arises, the task is best accomplished by constructing your own mannikin. A good example would be the bald eagle in this chapter.

Rarely mounted species such as the bald eagle outlined in this chapter, will "require" a handmade mannikin.

This majestic bird cannot be mounted without a permit; therefore, due to lack of commercial demand, there is no commercial mannikin available for an eagle. In such cases it becomes not just highly desirable, but a "necessity" to know how to accurately construct an anatomically accurate mannikin of your own. The primary thing to remember when doing so is to *build your mannikin to represent "exactly" (to the best of your ability) the actual bird's body.*

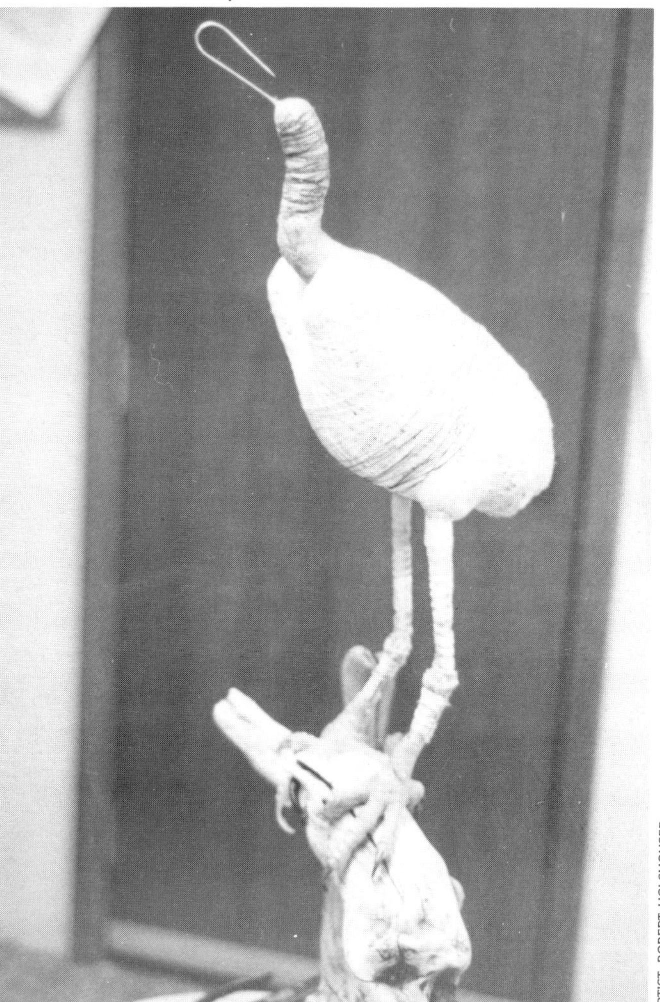

The Wrapped Mannikin Method

The wrapped mannikin method is probably the most popular method of bird mannikin construction. This usually entails wrapping a body using a combination of tow, excelsior, cotton batting, and cotton twine. The only tools necessary will be a pair of calipers, a 6" sewing needle, and a pump sprayer.

As this sandhill crane body (wrapped by master taxidermist Robert Holshouser) demonstrates, the artist must have a keen understanding of the anatomy of a specie when wrapping a mannikin.

At first, this method may seem awkward and difficult to accomplish; but, if each step is followed, it is possible to wrap an extremly accurate mannikin and achieve excellent results. Wrapping a body is time consuming and is not really economical when a commercial mannikin is already available. One of the main advantages and reasons that every taxidermist should master this method is, that wrapping a body from scratch with excelsior "forces" the taxidermist to study and to fully understand the anatomy or "inner body" of that particular specie.

Because you will not be able to rely on a commercially prepared mannikin for attitude and accurate anatomical detail, sketches made prior to, and after skinning, will prove to be invaluable aides in both wrapping and posing the bird. The carcass of the skinned bird is also a valuable reference for this process.

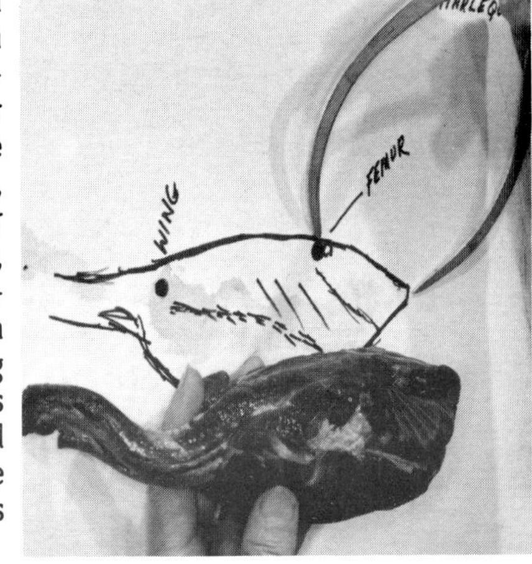

Carcass Sketch

Before the actual skinning operation begins, place the bird on its side and carefully trace the outline of the head, neck, body, and legs.

Pay particular attention to the location of the eyes, feather tracts, and the placement of the wings and legs. Notations of the color of the feet and bill should be made as soon as possible after the bird expires. These will prove to be helpful later when finishing the mount.

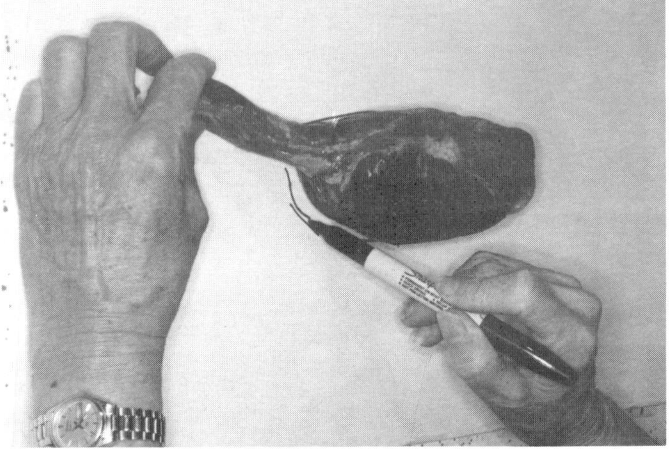

Once the bird is skinned, accurate sketches should be made of the side view (profile) and the top view of the carcass. Be sure to incorporate tracings of the neck and union of the neck to the body into these sketches. It is very helpful to actually position the neck of the skinned carcass into the pose that it is to be mounted in, and then to sketch the neck in that exact position.

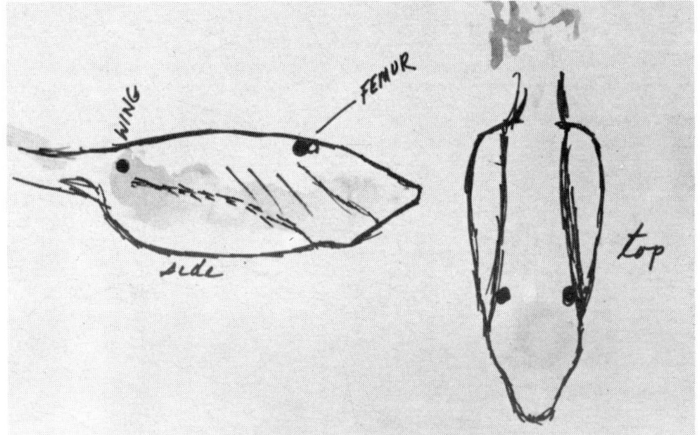

Be sure to note on the sketches where the leg attachment points are supposed to be, and whether the femur or tibia is to be attached to the wrapped mannikin. Also note where the humerus attachment points for the wings are located. Remember, the more information available on the carcass sketches, the less "guesswork" will be involved later when the bird is actually being put back together.

When the sketches have been completed, wrap the carcass in a plastic bag and place it back in the freezer. It will prove to be a valuable three-dimensional reference tool during the mounting (and/or wrapping) process. The carcass can be taken out of the freezer at any time for easy reference.

Once the bird is skinned and the sketches are completed, the bird should be defatted, washed, and tanned as previously instructed. The body can be wrapped, all of the wires can be cut to length and sharpened, and all materials and tools assembled before the actual mounting process begins.

Constructing the Bird Mannikin

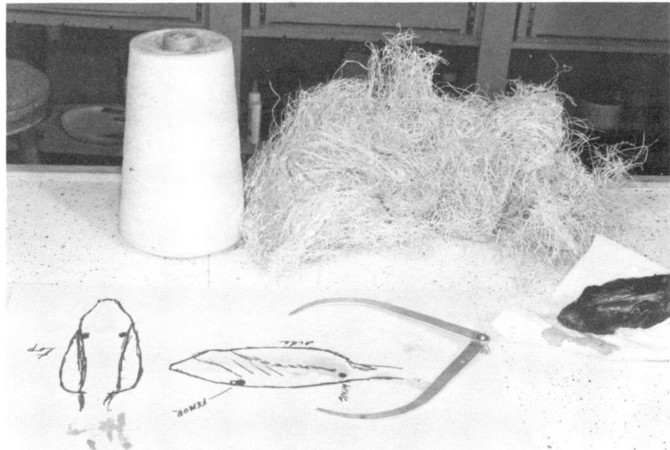

One of the best materials to use for the construction of the bird mannikin is slightly dampened excelsior.

If the excelsior is slightly dampened, it is much easier to shape and mold to the dimensions and contours of the carcass sketches. It can effectively be dampened by using a WASCO pump sprayer and Bacteria Stat (or Lysol) treated water.

To begin making the wrapped body, take an amount of excelsior approximately 1/4 the size of the carcass. Compress and "shape" the excelsior into the form of an oval cylinder.

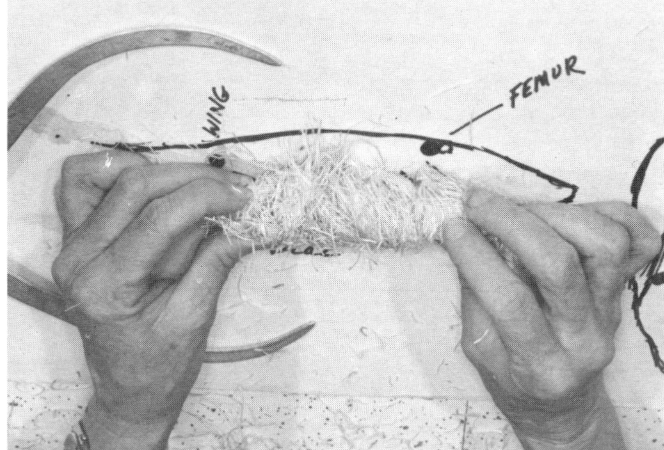

This core should be shorter in length than the carcass and about 1/4 the width of the carcass (as shown in photo above). Once the core is formed, begin wrapping cotton thread tightly around the core.

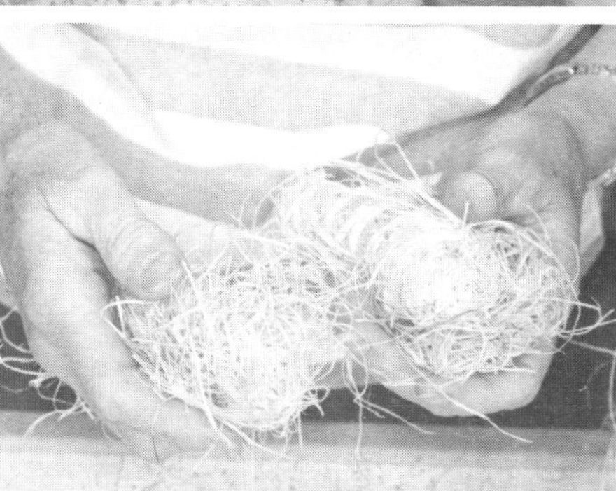

Now take several small tufts of excelsior and begin rebuilding the breast area to its original dimensions.

As the excelsior is added, compress it slightly with your hands and "simultaneously" wrap the excelsior with the cotton thread. Continue to add excelsior until the desired shape is achieved. Periodic measurement checks using calipers will prevent wrapping the body out of proportion. Just as it is helpful for a carver to work from side to side in order to obtain symmetry; likewise, it is helpful for the taxidermist to work from side to side as they add excelsior. This builds up each side of the mannikin equally and will help to achieve good symmetry.

Remember, the wrapped excelsior body should be no larger than the actual carcass of the bird. To put additional detail into the wrapped body, take a 6" sewing needle and heavy cotton thread and begin sewing approximately 1/3 of the way down the side of the breast. Pass the needle through the body and over the

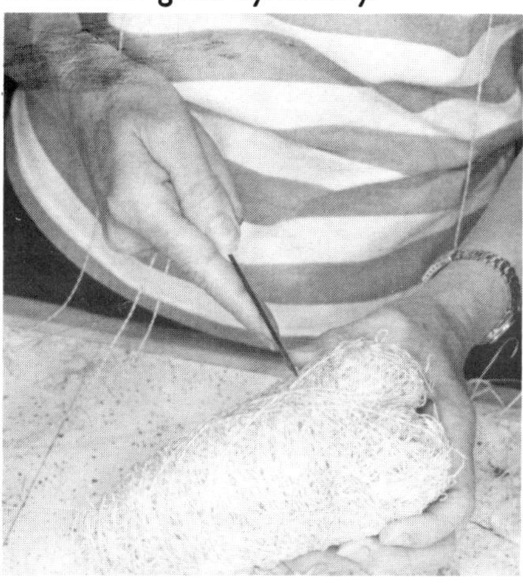

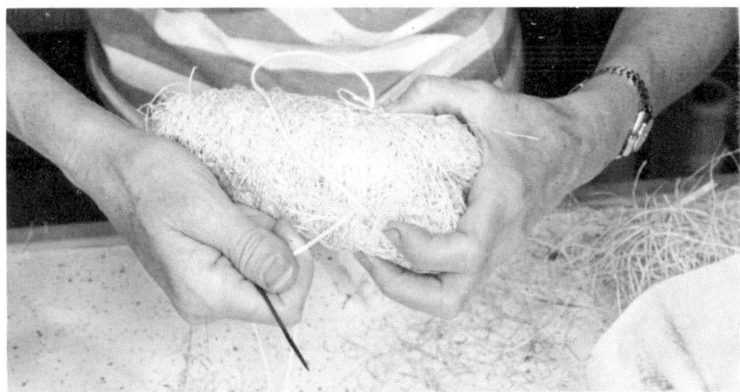

back. Pull the thread snugly across the back before beginning the next stitch.

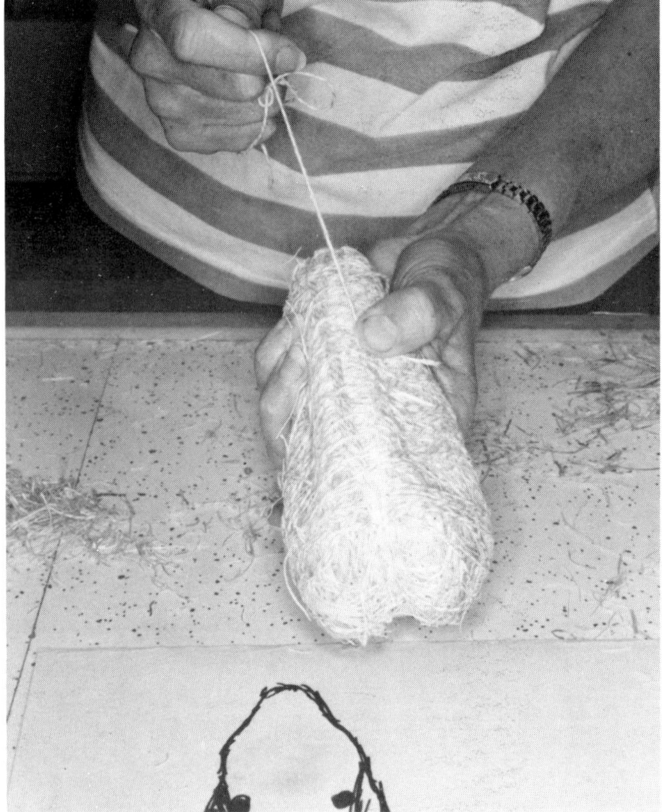

Repeat this process approximately two-thirds of the way back toward the rear section of the body. This will form a pocket for the wings (humerus, radius and ulna) and will facilitate align-ing the feathers of the side pockets.

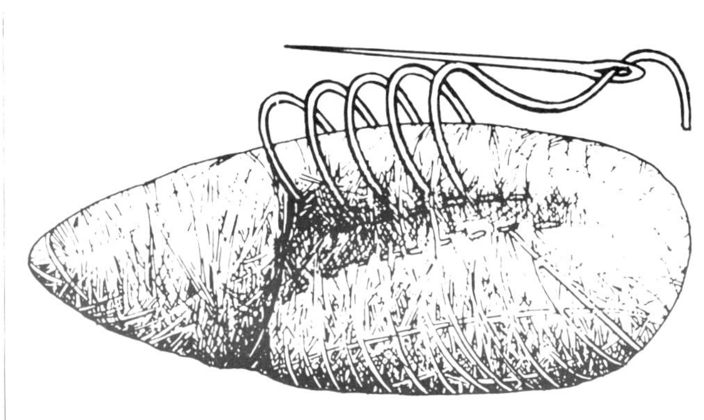

Several important things to remember about wrapping a body are:

1. The wrapped body should be "no bigger" than the actual carcass.

2. For standing birds, put detail into the wrapped body for the humerus, radius and ulna so that the side pocket feathers will easily track.

3. The wrapped body should be wrapped firmly enough to support the wings and legs, but not wrapped so tightly that it is impossible to pass a wire through the wrapped body.

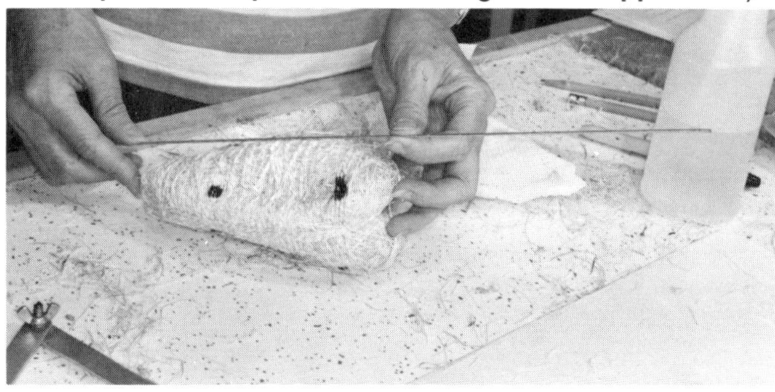

With the wrapped mannikin complete, insert a proper sized neck wire into the wrapped body. The neck wire should be of a heavy enough gauge wire to support the head, but be pliable enough to position the neck easily. (See the wire gauge chart on page 53.)

The wire should be cut approximately 3 times the length of the body and sharpened at both ends with a bench grinder or file.

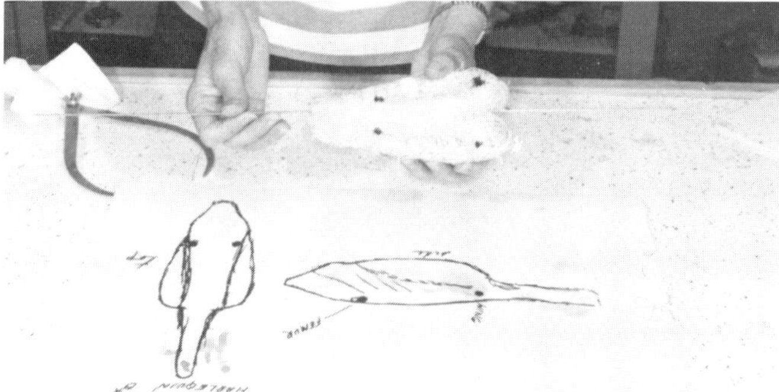

Refer to the sketches made of the side profile of the body where the neck joins the body and insert the wire into the wrapped body at that point.

Push the wire through the manni-kin, angling toward the base of the tail. Exit the wire at the base of the tail and pull approximately two-thirds of the wire through the bird. Bend the exposed two-thirds of the wire in half; then push the bent wire back through the mannikin toward the front of the body.

The wire should be pulled snug against the tail section. The short portion of the wire that protrudes from the breast should now be clinched back into the body with pliers (see illustration below). The neck wire will now be solidly anchored into the body.

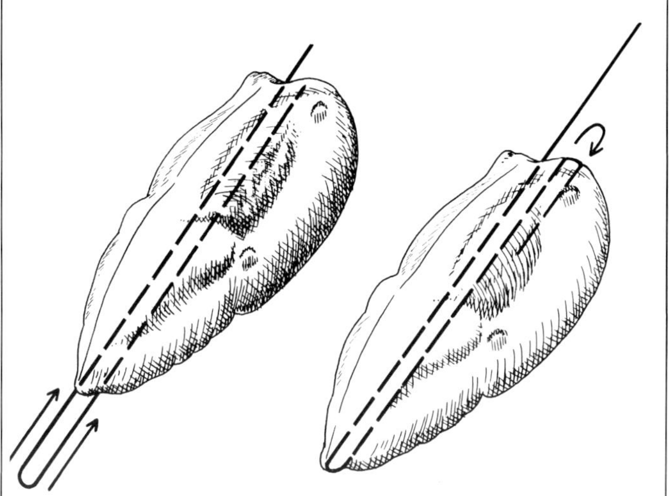

In order to prevent the cotton neck material from slipping and rotating around the neck wire when wrapping it, and to keep the proper oval configuration to the neck, another piece of wire must be secured into the main neck wire. Cut a piece of 20 gauge wire 2½ times the length of the bird's neck.

Insert this wire into the wrapped body slightly above the neck wire (see illustration below).

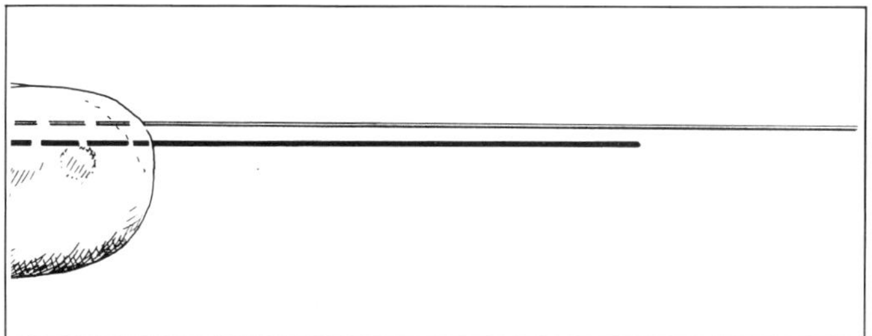

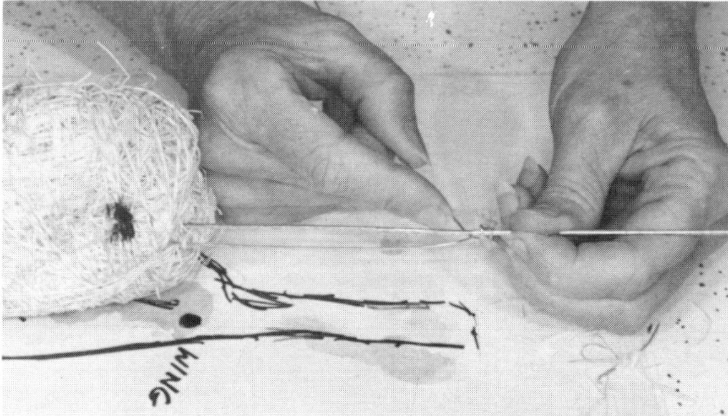

Run this 20 gauge wire on top of the neck wire down to a point that is the exact same length of the bird's neck. At this point on the neck wire, make several tight loops around the neck wire with the 20 gauge wire.

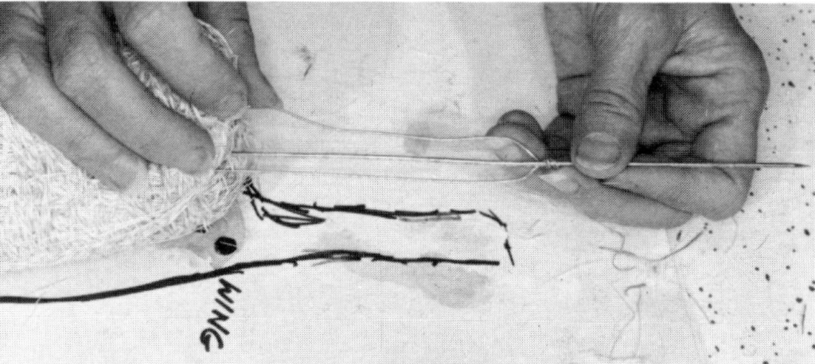

Proceed back to the body beneath the neck wire and insert this end of the wire into the body (see illustration below).

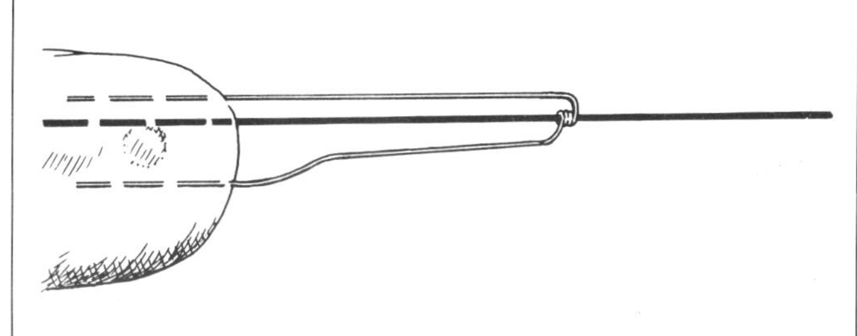

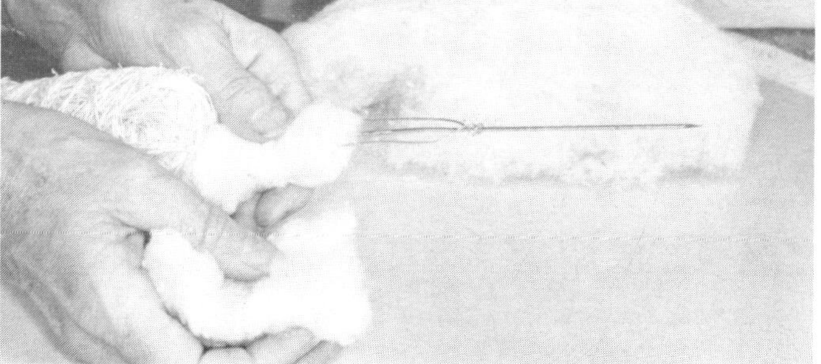

Now using the sketches to duplicate the length and shape of the original body, begin wrapping the neck with cotton.

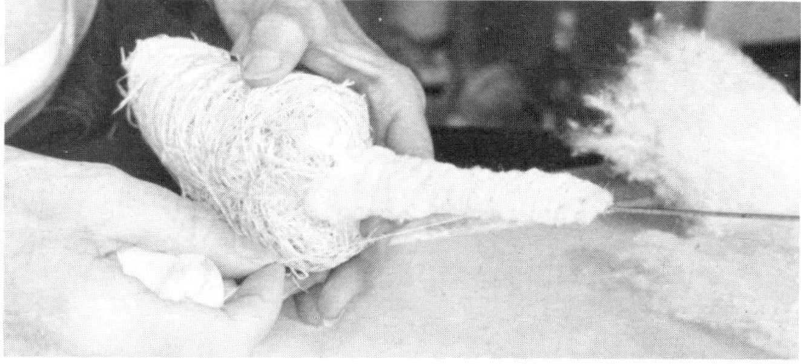

Secure the cotton in place with light cotton thread. Continue to build up the neck with cotton and thread until the desired shape and length is achieved.

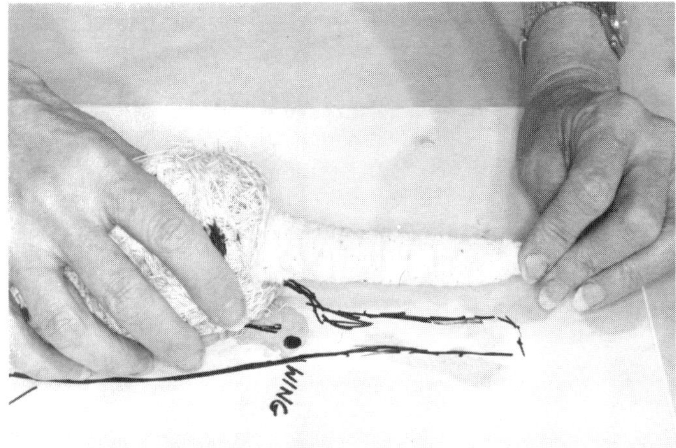

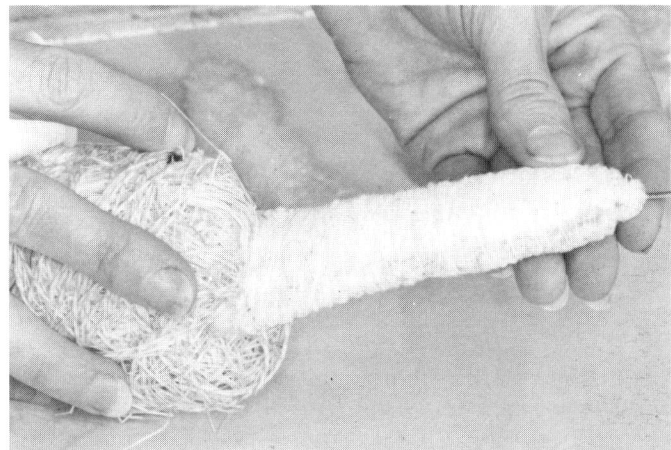

Making "tight," "close," wraps with the thread will keep the cotton smooth. Remember, the skin on the bird's neck is thin, and lumps on the wrapped neck will show on the finished mount. This means that the mannikin must be rebuilt with craftsmanship.

To duplicate the shape and contour of the neck, it is necessary to incorporate the trachea (wind pipe) into the wrapped neck.

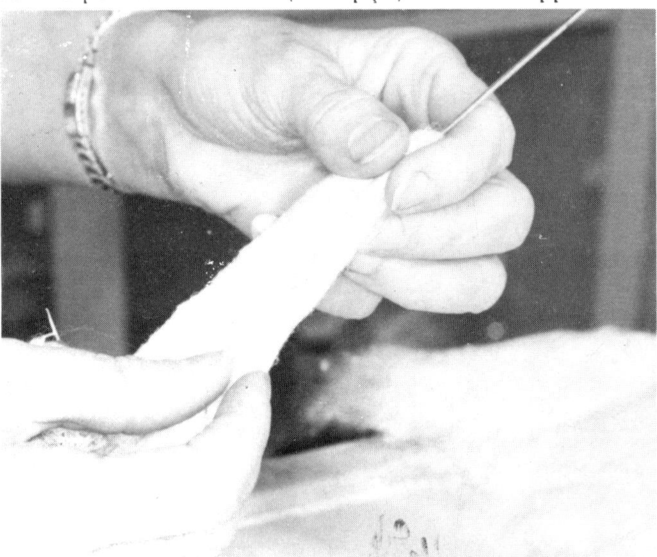

Simply take cotton and roll it into the shape of a round cylinder. Place this cotton cylinder under the wrapped neck.

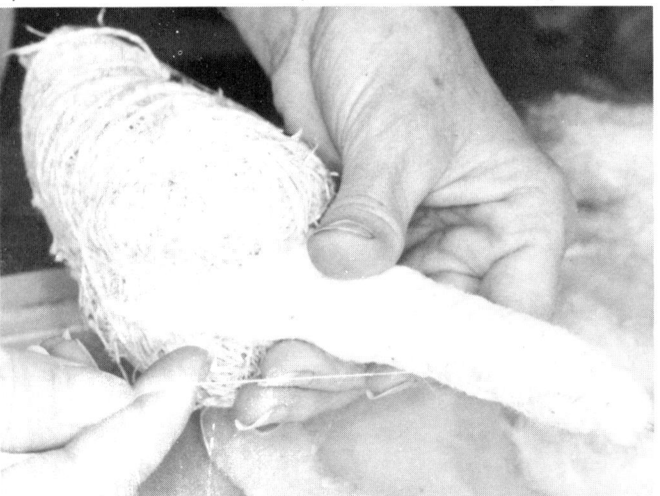

Begin wrapping cotton thread to hold the cotton cylinder in place.

Continue wrapping the neck with cotton thread and shaping the neck until the desired shape is achieved that matches the natural neck.

When the neck is completely restored to the original dimensions, it will be bent into the approximate shape that it will display on the finished mount. The exception to this procedure would be extreme or severe turns that would prohibit the skin from easily sliding down the neck. Check your references and sketches to make sure that the neck is bent into the proper position. "Presetting" the neck into its final position will eliminate unnecessary handling of the feathers and skin later in the mounting process and is highly desirable whenever possible.

With the wrapped body complete, support wires may be prepared and sharpened as described in Chapters 7 and 8. After assembling all the neccessary materials, the tumbled bird skin can now be mounted.

Mounting a Bald Eagle "Wrapped Mannikin" Style

Robert Holshouser is one of the world's leading bird taxidermists. He occasionally mounts bald eagles for museums and other scientific or governmental organizations. (Eagles must be accompanied by a valid permit in order to be mounted.) This step-by-step account for mounting this large bird should give you a good idea of how to prepare a mannikin for a large bird with a large wing spread.

by Robert Holshouser

In the years that I have lived in Alaska I have been very fortunate to have had the opportunity to mount bald eagles. I feel that it is a great honor to work on such a beautiful bird. With a body length of 30 to 43 inches and a wing spread of up to 96 inches, it makes one wonder at the incredible size, beauty, and "power" of these big birds. Only mature adults of 5 years or older have a snow white head, neck and tail.

I put "lots" of extra thought and planning into this difficult task. One of my greatest concerns with this project was to obtain "stability" in the mount.

Since this particular bird was to be an open wing mount, it created a tremendous challenge. Basically, the bird was skinned and prepared using a unique skinning procedure that employed detachable wings and legs. This was done because wiring and mounting a bird this large is best accomplished if done in sections.

The photo above shows the body of the completed mannikin which was hand-wrapped from excelsior. The reason I wrapped the mannikin for this bird was mainly due to the fact that I don't know of a commercially available mannikin for an eagle

anywhere. Hand-wrapping a body to recreate the actual body myself also gives me a much greater amount of satisfaction than using a commercial mannikin.

Constructing the Mannikin

The first step was basically to determine the position and pose that this bird was to be mounted in. The mount was to be an open wing, tabletop mount, with the eagle perched atop a piece of driftwood. So once the bird was skinned, prepared, and ready to be mounted, I made some drawings and templates of the skinned eagle's body structure. At this time, I also took extensive measurements and recorded them for later reference. From my measurements and templates, I started reconstructing the leg placement into the excelsior body itself. One must take into account that *good measurements are a "must"* and in order to get everything to work right, all measurements must be "accurate!" I wish to "stress" to you that *the finished product will only look as good as your initial measurements*. It is *very important* to have good measurements to refer to during the mounting process.

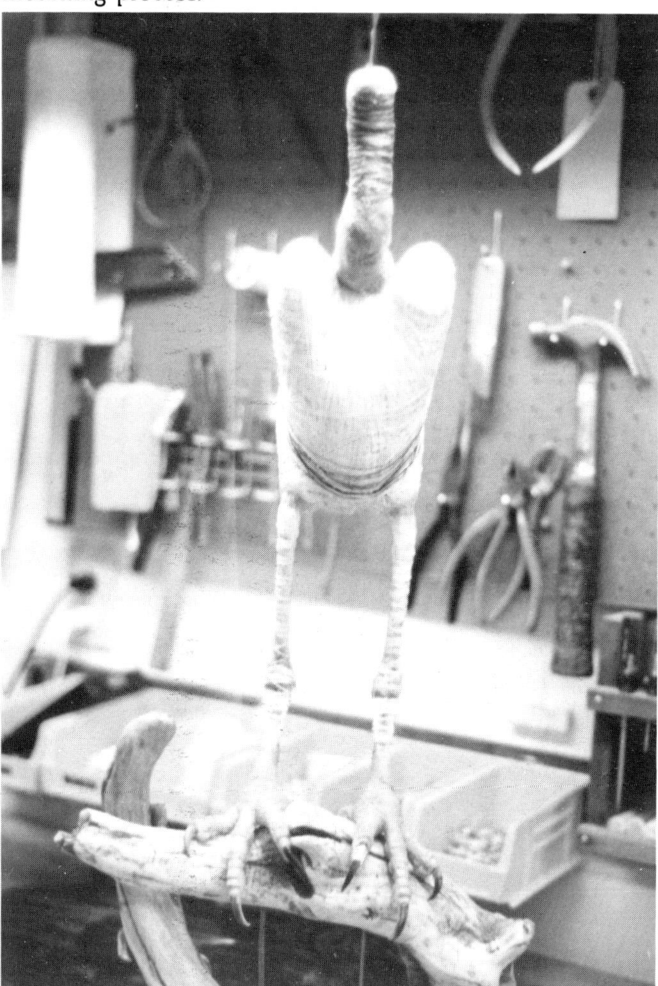

The legs were anchored into the excelsior body at the knee. The wire was run through the legs, anchored into the mannikin and then hot-melt glue was used at the joints to bond them to the mannikin. The hot-melt glue reinforced the joints to make them more sturdy.

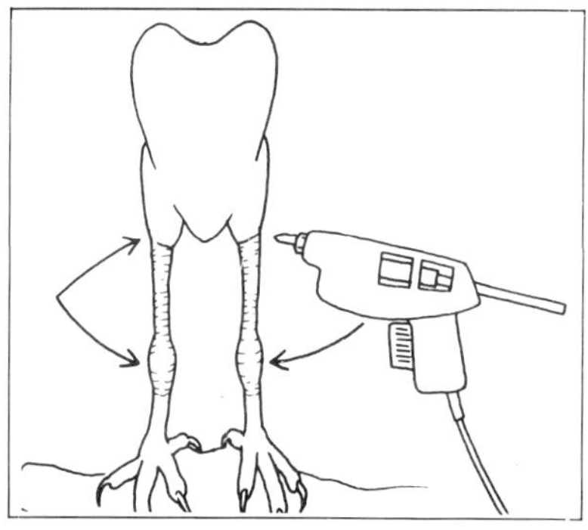

Finally, the legs were thoroughly and neatly wrapped with string. Hot-melt glue was applied over the top of the string for extra support.

This photo shows the muscles on the legs after they have been rebuilt and shaped back to the original size of the body itself. I used tow to wrap the legs, as it is a little easier to manage than excelsior when it comes to this type of wrapping. Tow is very easy to work with and it smooths out nicely.

Once the leg muscles were finished, the toes were injected with formalin to ensure that

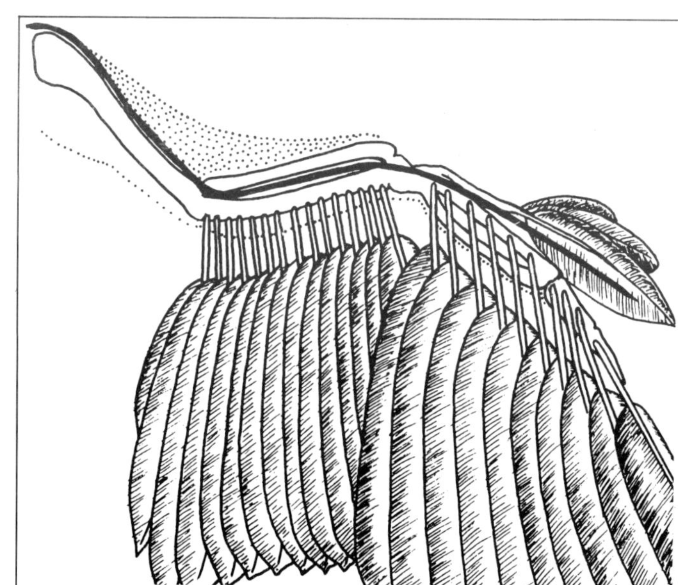

shrinkage would be minimized in the toes and feet. At this point the mannikin was allowed to set for about three days to firm up a little more. Once the legs started hardening up, I got a little more stability into the form itself; it also gave the toes a little "breather" and allowed time for the formaldehyde to dry out and evaporate so I could move right along with the rest of the mount.

The entire mounting process took three days (with the exception of the three day drying period for the feet and toes). It took me a little over a week to completely mount the bird. Big birds like this are almost impossible to mount in one day. To be able to mount them "right," one must take their time and not rush into the different steps.

The wing wire is inserted and clinched into the body itself and is securely tied to the humerus, radius and ulna bones. The end of the wire protruded through the alula (see illustration above).

All of the muscles in the wing were rebuilt to their original dimensions with tow and securely anchored into the body. At all of these anchor points (as far as where the actual bones were touching the excelsior body) hot-melt glue was used to securely fasten them to the excelsior body. This technique seemed to work pretty good. It really made the wings and these anchor points sturdier. Now with the wing rebuilt, I moved to the next phase.

This photo basically shows the wing being set into the body. The wings were detached and cut off from the main body at

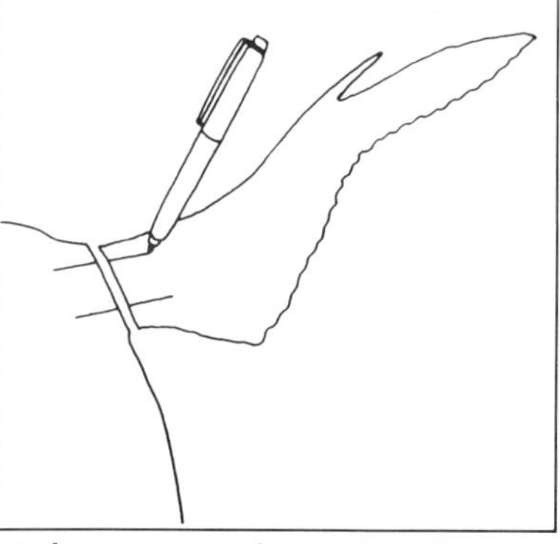

the humerus. When making incisions to remove different parts of the body, one must have good reference marks as to where they will begin sewing everything back together again. It is very important to make "good" reference points on the skin itself showing where the actual cut was made and where it will be sewn back together again later on. On the surface this process looks very difficult, but it will all go back together with ease. It is, however, "time consuming." It may "look" like a big mess at first, but you must have some patience.

The photo above shows the wing being inverted back over the wing muscle. Right before I inverted it, I applied a thin

The Breakthrough Bird Taxidermy Manual

coat of hide paste to the wing muscle of the mannikin. This actually helps to glue and hold down the feathers and also adds a "lubricant" to the surface of the mannikin for ease of pulling and adjusting of the wing feathers and skin. Once the glue sets, it really locks the primaries and secondaries in place. I also like the "feel" of the glue when taxiing the skin. I can move the skin and feathers around, positioning them just like I want them (and they will stay in place once the glue sets).

This photo shows the wings set and the wires coming out of the side. It has been "roughed in" and two support wires exit the side of the body. This helps to hold the primaries and secondaries up and out of the way while the main body portion of the skin is pulled over.

The body skin was then pulled over and hide paste applied. It was then sewn and adjusted. The feathered portion of the legs were cut down the back of each leg to the heel. Each leg was sewn up the back of the feathered portion of the leg and then sewn around to the breast incision in the front. Once the legs were sewn up, hide paste was applied to the area around the heel where the feathers actually meet the skin portion of the leg. Pins were used to hold it in place.

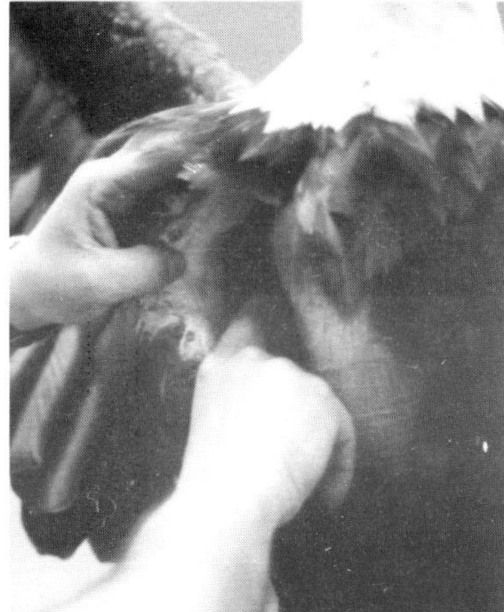

The main body was also glued with hide paste once it was pulled over the mannikin. Next, the tail was glued and the tail wire pinned into place.

It required quite a while to sew everything back into place for this mount. It is very difficult to describe how long it takes and how "aggravating" it is to sew something like this. Suffice it to say that it is very time consuming. So be prepared!

The head was mounted next. The eyes were set prior to placing the skin over it. The head was extremely difficult to model. Before I attempted to remodel it, during the skinning process, I made extremely accurate eye placement sketches of the skull itself. I then rebuilt the muscles to their proper proportions and the eyes were set into the head. To me, this was the hardest part of the entire mounting process. An eagle has an expression that is quite hard to achieve. The eyes are looking fairly straight ahead and they seem to appear "angry" or "fierce" looking. Once you look at them and study them for awhile, you will quite easily recognize this expression. It is a very difficult expression to achieve with a mount. It is one that can only be accomplished by intently studying a wealth of good reference photos. Good reference material is readily available for eagles, and there are many photos available of them (this is to your benefit).

Once the head and eyes were set and modeled back in, the head skin was pulled down over the neck wire. The end of the neck wire was run up through the mouth opening, and inverted back through the top of the bill. This gave me a very solid attachment without having to run the wire through the top of the head. (I always try to avoid running a neck wire through the head of a bird.)

The carding system that was used on this mount was quite extensive as there are many feathers and different tracks that must be carded on an eagle. These are also "time-consuming" steps that will take quite a while, yet the end result will definitely pay off! Anything that looks out of place or that is not laying just right, must be adjusted, carded, and pinned into its proper place.

Sewing takes up some time but the feather arrangement was the "real" time-consuming part. It took several hours once the bird was completely sewn up to arrange the feathers and position them into the positions that they should be in. Thousands of pins (it seems) were used.

I used the carding method which consists of two strips of cards; one on each side of the wing itself and pinned with insect pins to minimize damage to the feathers. I don't like stapling which tends to cut into the feathers.

In some places where the primaries were carded, I used masking tape underneath the cards which gets the feather arrangement in the "ball park," and also helps to hold the feathers and prevent them from pulling apart by themselves. Masking tape was used extensively in the longer run carding areas. This did help to hold them together much better.

The photo above shows the finished bird once it was completely dried and painted. The end results were quite satisfying.

The wing tips were actually bent a little to give a little more motion and realism to the mount. I bent the outer primaries with a hot tool which must be used very carefully. I start on the back side of the quill and lift the wing tips from the tip out, actually bending the primaries back to give a little more wind resistant effect (which gives a natural appearance).

I used Tohickon eyes which were really beautiful. The eyelids and the tissue around the beak itself were rebuilt to their original size with sculpall. I had quite a bit of shrinkage in this area. This was all rebuilt to size "after" it had completely dried. It was then repainted using good color photo references of a live eagle.

Basically, that is the process that I used to mount this bald eagle. This was a time consuming mount, and I found that doing it this way may not be the quickest way, but it is definitely a good method for mounting big birds. (Especially big birds that must be very stable.) There is always going to be a problem with big birds as far as stability is concerned. They are hard to work on, and trying to mount one using the standard method (without detaching the wings or legs) will create all kinds of problems, especially when bending the heavy wires which are required for support. It is almost impossible to get inside the bird to work. It is similar to working on a ship inside of a bottle when trying to get inside and bend heavy wires without ripping the skin and tearing it up completely.

At any rate, in what I was trying to produce, I found this method to be quite satisfactory. It is a little "old timey" but it seems to work for me. As far as hand wrapping bodies goes, I get a lot of enjoyment out of it. As far as making a living and doing it this way, you better be charging a "bunch" for a mount done like this.

The "Carved Mannikin" Method

Some taxidermists prefer to carve their own mannikin. A variety of substances are used to carve bird mannikins with, but urethane foam and open cell styrofoam are probably the most popular. Some taxidermists use wood and even cork to carve their mannikins from, but foam seems to be the quickest route for now.

The bird mannikin that we carved was carved out of urethane foam. It is readily available from altered forms, etc., and holds wires extremly well. Perhaps the most important thing to remember is to take some off of "each" side as you carve. If you take a small slice off of the left side, do likewise with the right side! This will ensure good symmetry.

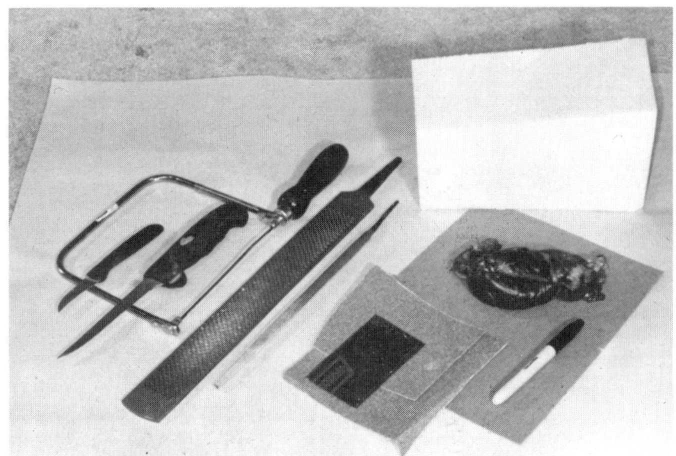

Once the bird skin has been removed, assemble the tools and materials necessary to sketch and carve the mannikin. This would include cardboard, felt-tipped pen, sandpaper, sanding screen, farrier's rasp, file, Perfect Knife, "All-Pro" skinning knife, coping saw and block of foam.

To begin, make a detailed outline of the side and top view of the carcass. These sketches will be an important aide in recreating an accurately carved mannikin.

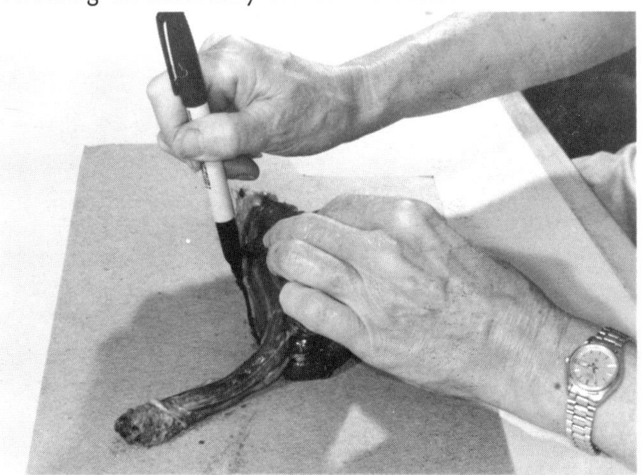

Lay the carcass on its side on a piece of cardboard and carefully trace the outline with a fine, felt-tipped pen. The pen should be held perpendicular to the surface of the paper and should be pressed snugly against the carcass to ensure duplicating the exact contours of the body.

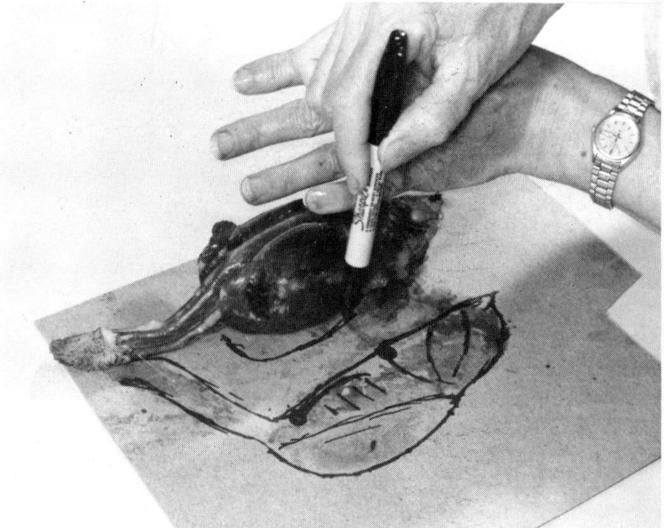

With the side sketch completed, place the carcass on its breast and trace the top view of the carcass.

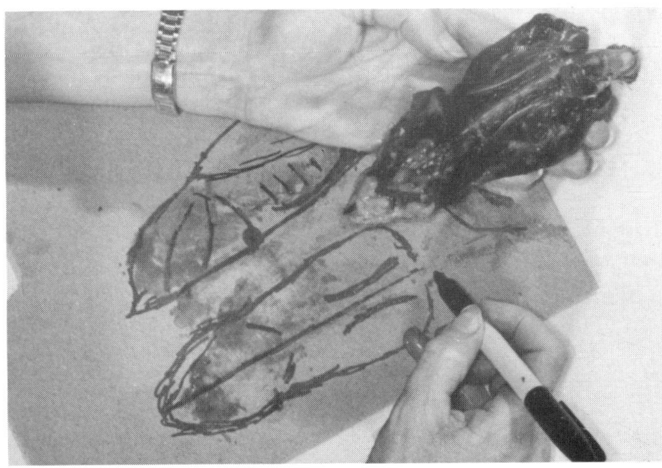

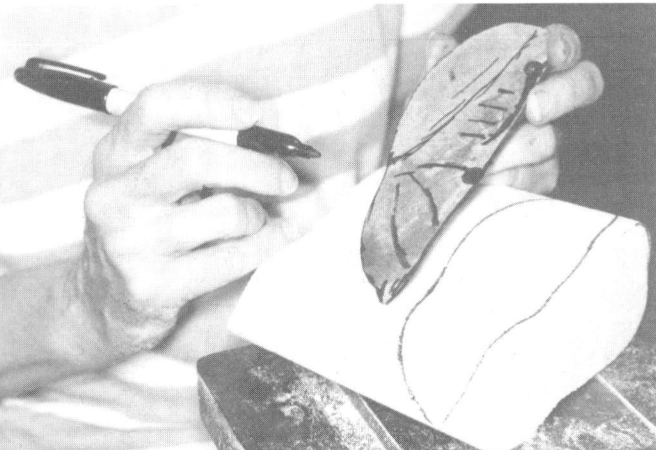

Referring to the carcass, sketch in the surface detail on both the side and top view drawings. Remember, the more detail placed in the sketches, the less "guesswork" will be involved later in the carving process.

Once this cut is complete, place the side view carcass pattern on the cut block of foam and trace the outline.

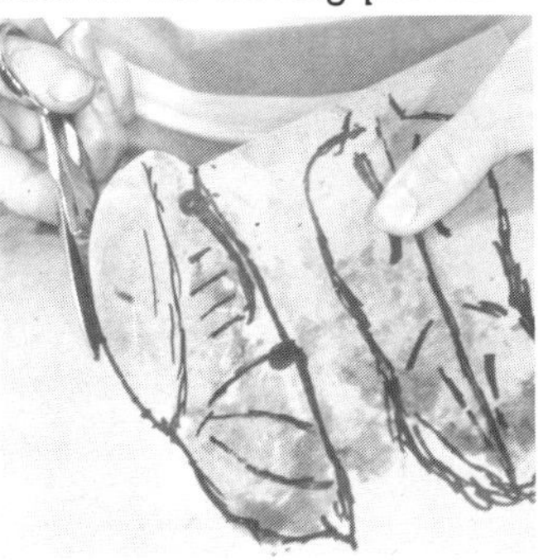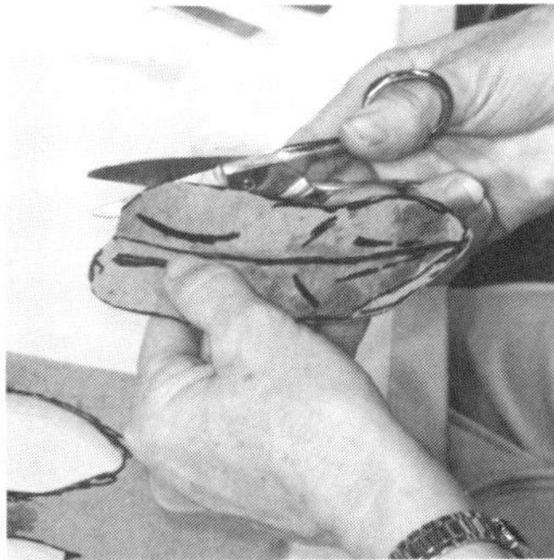

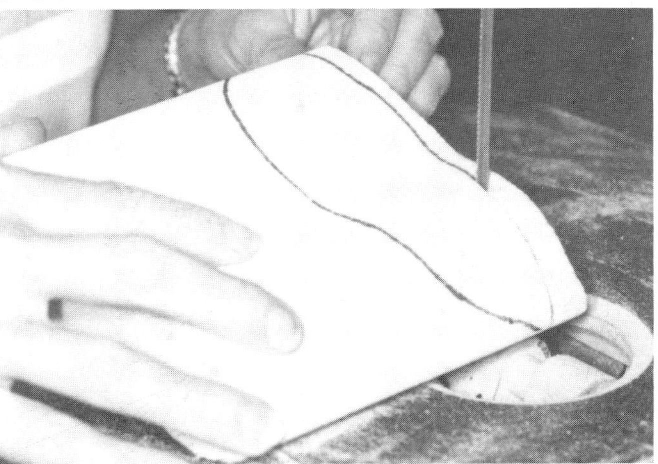

Check the tracings for symmetry and make any necessary adjustments to the drawings. With the carcass sketches complete, carefully cut out the drawings with a pair of scissors. The block of foam should then be checked for air pockets or stress cracks to determine the best place to make the cuts for the mannikin.

Again, use the bandsaw (or other suitable saw) and cut this side view profile from the foam.

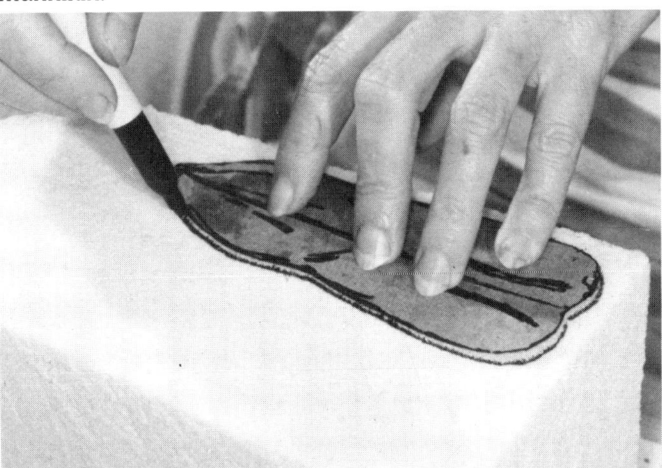

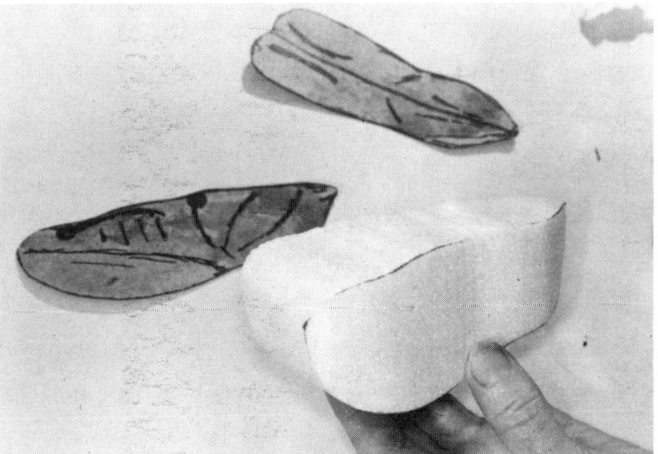

The foam block should now begin to resemble the basic shape and contour of the carcass.

Once this is determined, place the top view carcass sketch on the foam block and with a felt-tipped pen, trace the outline on the foam block.

Carefully cut out the top view outline with a band saw. If a bandsaw is not available, a hand-held coping saw will work well. Remember, take your time and make the cut exact.

With a Perfect Knife, begin to remove the square edges of the foam. There are two very important techniques to remember when carving foam. First, always remove foam from "each" side as you carve. This will help to maintain symmetry in the carving. Secondly, remove only small slices each time. (Once a large slice of foam is cut away, it is gone *forever*.)

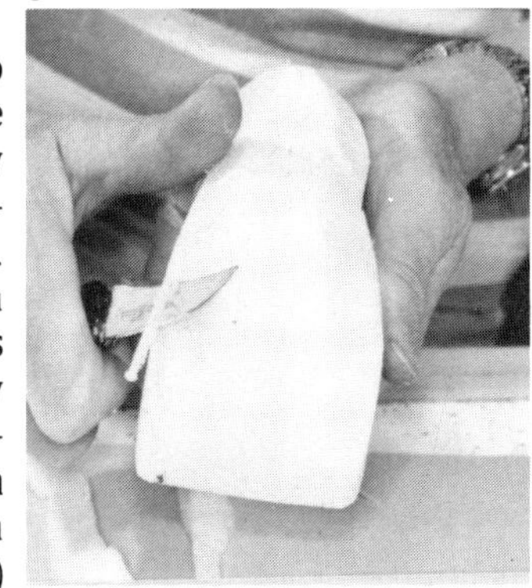

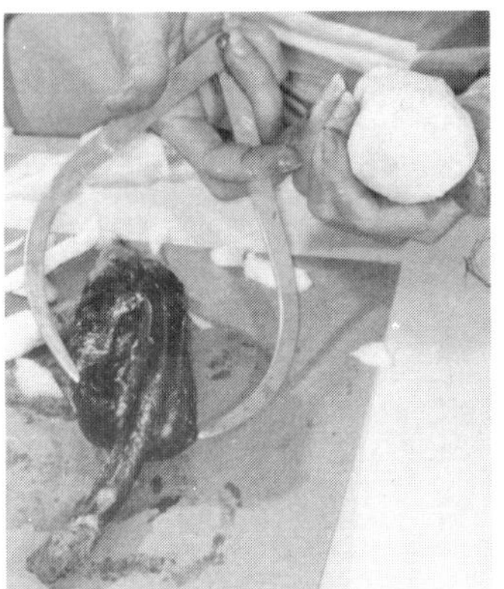 

With the rough edges removed, it is important to use the calipers to ensure the carved mannikin is conforming to the proper dimensions of the original carcass.

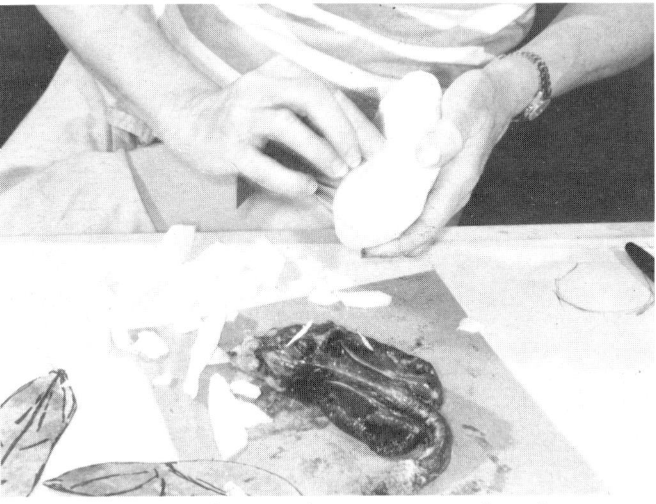

Once the proper shape and measurements have been incorporated into the carved mannikin, use the actual carcass for three dimensional reference and begin placing the detail of the side pockets, thigh muscles, clavical and caudal area into the carved mannikin.

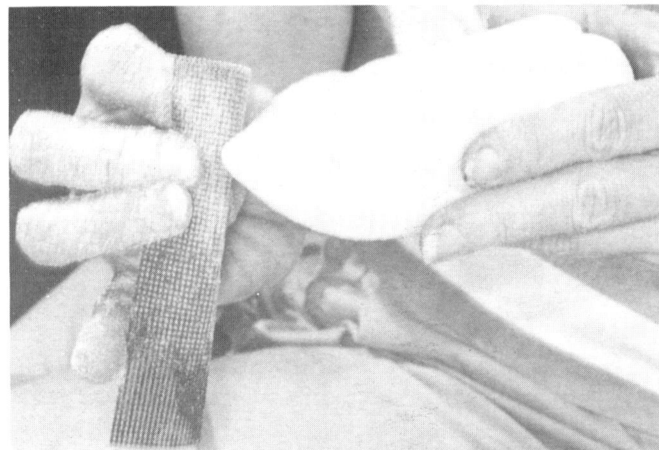

This anatomical detail is easily put into the carved mannikin by using sanding screen. Again, use caution when removing the foam with the sanding screen. It is very easy to apply too much pressure and remove a large portion of the foam. If it is neces-

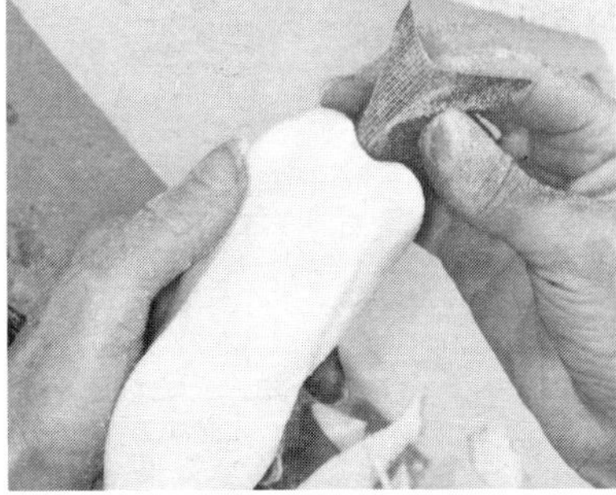

sary to carve a large mannikin, remember that a farrier's rasp works well for this type of project.

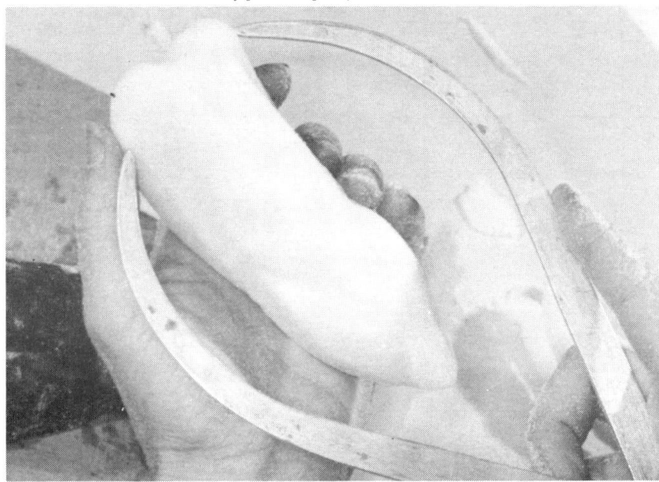

With the detail in the carved mannikin completed, make one last check with the calipers to make certain the dimensions are the same as the carcass.

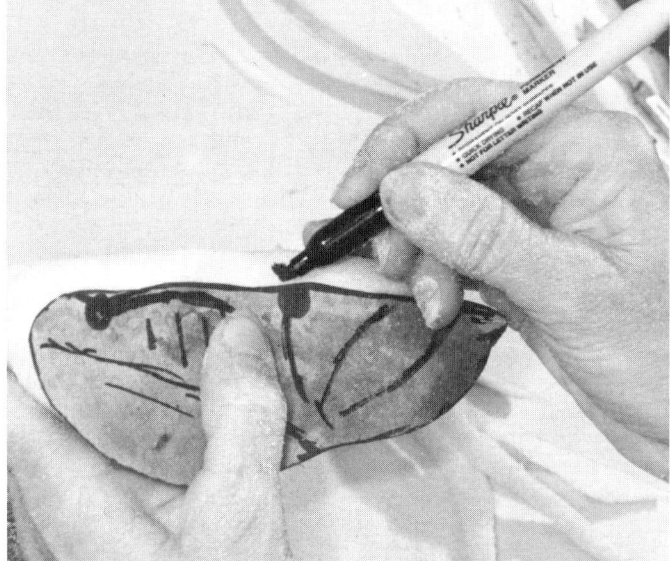

Refer once again to the actual carcass and the carcass sketches and mark the neck, wing and leg attachment points.

The carved mannikin is now ready for either the installation of a wrapped neck or the insertion of a Sportsman Series Flex-Neck.

CHAPTER 10

Mounting Wild Turkeys

World Champion Ed Thompson: Mounting a Strutting Turkey

Perhaps no other game animal or bird has caused so many taxidermists and hunters alike to throw up their hands in utter despair and frustration than "his majesty," the mature wild turkey. The wild turkey is considered to be a most worthy opponent for even the most experienced hunter; and to the taxidermist, when it comes to the degree of difficulty, mounting a "strutting" gobbler is definitely the "ultimate" challenge.

While most waterfowl and other migratory birds present some unique problems of cleaning and defatting, these problems simply do not compare to the problems of individual feather placement, the large size, and the difficulty of mounting and re-coloring the fleshy head of a turkey.

Many taxidermists feel that they "have truly arrived" when they can mount a satisfactory strutting turkey; one that portrays the pose, feather placement, and head coloration accurately enough to win a ribbon in a major taxidermy competition. In this chapter, World Champion bird taxidermist Ed Thompson shows how you can do just that!

Ed has won numerous competitions with his favorite subject, the wild turkey, and is well qualified to teach this difficult art. He is one of the most accomplished bird taxidermists in the United States. Currently the curator of the Fernbank Science Museum in Atlanta, Georgia, he has won nearly every important award ever offered in taxidermy competition, including over 50 First Place awards, Best of Shows, Judge's Choice and People's Choice awards. He won the coveted Carl E. Akeley Invitational Award at the World Taxidermy Championships in 1986, and most recently won First Place, Best of Category, and Best of Show at the National Wild Turkey Federation Convention. Ed is in constant demand as a seminar speaker and judge at the various taxidermy competitions.

Tools and Materials:

The following list of tools and materials are what Ed used to mount the strutting turkey in this chapter:

Tools: Scalpel, blades
Bone cutting scissors or pliers
Ultimate Scissors

Needle nosed pliers
Powered wire wheel
Measurite (MR100)

Sewing needles (SN104)
Paasche VL-3 Airbrush
Skinning knife

Loop Head Pins   Apron (APR1)
Jumbo Head Pins   Bird Skinning Gambrel (JHT102)
Tack Hammer (MR140)   Syringe (SY) and Needles (SYN)
Feather Duster (FD)   Skinning Shears (MR112)
Calipers (MR30)

Supplies:

Turkey Mannikin (FN20)   Fur Dresser's Sawdust (CT99)
Powdered Borax (PB10)   Ultra Lite Filler (UL32)
Puffed Borax (PBX10)   8 gauge wire (GW8)
True-Tan Skin Prep (CT30)   10 gauge wire (GW10)
True-Tan Bacteria Stat (CT10)   Super-Pro Thread (STS)
True-Tan Bird Tan (CT80)   Tohickon Bird Eyes
Grease—Buster (GB32)   WASCO Clay (PC5)
Polytranspar Degreaser (FP1005)
Instant Bonding Glue (AG7432)
Bird Photo Reference Set (FNR112)
Taxidermy Business System (WL100)
Strutting Turkey Head (FNB201)
Sallie Dahmes Hide Paste (SDHP)
Polytranspar FP or WA Paints

General Field Care of the Turkey

As with all forms of quality taxidermy, the "care" given to the wild turkey in the field is of prime importance. All sportsmen desire as clean of a kill as possible, and when a bird is down, it is important to get to it quickly. If the bird is still alive, it should be pinned to the ground until it succumbs to prevent the loss of feathers caused by the turkey beating his wings upon the ground. Many times the bird will be bleeding from the mouth and it is a good idea to support it, head down, to permit as much blood as possible to drain. Then the mouth, vent and any large shot holes should be plugged with pieces of paper towel, toilet tissue or rag to keep blood from the feathers. If a bloody spot does appear on the feathers, it can be quickly rinsed with cold water and then dried with paper towels. Carry the bird by the feet, which will permit good air circulation to remove some of the body heat. The turkey should then be placed in a freezer as soon as possible. It should definitely be frozen the same day it is collected, to further remove all body heat. If the bird cannot be frozen right away, it should be eviscerated neatly and cleanly, and the body cavity packed with dried grass or paper towels to absorb any remaining fluid. Always suspend the bird by the feet until it can be frozen.

Skinning the Turkey

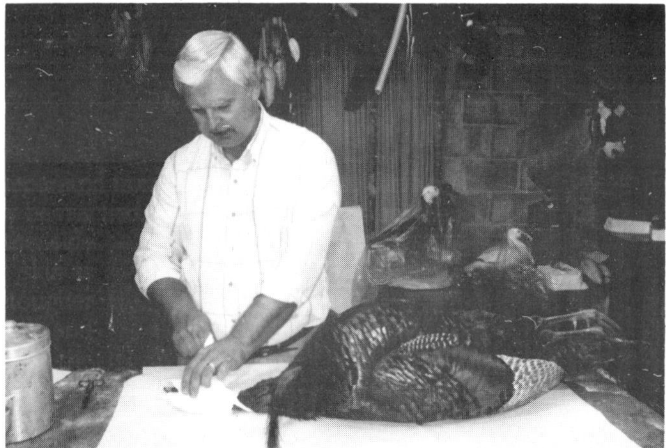

The turkey must be thawed completely prior to skinning. Ed places a piece of paper towel around the head to contain any fluids that may drain. A clean piece of paper is used to cover the surface of the work table in order to keep the feathers clean.

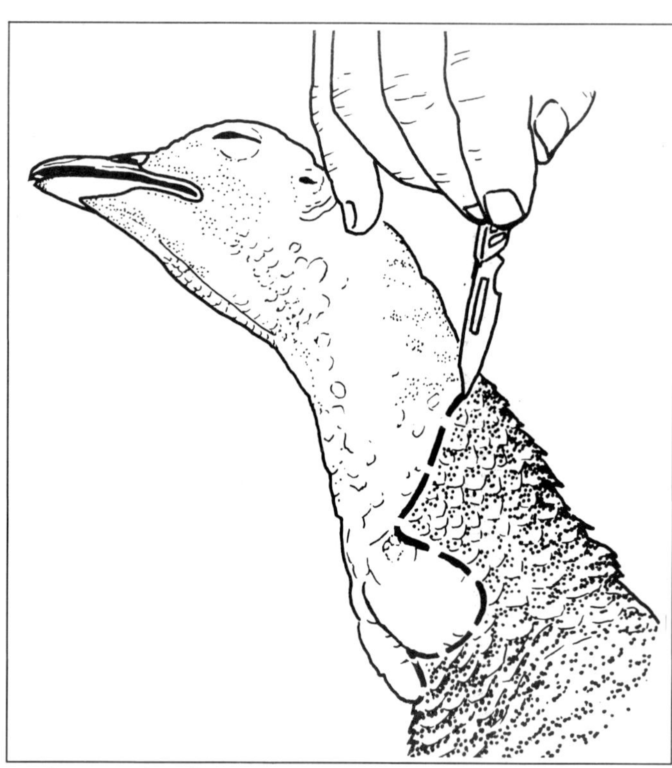

Ed starts the skinning operation by using a scalpel to cut through the skin where the head joins the feathers. (Ed cuts "just" **deep enough to cut through the skin—not any deeper.**)

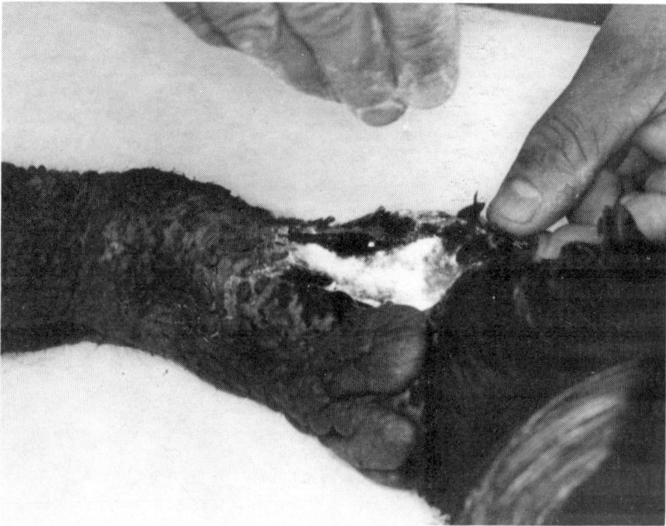

Powdered Borax is used to absorb blood and fluids as he proceeds.

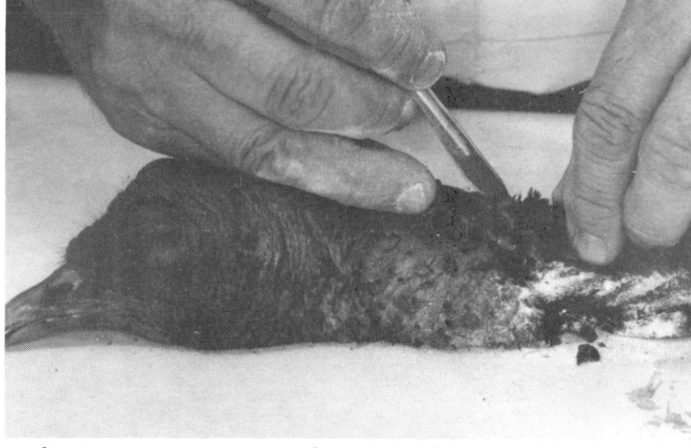

The incision is continued completely around the head as shown.

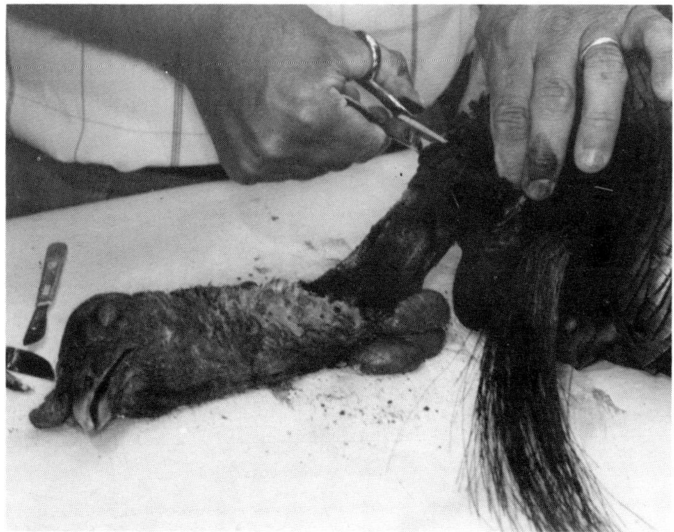

The neck skin is then peeled down the neck for a distance of about 3 inches, and the bone is cut using a pair of Ultimate Scissors. The neck skin is then packed with paper towels to prevent any blood leaking from the body.

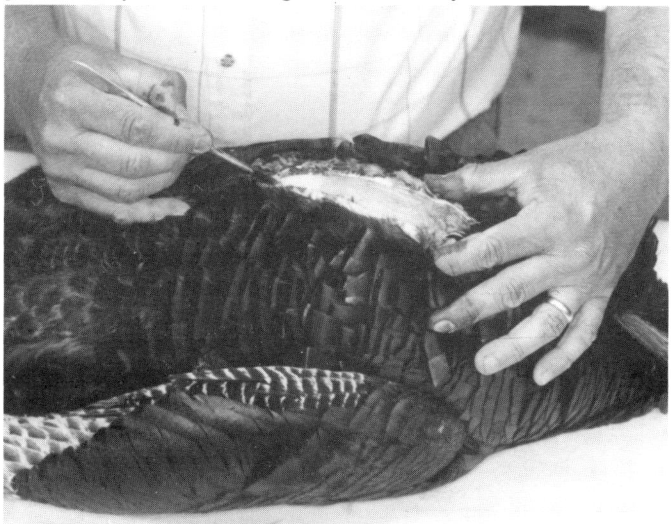

The breast incision is then made from the point of the brisket to the leading edge of the vent. By stopping at this point, a "pocket" is created to later receive the mannikin.

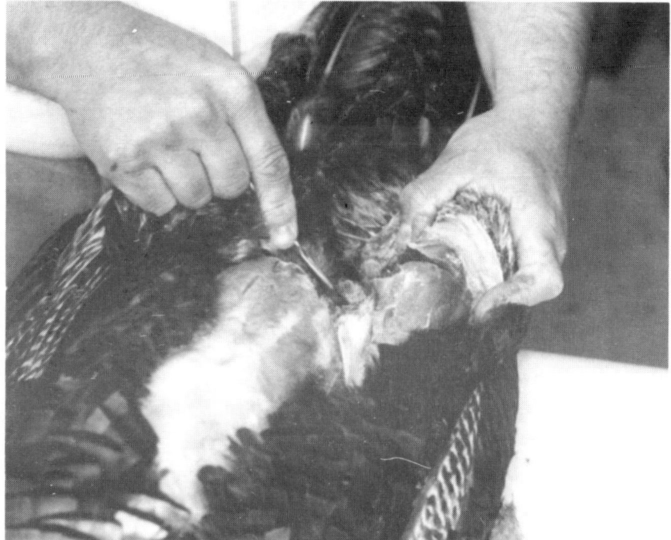

Next, the skin is worked loose from the body and over the legs with the fingers and careful use of a scalpel. Ed uses a

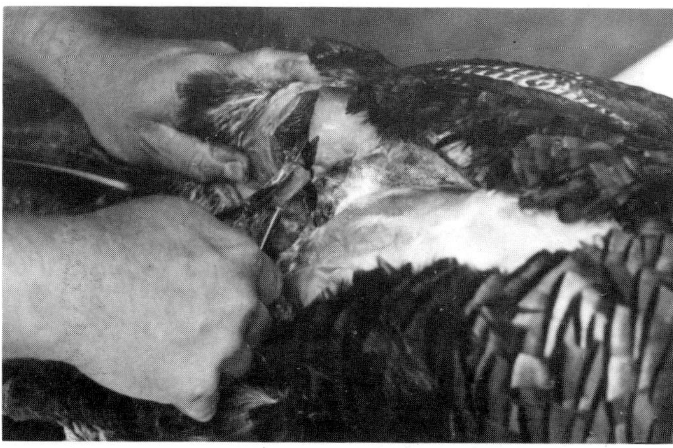

pair of bone cutting pliers to sever the leg bone just below the femur-tibia joint. Be careful not to cut through the skin on the other side with the sharp pliers.

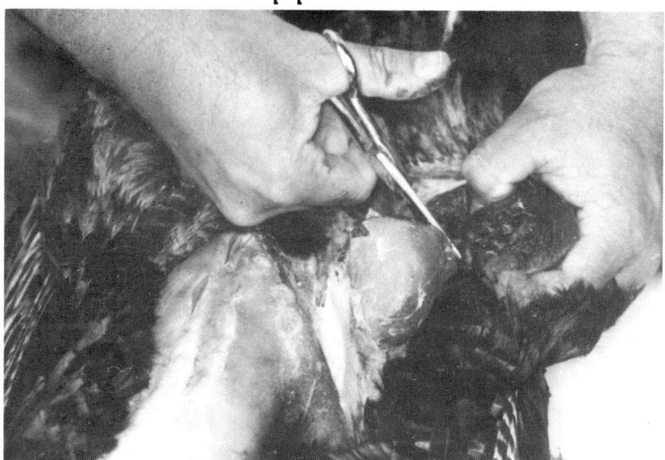

Ultimate Scissors are then used to cut the remaining tendons and flesh.

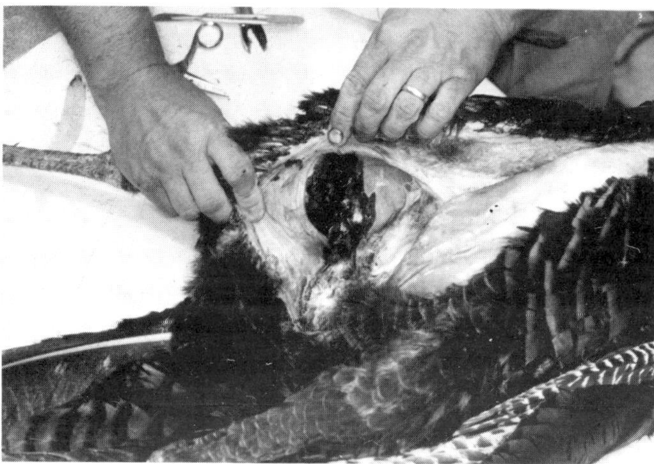

The skin is then separated from the body down to the back as far as it can easily be reached. The same operations are then completed on the other side of the bird. Turkey skin is much heavier and tougher than migratory birds and when skinning turkeys the hands can be used more often to separate the skin from the flesh.

After the vent is severed, the skin is worked free of the abdomen to the base of the pygostyle area.

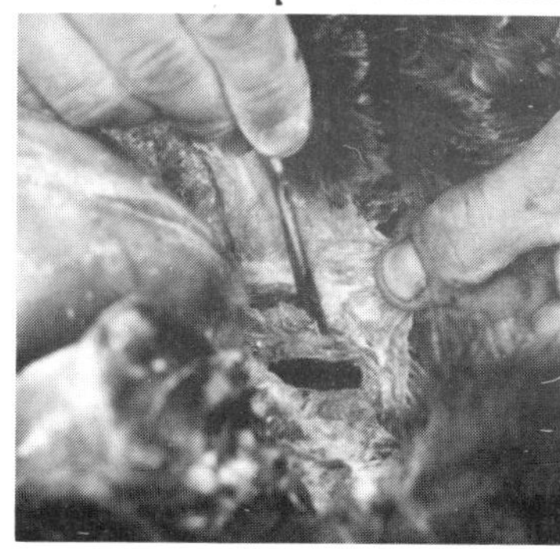

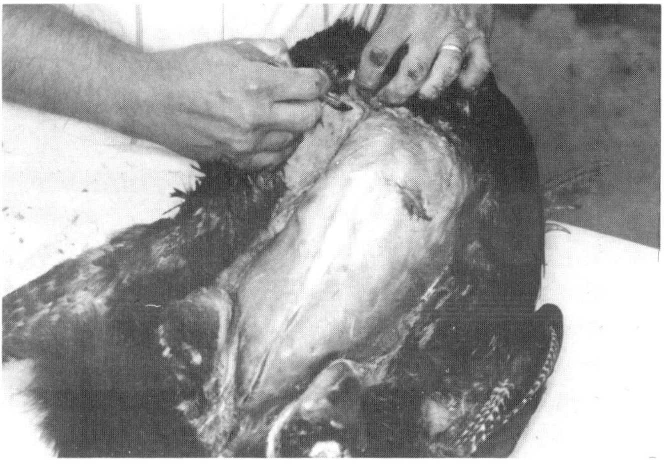

Ed next turns his attention to the front of the breast and carefully separates the skin down to the crop area.

As the wing bone is reached, it is severed with the bone cutters and the skin is removed down over the shoulders. This same operation is then completed on the other wing.

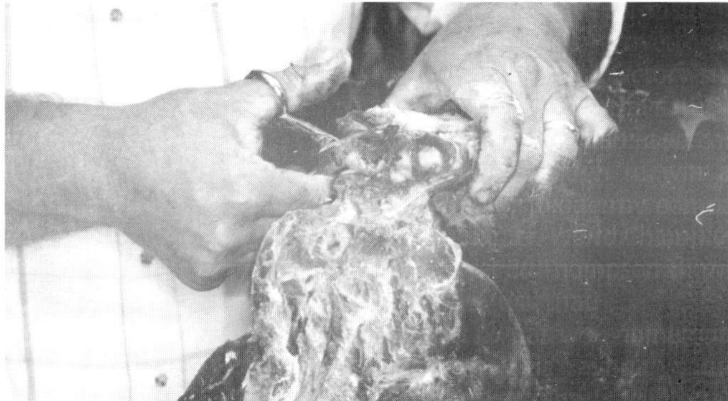

Next, the tail is severed at the joint of the pygostyle with a pair of Ultimate Scissors (above), and the skin is pulled down over the back to the shoulders (below).

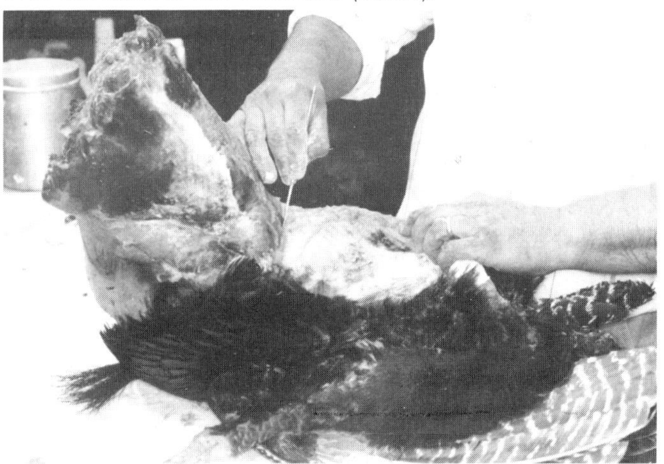

The skin is carefully removed from the breast area up to the original neck cut (by working close to the skin with a scalpel),

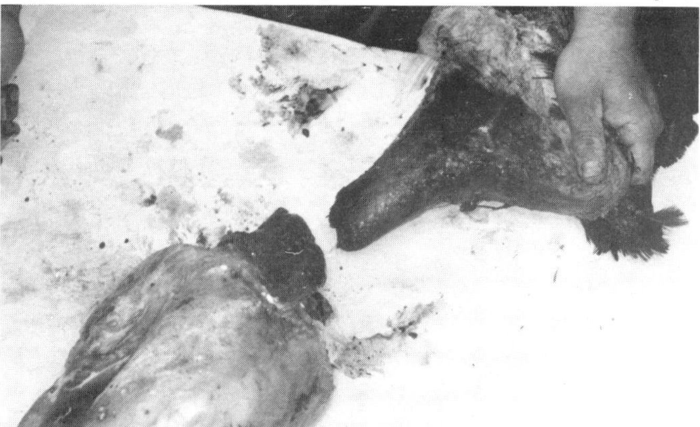

and is then completely removed from the carcass.

Ed recommends skinning all fresh turkeys up to this point and then storing them in a freezer until needed. In this way, freezer storage space is minimized and the bird will thaw faster at the time of the mounting.

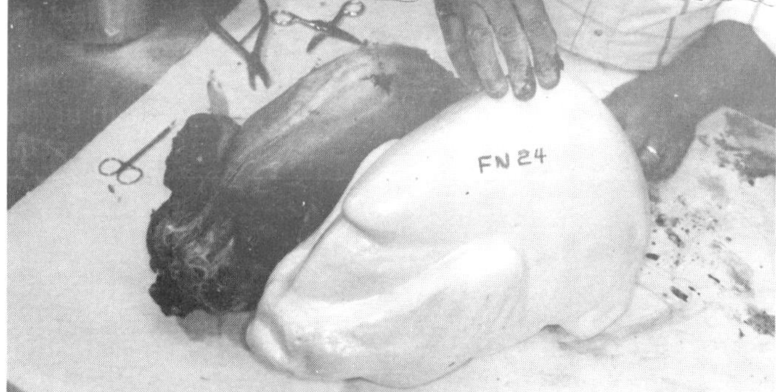

To select the proper sized mannikin for this turkey, Ed compared the actual carcass to a Sportsman Series FN20 and FN24.

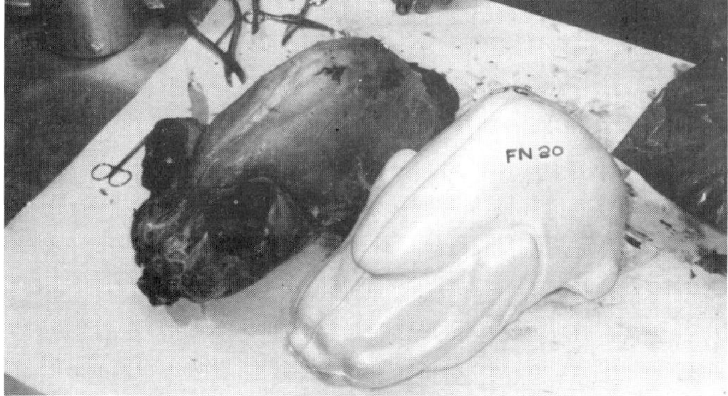

As can be seen in the photos above, the FN20 is the more correct size and was the mannikin that Ed used.

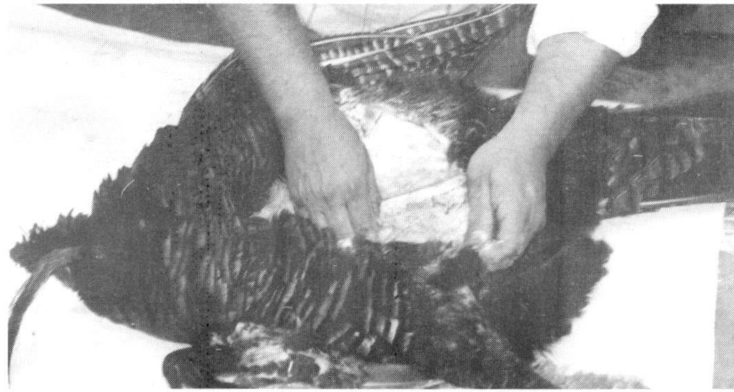

After a clean piece of paper is placed beneath the turkey skin, the inside surface of the skin is covered lightly with Powdered Borax.

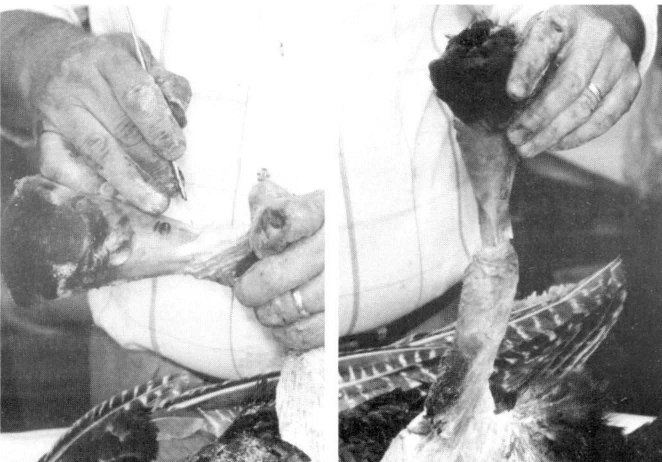

Each leg skin is then worked down to the unfeathered portion of the leg (heel).

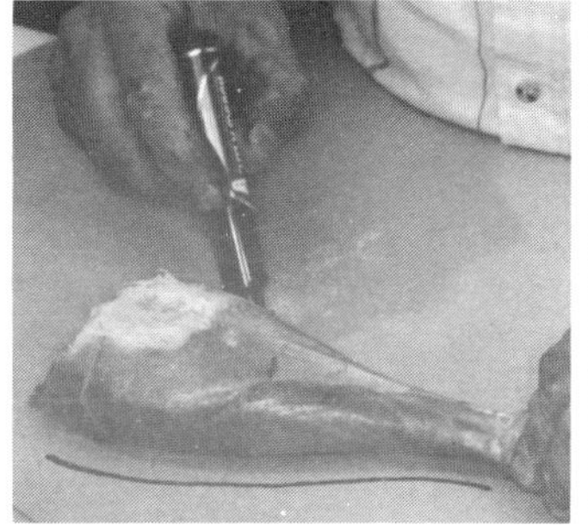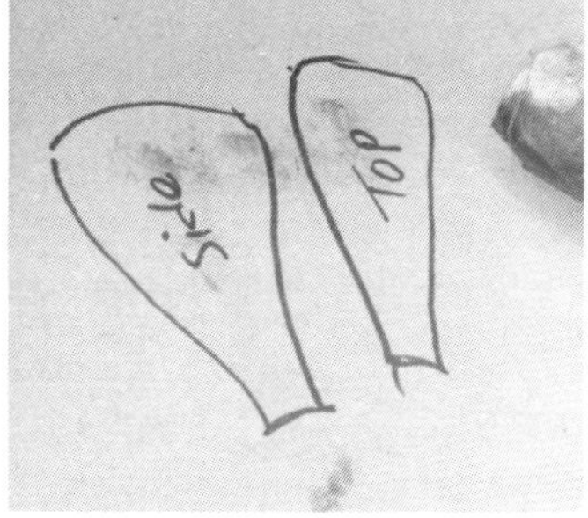

At this point, Ed makes a top and side view tracing of the "drumstick" section of the leg to be used as reference later. Note how the drawing demonstrates that the thigh is not round, but oval in shape. (If so desired, the whole body can be traced, side and top view, to be used as reference and to keep track of the exact size of the carcass. A Sportsman Series mannikin can be compared to these tracings to determine the correct size for mounting.)

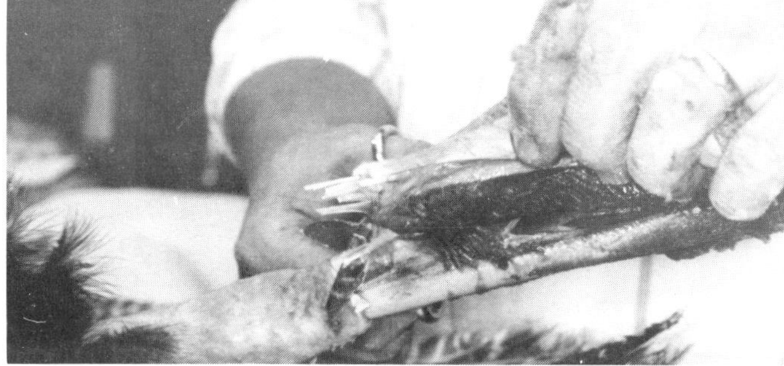

Now using the fingers and a scalpel, the tendons are cut and the flesh is removed from the leg bones.

Next, a piece of 9 gauge wire is inserted into the center of the leg bones to force out the marrow.

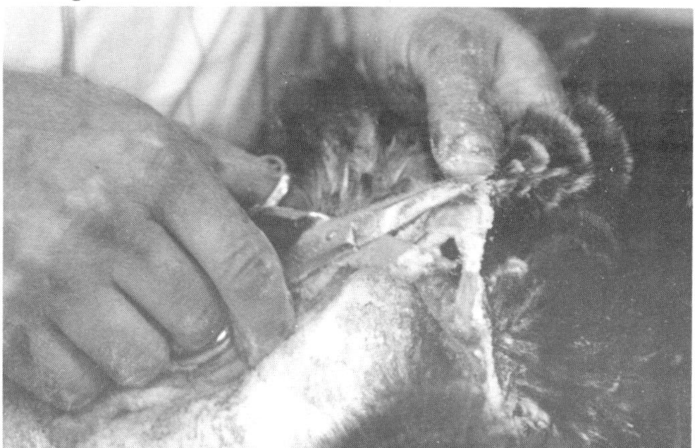

The Ultimate Scissors are very useful in trimming the excess flesh from the skin. Much of this flesh and membrane can also be pulled off with the fingers.

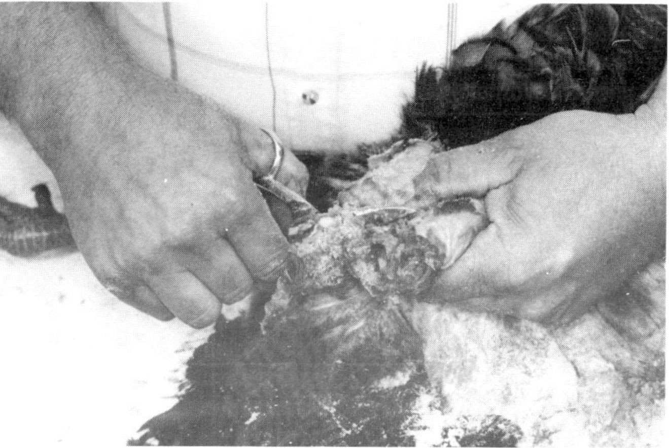

Ed then continues skinning up to the tail (or pygostyle) to the feather line, and the excess flesh is removed with Ultimate Scissors.

A wire grooming brush is used to clean between the feather quills.

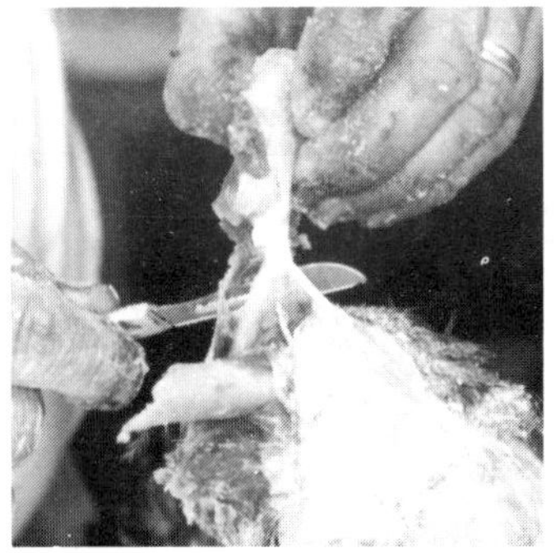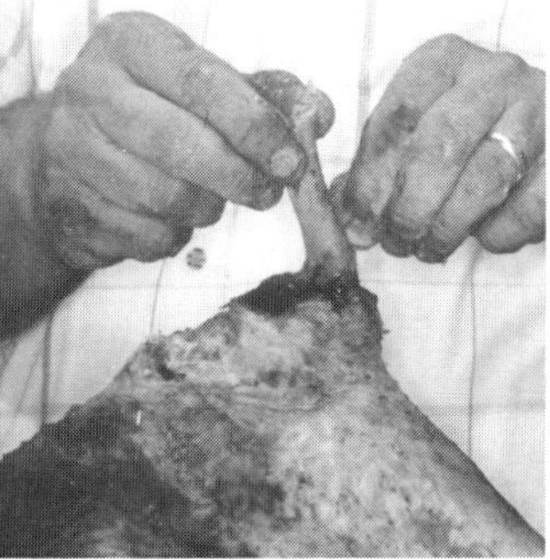

The next step is to prepare the wings. The meat and tendons along the humerus are removed as shown in the photographs above.

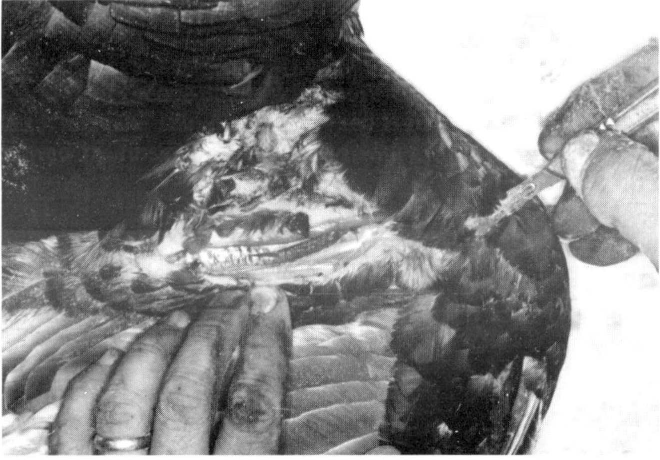

On the underside of each wing, an incision is made to expose the flesh and tendons located along the radius and ulna (see illustration below).

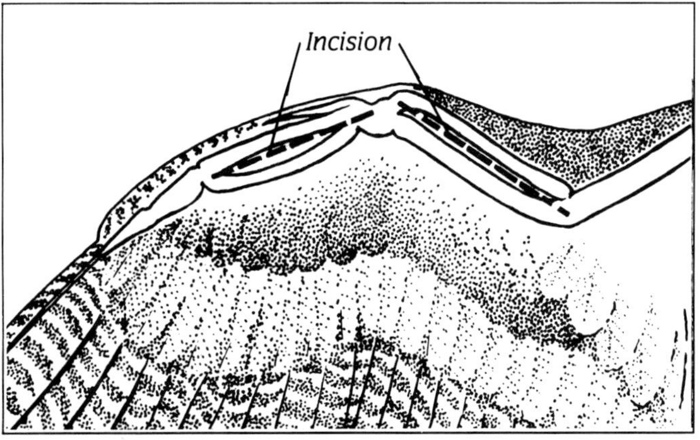

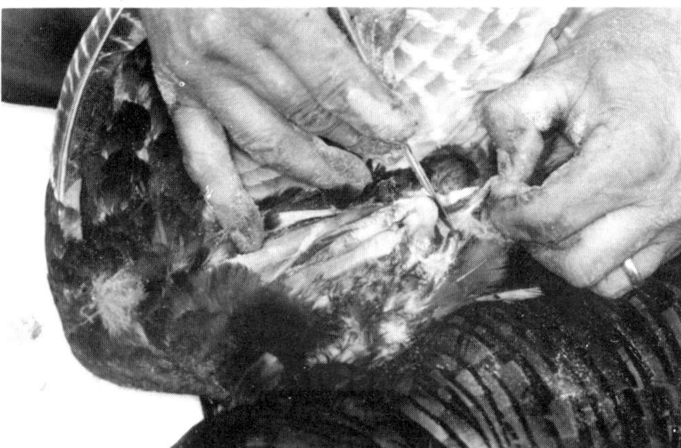

This area is then thoroughly cleaned with the use of a scalpel, scissors, and Powdered Borax, especially on the underside of the bones and joints.

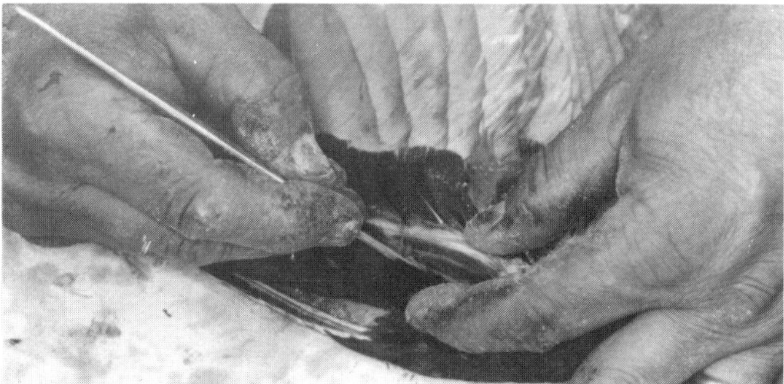

The next section of each wing to be cleaned is the "hand" (carpal/metacarpals), and is done in the same manner as the radius and ulna.

The photograph above shows the prepared wing section.

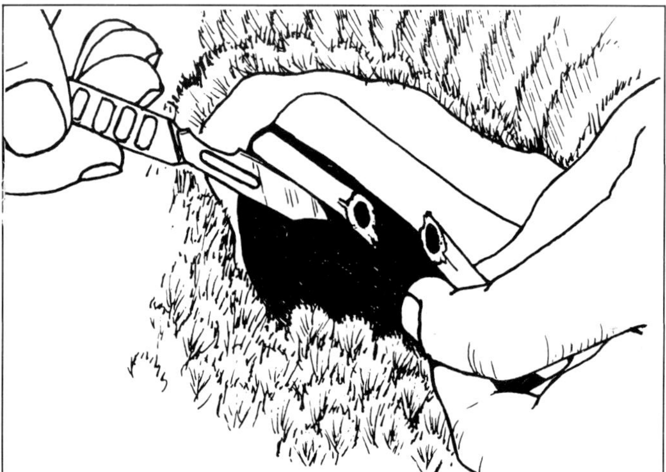

One wing of this turkey has a broken radius bone.

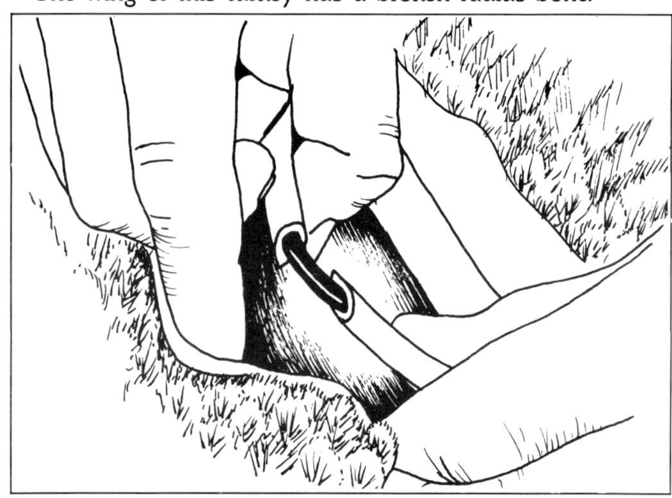

To make a very quick and serviceable repair, Ed first cleans the marrow from inside the broken bone halves, and then installs a short section of 9 gauge wire.

He then bends the wire straight again, the bone ends are positioned correctly, and the repair is complete.

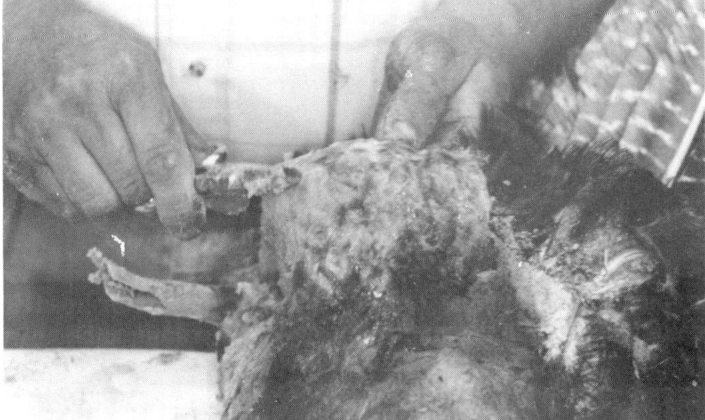

The breast area of the skin contains a spongy-like flesh sometimes referred to as the "gobble box." Ed carefully fleshes this area with the Ultimate Scissors.

Next he works out any remaining flesh between the feather tracts of the skin.

The final fleshing is accomplished with the use of a powered wire defatting wheel.

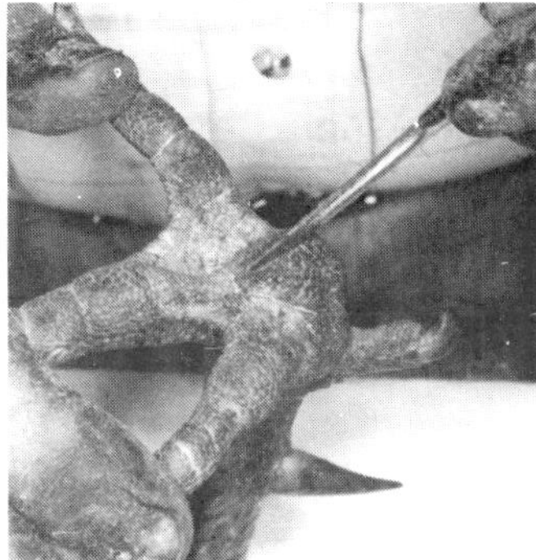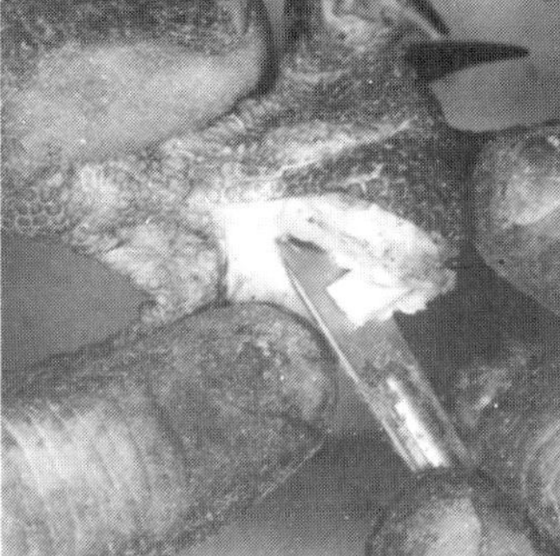

In order to allow the heavy gauge wire to be inserted and easily run through the leg, it is necessary to remove the tendons from each leg. An incision is made at the pad of the heel and the scalpel is worked around the sides of the tendons.

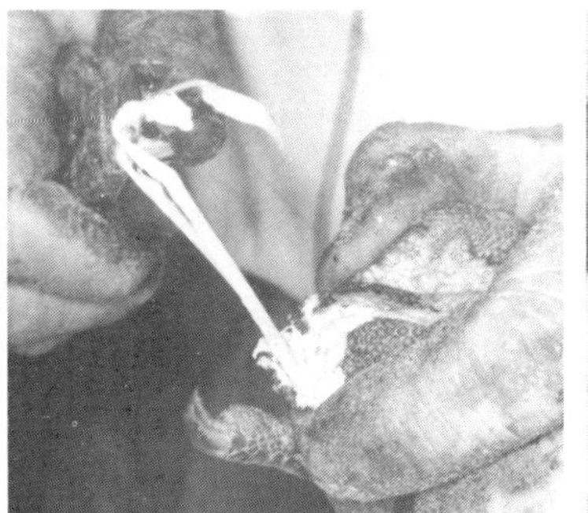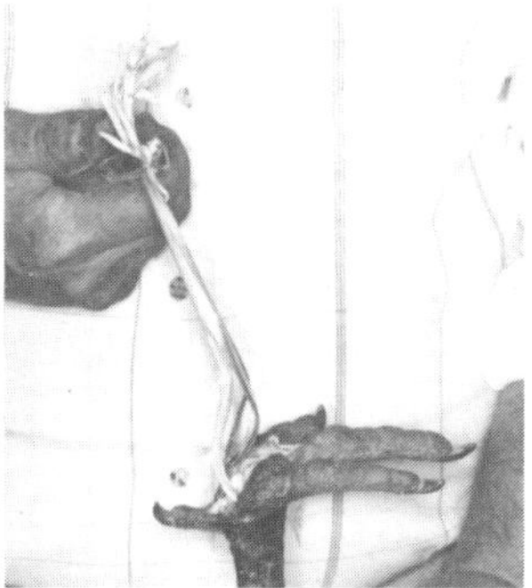

The tendons can then be easily pulled from each leg with a pair of needle-nosed pliers.

After "mechanically" cleaning and defatting the skin, it is "chemically" cleaned and defatted with chemicals that remove both the "surface" fat and the "inner" concentrations of fat. First, we use *Polytranspar* Degreaser (solvent) to remove the outside fat deposits and then Grease-Buster to remove the inner fat that is located within the skin itself. Skin Prep is used as a final wash, followed by rinsing in cold, clean water.

Tanning

After chemically cleaning the skin, it may be tanned with the *True-Tan* Bird Tanning solution according to the instructions in Chapter 6 or dry preserved if you are so inclined. After tanning the skin, it is gently squeezed, drained, and "tumble-dried" in Puffed Borax. Finish drying the turkey skin using compressed air at a low setting.

Wiring the Wings

Since this turkey is to be mounted in the closed wing position, Ed selects 12 gauge wires to support the wings. (For flying turkeys, 10 gauge wire is required.) Two 36 inch lengths of wire are cut, straightened, and sharpened on one end of each wire on a bench grinder.

The sharpened end of the wire is inserted into the "outer" or last section of the wing (metacarpal). The wire is then pushed past the radius/ulna and the humerus. A sufficient amount of wire will be pushed past the humerus into the body cavity to allow the wire to be secured to the mannikin.

To permit greater flexibility when positioning the wings along the body, the wire is *not* fastened directly to these bones.

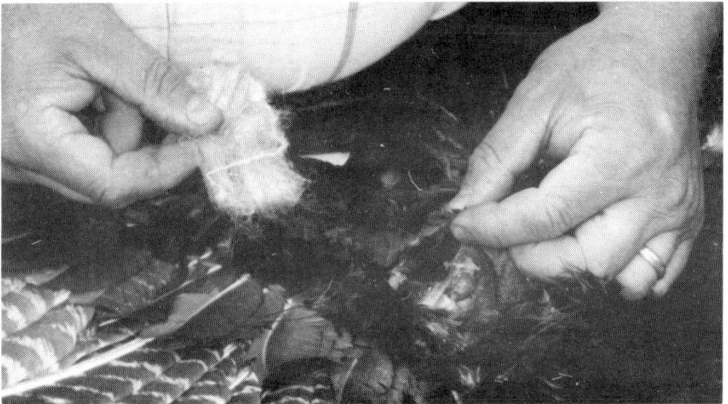

Ed then begins to close the wing incisions with a three-cornered needle and thread. When each of the wing incisions are nearly closed, he inserts fine tow filler material to recreate the muscle and cartilage removed during the skinning operation.

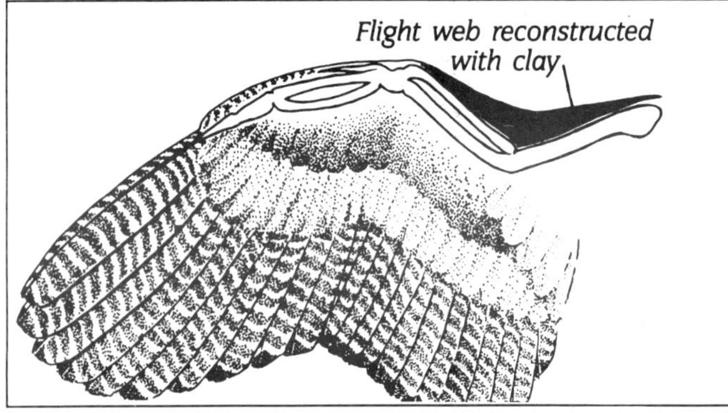

The flight "web," or the portion of skin between the second joint of the wing and the body, is reinforced by adding a small section of WASCO clay and shaping it into this area.

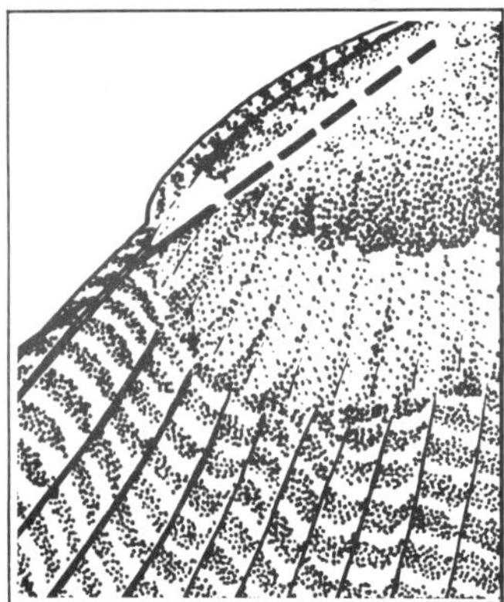

The first primary wing feather is then placed over the protruding wing wire to offer additional support to the wing as it is being mounted.

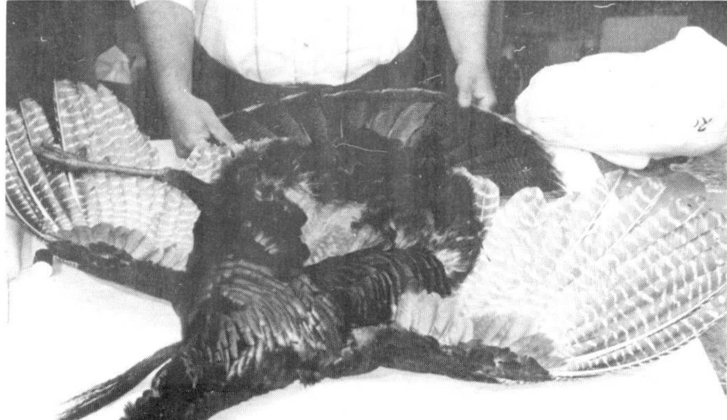

The other wing is wired and processed in the same manner. At this point, the wings are able to "slide" along the wing wires to aide in the proper positioning of the wings at the side of the body.

Working from inside the body cavity, Ed then wraps the humerus bone with fine tow and thread to recreate the muscles that were removed. Note that the wing wire is *not* included in this wrap; this was done purposely to provide easier wing positioning later. Once the wings are wired in this manner, Ed proceeds to mount the tail.

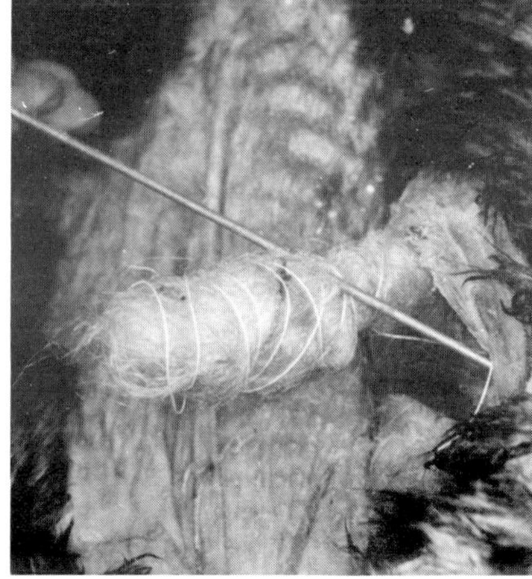

Mounting the Tail

One of the most important aspects of a strutting turkey is, of course, to properly mount and position the prominent fanned tail. Contrary to many taxidermist's beliefs, the turkey will often angle its tail from side to side to better display himself to hens. The tail does not "have" to be in the more conventional "centered" pose; however, it probably looks best centered.

A new technique for simplifying holding the fanned tail in its position has been developed that is called the "bonded tail method." This method permanently locks the tail feathers together in their proper positions from the *inside* and makes displaying the fan much easier.

To begin this process, the tail is first "fanned" out and placed over a sheet of cardboard that has been formed into a slight curve. Before "bonding" the tail, Ed first adjusts the curved

surface of the cardboard and the feathers until he is completely satisfied that the shape is exactly in accordance with his
reference materials.

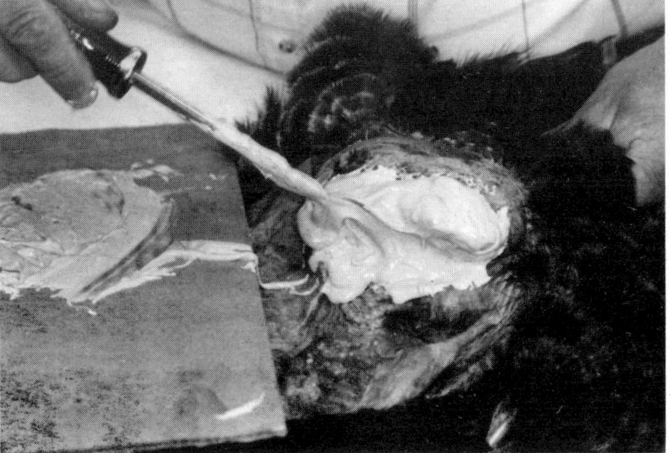

He then mixes a quantity of Ultra-Lite Filler and catalyst and
applies a liberal amount to the pygostyle area of the inside
of the tail. He makes sure that the filler thoroughly surrounds
each tail feather root.

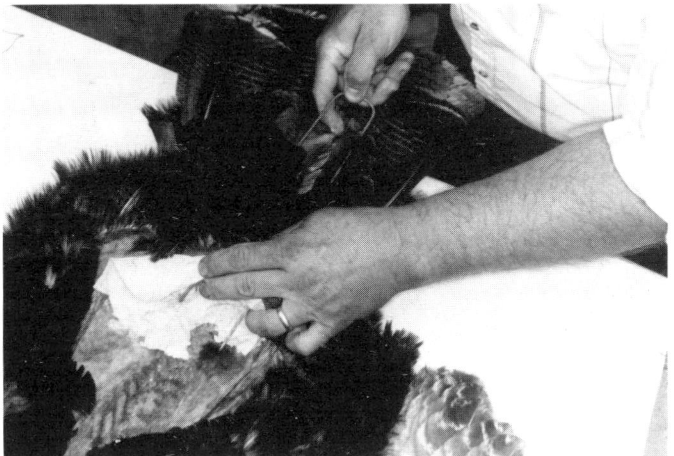

Next, a "U" shaped 12 gauge wire is inserted through the
skin on the underside of the fan as shown. A piece of paper
towel is placed around the Ultra-Lite Filler to prevent it from
touching any of the feathers.

The tail is then placed so that the fan can be spread and
positioned over the curved sheet of cardboard. Pieces of masking
tape hold the fan in position until the Ultra-Lite cures. Notice
that the fan is spread to nearly 180 degrees.

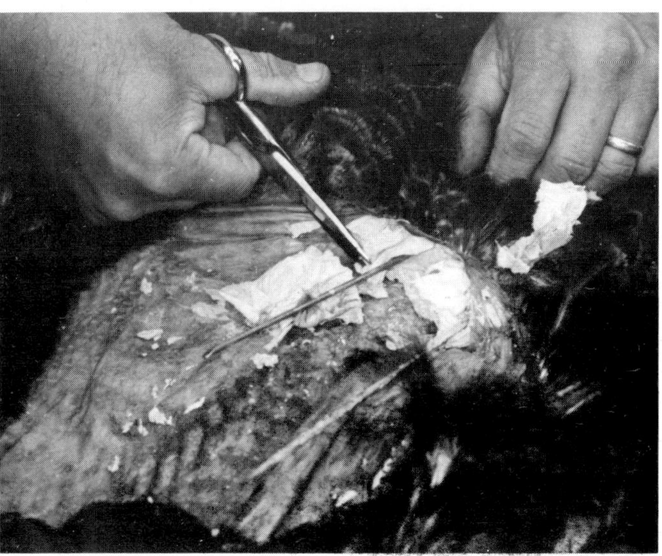

It requires about 15 minutes for the Ultra-Lite to completely
harden, after which Ed trims the excess filler that is not needed
from around the base of the tail. The photo below shows how
nicely the tail is held in the fan position with the use of this
fast and easy method. Perfect, permanent feather placement
is achieved in minutes!

Mannikin Preparation

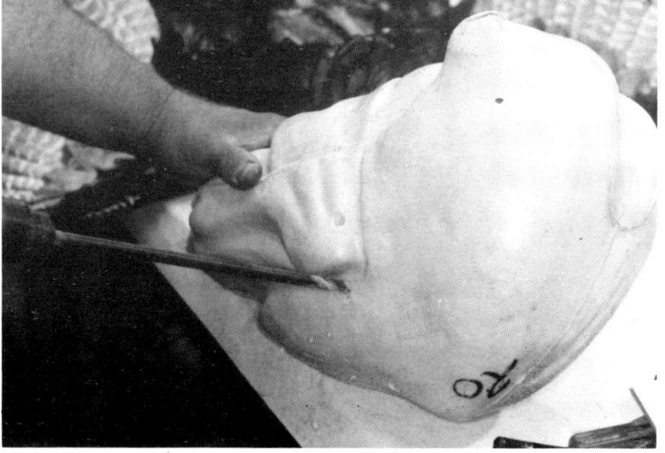

To make it easier to insert the leg and wing wires into the
mannikin, the holes for the wires are first pre-drilled with a 1/4"
extension drill as shown. All Sportsman Series mannikins have
the exact reference points for attachment of the legs, wings,
and neck which makes accurate placement very easy.

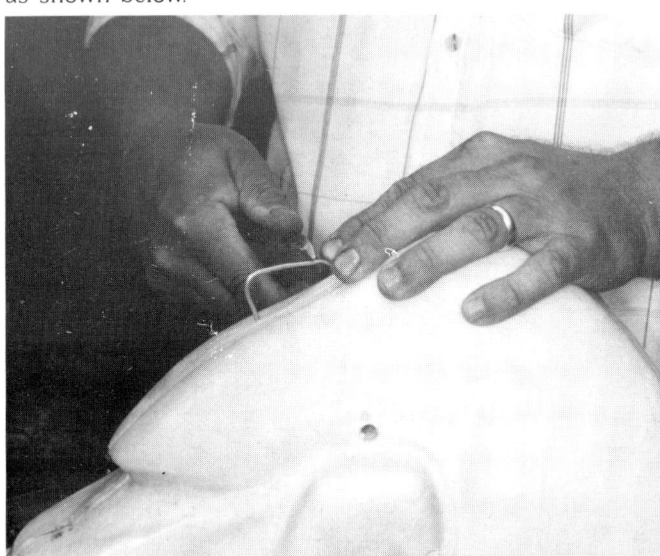

Ed then installs an FNN10 turkey Flex-Neck by inserting the neck wire into the mannikin, and then clinching the end in place as shown below.

The mannikin, with the neck attached, is then test fitted into the skin. Notice how easily the skin can be positioned about the mannikin. A bird skin should never be tightly pulled or stretched to "make" it "fit." Such action will prevent the taxidermist from being able to properly "taxi" or adjust the skin and feathers.

Mounting and Securing the Wings

The 12 gauge wing wires are then inserted through the pre-drilled holes in the mannikin from each side. About four inches of wire is bent back into a "U" shape and forced back into the mannikin by tapping with a hammer. This securely anchors ("clinches") the wire into place.

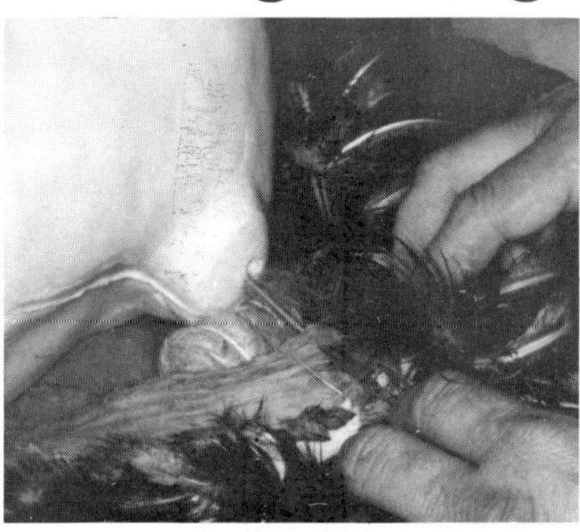

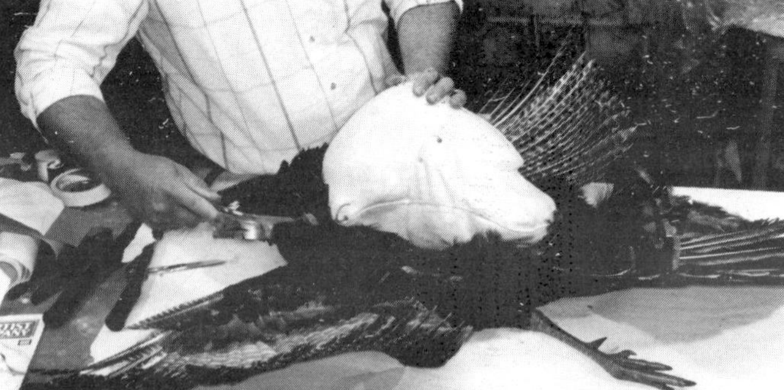

Next, a layer of "stiff" Sallie Dahmes Hide Paste is spread along the back and sides of the turkey mannikin. This paste really provides two functions. One, it bonds the skin to the mannikin, which is important on a large bird, and two, it also provides a secure base to support each individual feather root. This is especially important with turkeys as their feather patterns are so intricate.

Mounting the Feet and Legs

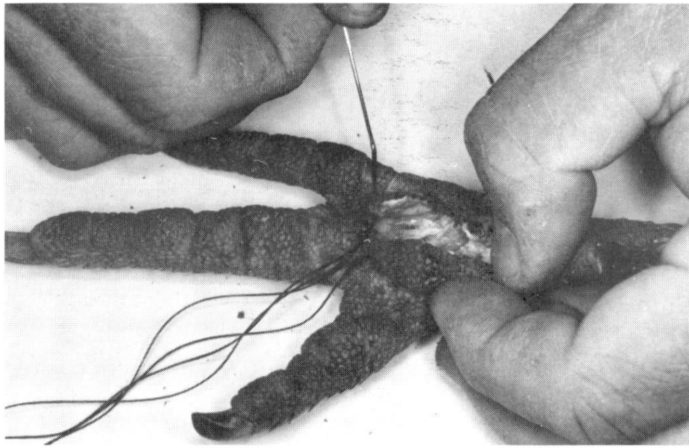

To begin the feet and leg mounting procedures, the incisions in the ball of each foot (made earlier to remove the leg tendons)

are sewn closed, using a fine needle and thread. Next, two 30" long 8 gauge wires are cut, and one end of each wire is sharpened for insertion into the middle toe and through the leg.

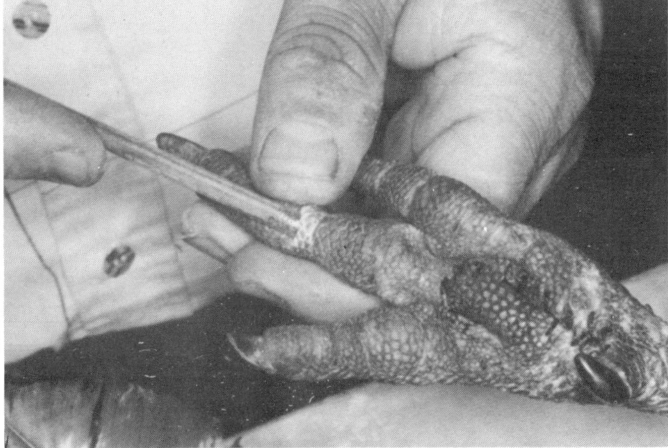

The legs are then wired to the mannikin in similar fashion as the wings. In order to hide the support wire and to balance the bird correctly, Ed inserts the 8 gauge wire so that it enters the foot half way up the center toe, past the ball of the foot, up through the back of the leg and into the body cavity of the skin. (For very large birds, he recommends 6 gauge wire.) Once in place, the wire is pushed back and forth several times to loosen the fit and to help make any final adjustments somewhat easier.

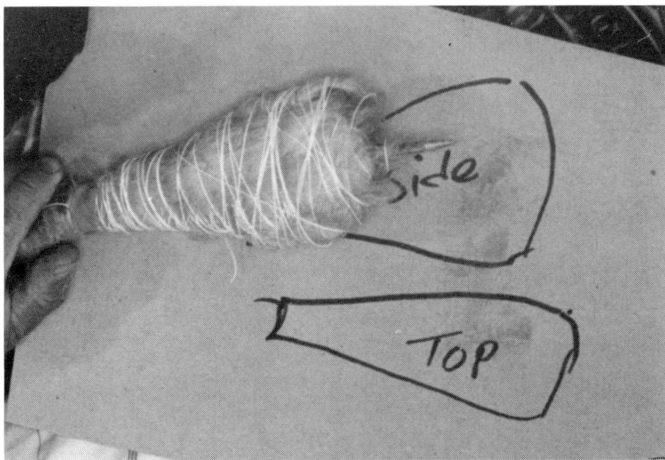

Fine tow is then placed around the bone and leg wire and wrapped with thread. Ed refers to his top and side view sketches of the skinned out leg to rebuild the leg musculature to its correct shape.

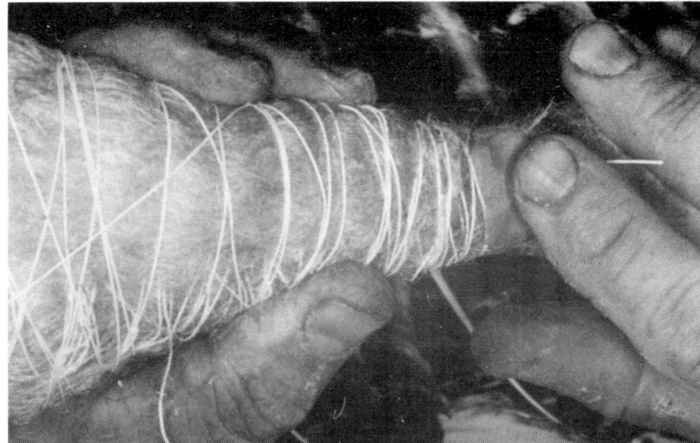

He uses a small quantity of WASCO clay to form a smooth transition between the wrapped leg and the bone at the ankle area.

The leg skin is drawn up over the wrapped portion, making sure that it fits easily and is not twisted.

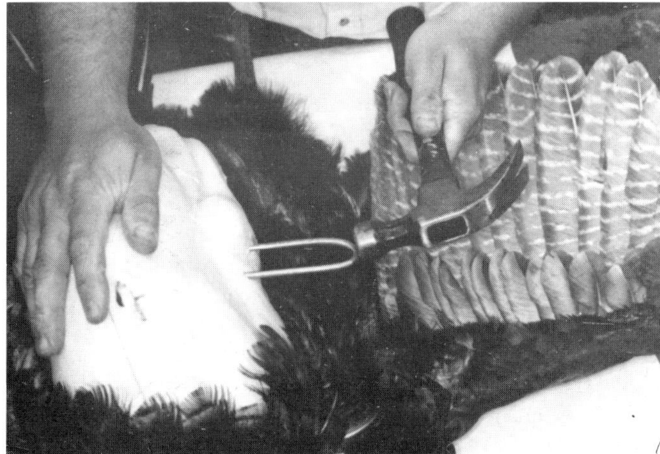

The leg wire is then inserted through the pre-drilled hole in the mannikin, reversed into a "U" shape, and pushed back into the mannikin so that it can be "clinched" in the same manner as the wing wires were done. (A hammer may be used to easily tap the wire in, if difficulty is encountered pushing the sharpened wire back into the mannikin.) The other leg is prepared with the same procedure.

Securing the Tail and Positioning the Legs

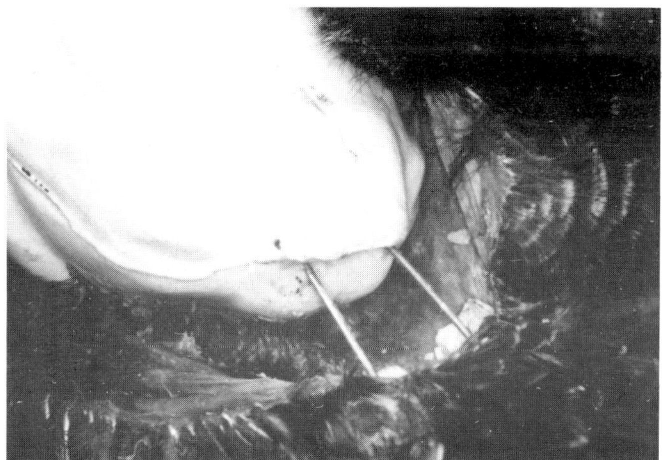

The pygostyle area of the Sportsman Series turkey mannikin is pre-notched to accept the stub section of the tail feathers. With the tail correctly located in this notch as shown, the "U" shaped tail wire (already installed) is pushed into the mannikin to hold it in place.

Next, the legs are bent into their approx-imate positions along the side of the body to make it easier to sew the incision. Heavy pliers are used to bend and shape the wires where they penetrate the toes.

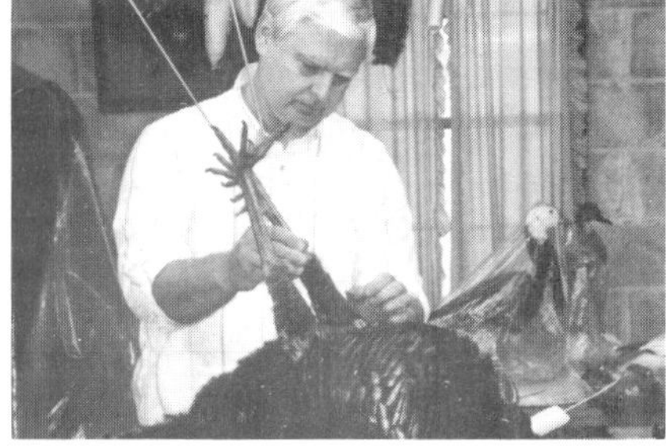

Initial Shingling and Positioning of Feather Groups

The "foam-like" adipose tissue and fat layer that was removed from the breast area of the skin is now recreated with tow.

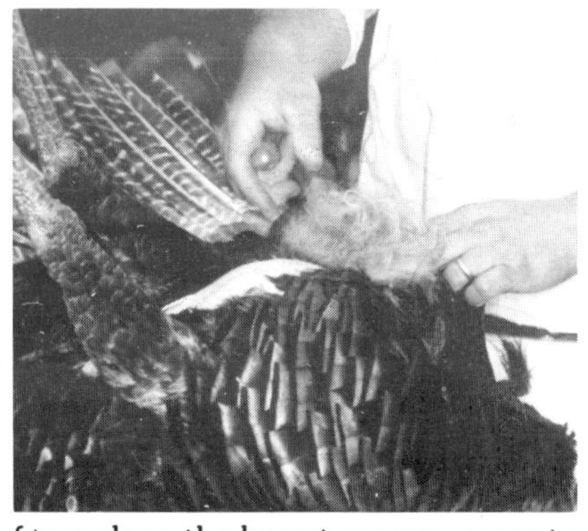

This step is easily ac-complished by placing loose pads of tow along each side of the mannikin at the front of the breast. When an old gobbler "struts," he puffs and fluffs himself up and actu-ally "bristles" the feathers out along his sides and back. The proper term for this is "shingling." The pads of tow along the breast are necessary to begin the "shingling" effect. By first placing a pad of tow in these areas and then pulling the skin into its proper place, Ed is able to check and make sure that the proper "shingling" effect is going to be obtained.

Some feather groups are *not* supposed to be shingled. Notice in the previous illustration and following photograph the three areas of feather patterns of a strutting turkey that are not sup-posed to be fluffed or "shingled." These areas are located along each side of the front of the breast, and the area just behind the head. These feather tracts present a very smooth appearance of the feathers.

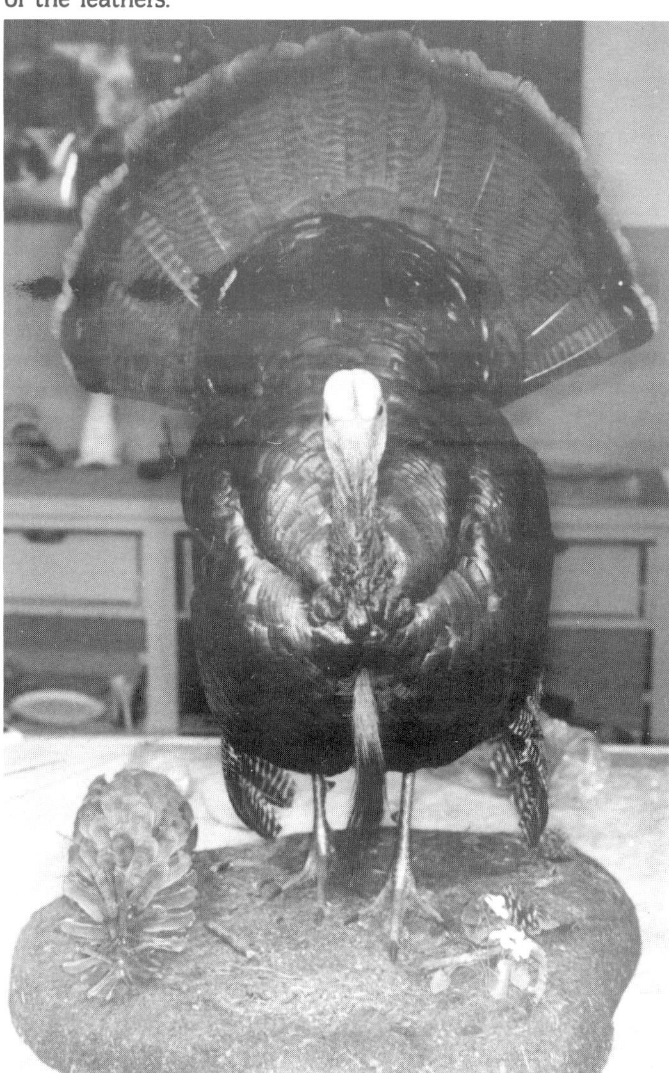

Closing the Incision and Positioning

Once the additional tow padding is in position, the ventral incision is sewn, beginning at the point of the breast and con-tinuing down to the vent.

Next, the excess length of wire protruding from the end of each wing is folded beneath the wing to provide additional support to the heavy wing. The wings are then folded into their approximate positions along the sides of the body.

The turkey is placed upright on a suitable sized base and positioned so that the leg wire holes can be marked and drilled.

A groove is cut adjacent to each leg rod hole on the bottom of the base. Then the bird is firmly seated on the base, and each leg wire is cut to length and bent to fit into these grooves. The wires are then tapped in with a hammer and stapled into place.

Posing the Turkey

With the turkey placed in the upright position, Ed begins to bend the leg and wing wires into their final positions. He refers

constantly to his reference materials to ensure that the position (or attitude) of the bird is accurate.

The fanned tail is then bent upright to the correct position, again referring to actual photographs of live birds. The breast skin is adjusted forward and upward so that the neck skin can be "taxied" into its correct position at the upper end of the neck.

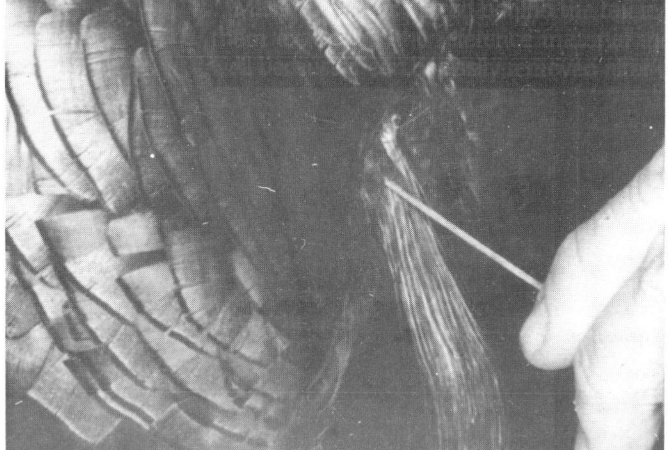

With the breast skin positioned correctly, the beard is pinned in place with a piece of sharpened 14 gauge wire. The head of the wire is buried just underneath the skin of the bird.

Installing the Head

The artificial head selected for use on this turkey is a FN201 strutting turkey. The head is painted prior to installing it on the mount. Since part of the neck area is already formed on this head, some of the material on the neck of the mannikin must be removed before installation. The amount removed is determined by holding the head into position and then marking the neck. The foam neck material is removed from the wire to the mark, and the support wire is trimmed to about two inches above the neck.

A 3/8 inch access hole is then drilled into the base of the artificial head to a depth of about one inch. A "U" shaped loop is bent in the end of the neck wire to fit up inside the head, and the assembly is test fitted. When the "fit" is satisfactory, Ed "hot-melt" glues the neck wire into place, and then allows the glue to cool.

Final "Shingling" of the Turkey

The final detailing of the feathers, or "shingling," is started at this point. While shingling is not particularly difficult to master once you have seen it done, it does require a great deal of tedious work and "time" over a period of several days. Ed states that he "easily" spends as much time grooming and "adjusting" the feathers on a strutting turkey as he does the entire process of skinning and mounting the bird, but the time is necessary if one is to "perfect" the attitude of the strutting turkey.

Ed gently "teases" the body feathers forward and upright with his hands.

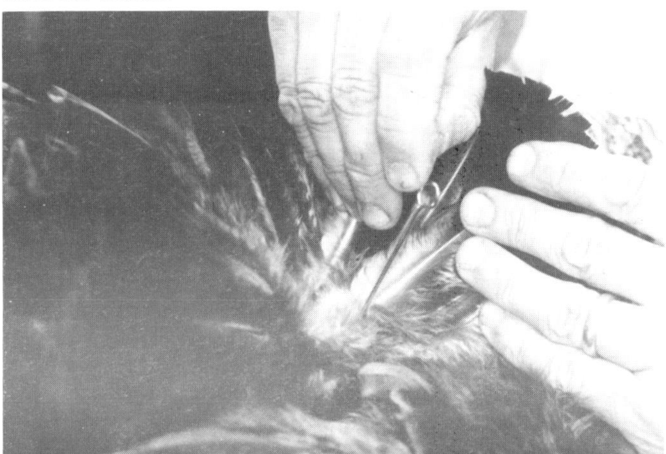

He then begins placing Loop-Head Pins behind various groups of feathers to roughly establish their positions.

When this is done, he places small cotton balls behind most of the feathers to further define the position of each one. As the bird continues to dry over the next few days, the pins and cotton balls are constantly moved and repositioned to obtain

Next, a small layer of WASCO Clay is placed around the union of the artificial neck and head to make a smooth transition.

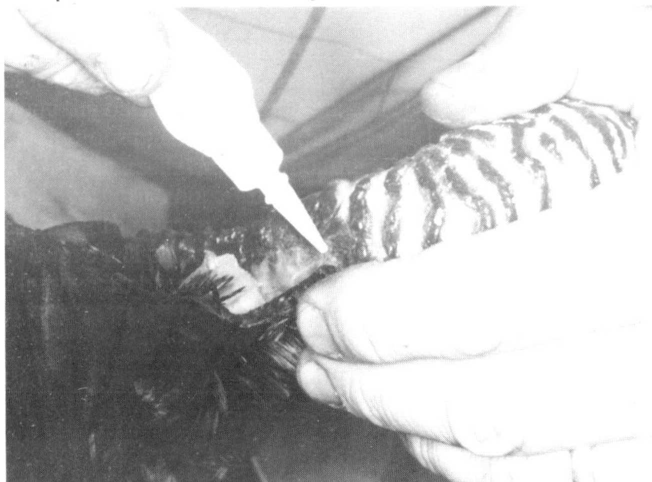

The back of the neck skin is then pulled up behind the head and pinned into the recess provided in the artificial head.

When Ed is satisfied that everything looks "just right," he glues the skin into position with Instant Bonding Glue (above).

Small insect pins are used to hold the skin in the exact position that Ed wants it in until the glue dries (right).

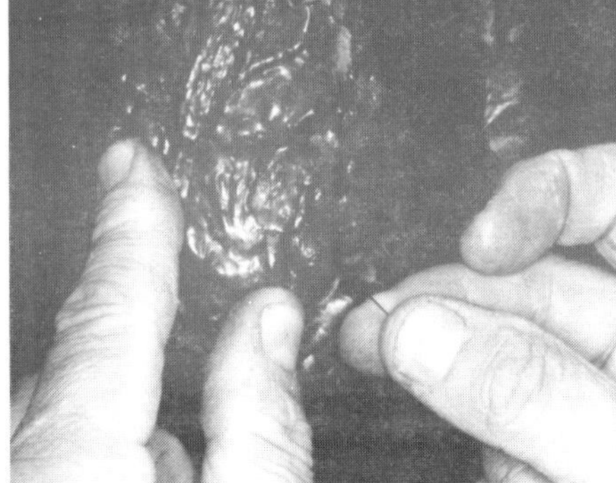

a more uniform appearance to the raised feathers. The feathers are not raised in rows, but rather, in more of a random pattern.

The large black feathers located over the side pockets are "shingled" in the same manner as the back feathers.

Sharpened wires and Loop-Head Pins are used to hold each wing in position, so individual feather adjustment can be made.

The side view of the bird illustrates the dramatic change in the visual shape of the strutting turkey. Notice how the "shingled" feathers along the back rise up gently from the back "shield" area to a peak in front of the tail, and then dip slightly before blending into the tail itself. The length of the feathers on individual birds will vary, of course, but this is a good "general" profile. Also, note the droop of the wings. When a turkey is strutting, its wings will actually drag on the ground, causing the primary feathers to become very worn, especially

on the more mature males. Also note how the tail covert feathers located beneath the tail actually "hang" down as shown.

Masking tape is placed behind the tail feathers to further position them exactly as the reference material indicates. This tape works well because it can be easily removed after the bird is completely dry.

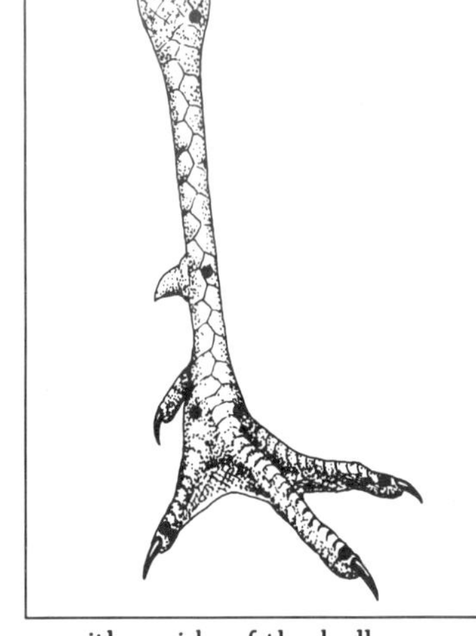

The feet are injected with Formalin (37% formaldehyde) at seven locations; at each toe, once on either side of the ball of the foot, and in two locations between the ball of the foot and the knee. At each location, sufficient Formalin is injected to puff the area up "slightly" larger than actual size, so that as the foot dries, it will shrink back down to its original dimensions. *Remember that Formalin is an extremely dangerous product to use.* "Always" wear rubber or surgical gloves and a pair of safety goggles when injecting this substance, and treat it with great caution. Perform any injecting operation in a well ventilated area, preferably in a ducted spray booth system, and place the bird (or mammal) where no fumes can be detected while it is drying.

Turkeys require several days to dry in the strutting position, and demand a great deal of attention. The taxidermist must adjust feathers one at a time, and refer constantly to the available reference material as the skin continues to set and "lock" the feathers into place.

The key to success in mounting a superb strutting turkey, to a large extent, is dependent on how much "feather adjustment time" is allotted to the bird. Four to five hours on the first day is about average, with additional hours spent on the second, third, and fourth days, until the bird is dry and the feathers are securely "locked in."

When the turkey is thoroughly dry, all of the visible pins and cotton balls can be removed, leaving the "shingled" feathers exposed in a gorgeous upright position.

The completed strutting turkey. Nice job, Ed!

Tommy Knight: Flying Turkey

Dr. Tommy Knight of Pensacola, Florida, is considered to be one of the best turkey taxidermists in the country. He has judged numerous taxidermy competitions as well as received several awards on the state, national, and World level for his outstanding turkey and waterfowl mounts. As a result of these endeavors, Dr. Knight is generally recognized by his peers as an expert in this field.

The following is Tommy's step-by-step mounting process of an Eastern wild turkey in the flying position. It is our intention that this chapter will help taxidermists to overcome and prevent many problems that may arise in mounting a flying bird of this large size.

Equipment Needed:

Scalpel	Edolan-U
Diagonal Pliers	Clay
Adjusting Probe	Excelsior
Needle Nosed Pliers	Sculpall
Wire Wheel Degreasing Machine	Masking Tape
Cotton or Poly Fiber	Dry Preservative

Skin Prep or Low pH Liquid Detergent
Polytranspar™ Degreaser
Wire (#10, #12 and #7 Galvanized Annealed Gauge)
Waxed Sewing Thread, #3 Needle, #2 Needle
Formalin (optional—extreme caution must be used)
Sportsman Series Medium Turkey Mannikin and Flex-Neck
Bondo or Ultra Lite Filler
Sallie Dahmes Finishing Wax

Mounting Process

On this particular turkey, Tommy is re-creating the bird by posing it in a very typical pose for a flying turkey, the banking/gliding position. The wings are extended, the tail is fanned, the neck is outstretched, and as in most upland game birds when posed in a flying position, the feet are pulled up underneath the body (rather than pointing away from the body like waterfowl). Most turkey hunters have seen many a turkey in this position. As with any mount, good reference photos are absolutely essential for re-creating an accurate pose. Tommy uses *Turkey Call Magazine*, published by The National Wild Turkey Federation and *Turkey Magazine* for many of his reference photos. Do not attempt to mount anything without reference photos. The skinning and skin preparation is essentially the same as for a strutting turkey; therefore, we will begin with the mounting procedures.

Wings

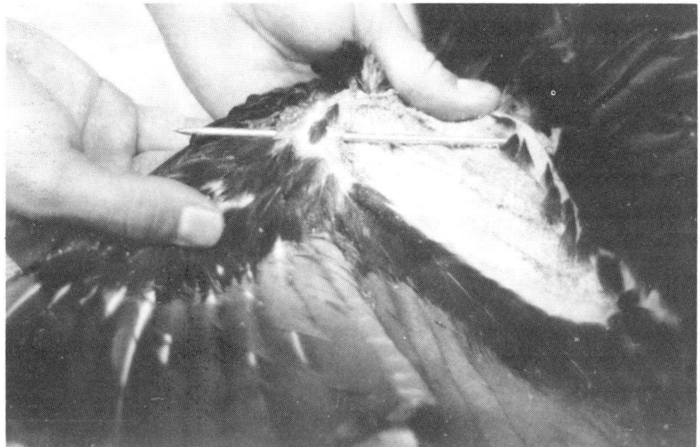

The first step in the mounting process is to wire the wings. This is done by inserting a 10 gauge wire starting at the

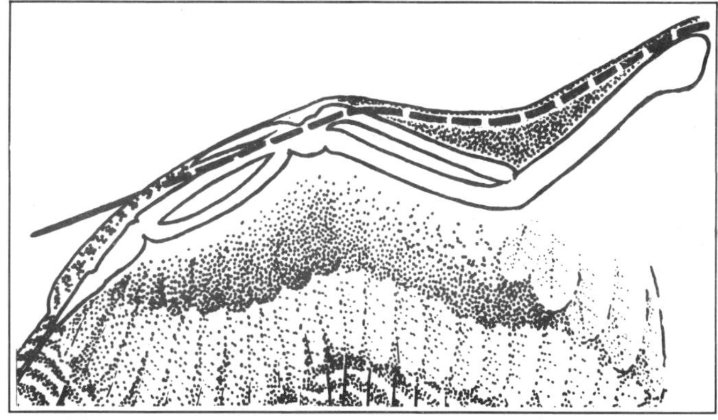

shoulder joint of the wing (humerus). Run the wire down the forearm (ulna and radius), go through the joint, and into the outer portion of the wing exiting the wing between the "fingers" (metacarpal).

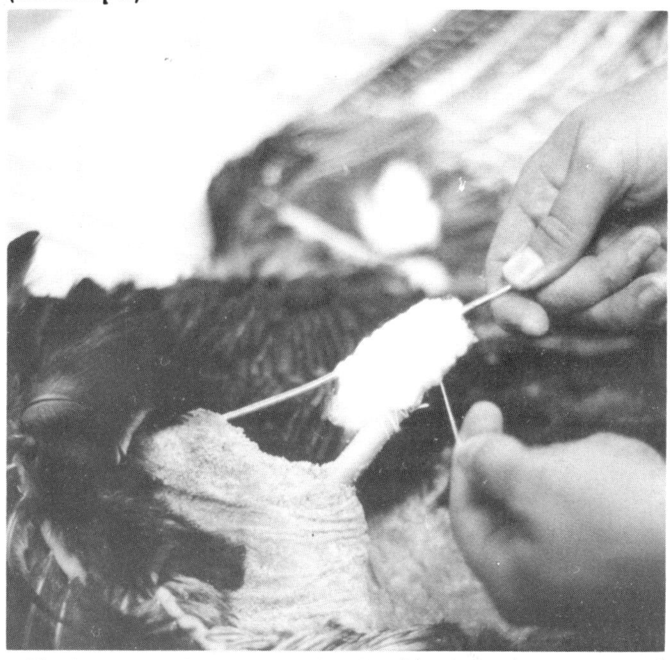

Next, wrap twine around the shoulder joint (humerus) and the 10 gauge wire, locking it into position. During the skinning process the muscle was removed and now must be rebuilt using cotton or poly fiber. Use reference pictures or a study carcass for reference when reconstructing this area. Turkeys have big wings and they should be rebuilt to their original proportions in order to look right.

Legs

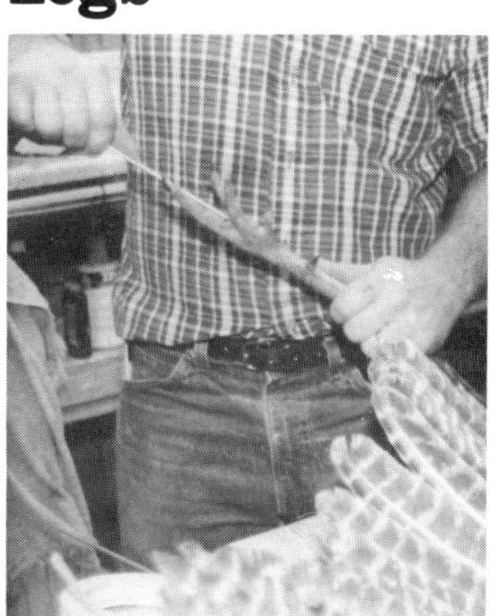

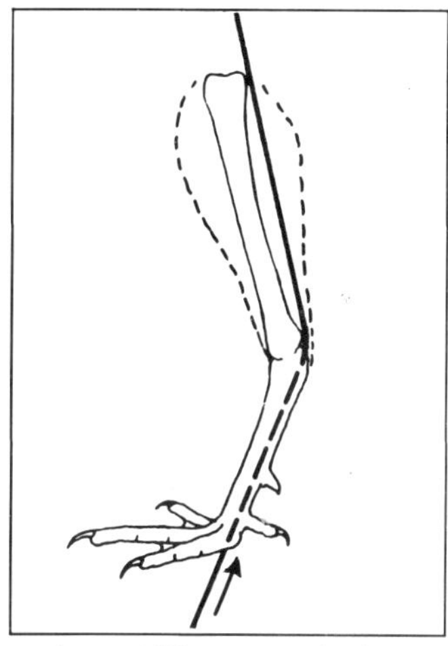

To begin the leg mounting procedure, a 10 gauge wire is run through the incision in the pad of the base of the foot up the inside of the leg (tarsus), past the ankle and up the tibia.

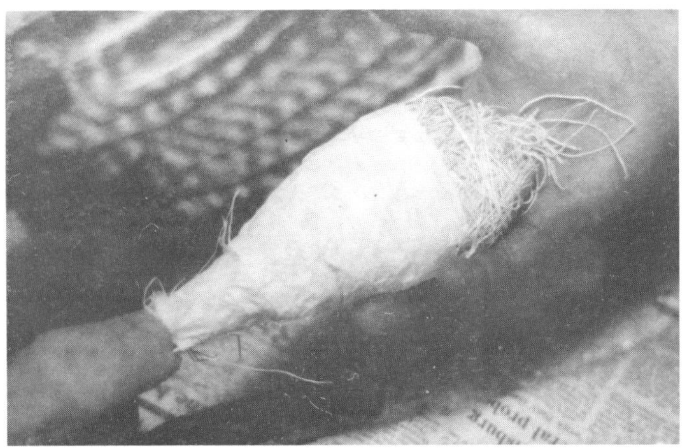

Once the wire is in place, the muscle is rebuilt to its original anatomy. (Again, good reference materials are invaluable for this step.) Excelsior is carefully wrapped with twine to rebuild this area. Clay is also used to rebuild the joint around the ankle (hypotarsus). Finally, masking tape is wrapped over the excelsior to make it smoother. Once the legs have been wired and wrapped, proceed to the tail preparation stage.

Tail

A wild turkey's tail is one of its most prominent features. It pays extra dividends to take the necessary time to properly display it. Tommy begins by running a small wire (12 gauge) approximately 10" in length through the flesh from the outside of the tail.

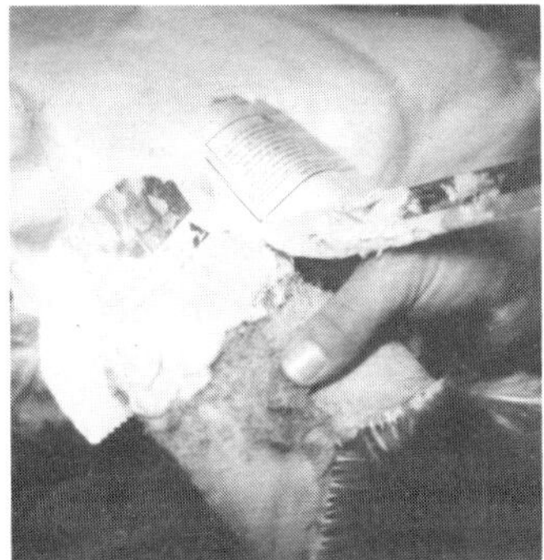 

Next, he mixes a small amount of Ultra Lite Filler to a fairly "hot" consistency. Apply the Ultra Lite Filler to the feather tracts of the tail and place newspaper on the Ultra Lite to prevent it from touching the skin of the bird. Now place the tail on a sheet of plywood and arrange it in its final fanned position. It should take approximately fifteen minutes for the Ultra Lite to harden. When it has completely cured, trim any excess from the roots of the feathers with a knife or scissors and the tail will be permanently locked into place.

Mannikin

The skin is now ready to be taxied onto the mannikin and mounted. First, attach the flexible neck to the turkey body mannikin. Sharpen and then insert the neck wire through the foam mannikin, bend it into a "U" shape and secure it into the foam. The Flex-Neck is inserted into the neck of the bird skin.

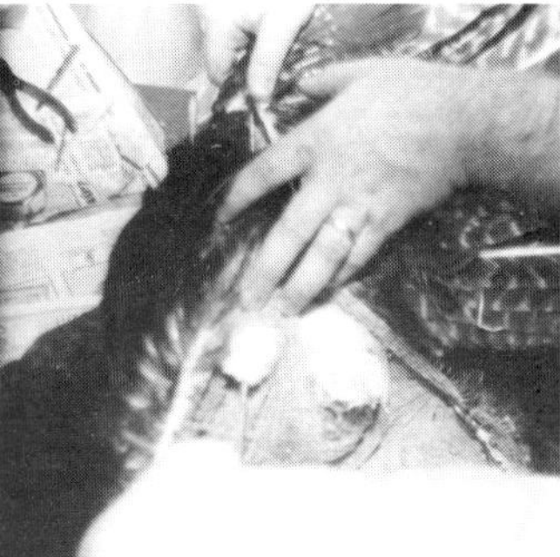 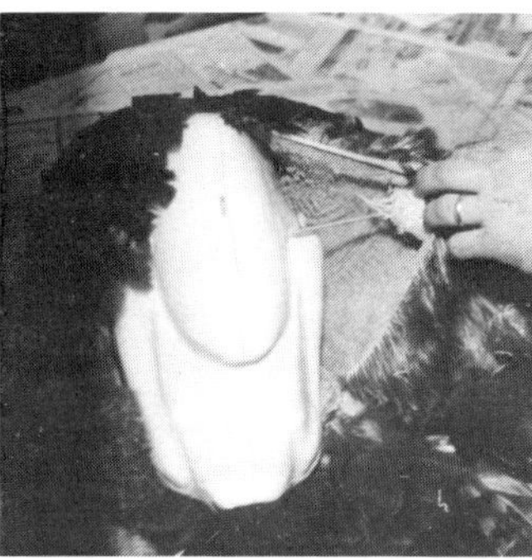

Attach the wing and leg wires to the mannikin using the reference points on the mannikin for points of attachment. Insert the sharpened wires through the mannikin, bend into "U" shapes and firmly clinch and secure them into the foam body.

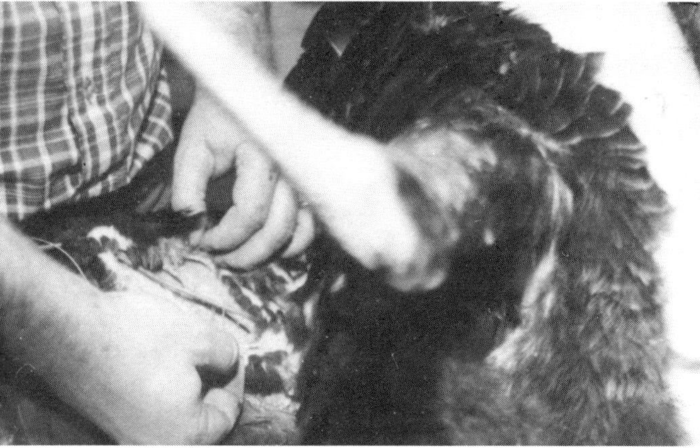

Once the wing and leg wires have been secured into the foam mannikin, tie a small piece of string around the forearm (radius) of the wing and the wing wire and lock both wings in place.

At this time, the muscle area between the radius and ulna of the wing must be rebuilt with clay and the incision made in the skinning process must be sewn shut. Once the wings have been restored to their original size and shape (use reference), 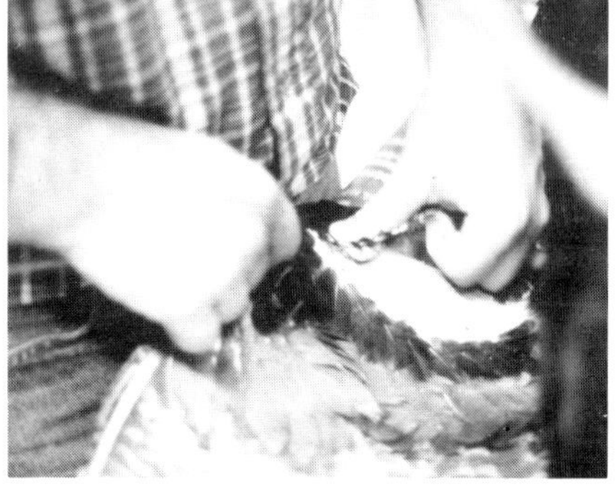 then the incision should be sewn shut using a no. 2 needle and fine waxed sewing thread.

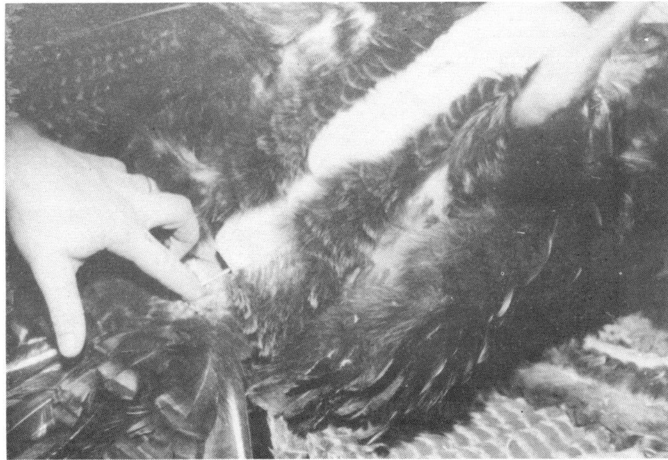

Place the tail into the pygostyle slot of the mannikin and push the 12 gauge tail wire into the mannikin to lock it in place. Now take small pins and lock the scapular feathers of the wings to the shoulders of the mannikin by pinning them in place.

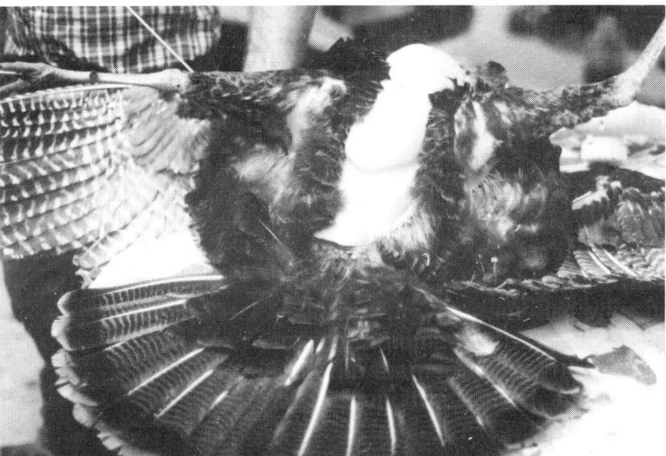

With these reference points locked in place, the ventral incision may be sewn shut. To sew the skin together start at the upper part of the breast and work toward the tail. Tommy suggests sewing from the top of the breast down as it makes the feathers lay into place better. Use a no. 3 needle and waxed sewing thread to sew the incision closed. A baseball stitch was used on this mount. It "may" be sewn from the bottom to the top if it is more comfortable for the individual.

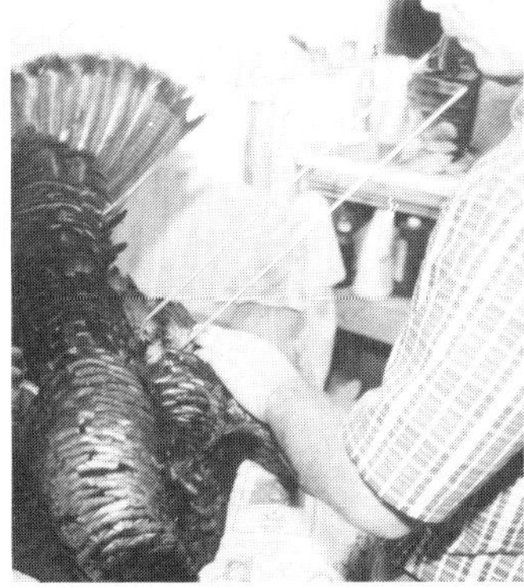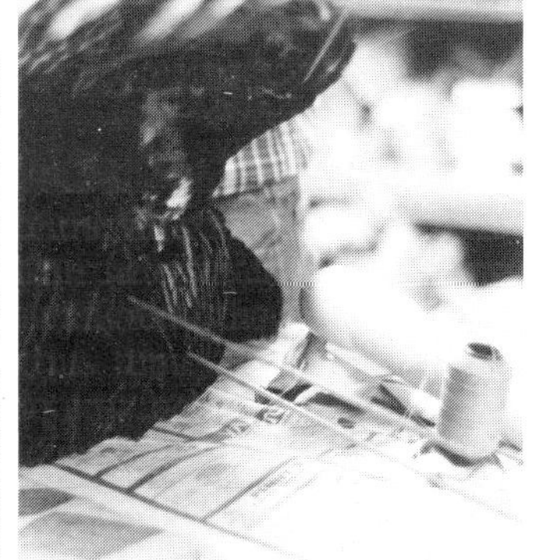

After the bird has been completely sewn, take a 7 gauge wire sharpened on both ends that is approximately 6 feet long and bend it into a "U" shape. Insert this wire under the left wing, pushing it through the mannikin directly under the right wing, and attach it to a plaque or piece of driftwood. NOTE: This may be done prior to sewing if the wire showing "underneath" the mount will be a problem. It is not readily visible when viewed from the front but rest assured that a "judge" would surely fatal flaw the mount if the wire was not hidden.

Now take a small gauge wire (14 or 16 gauge) and place it into the mannikin under the secondary and primary wing feathers. Clamp this temporary support wire to the wing wire where it exits the wing. This keeps the wing feathers from sagging and drying unnaturally. Finally, take a piece of cardboard, bend it, and place it underneath the first two primary feathers and clamp it into place with a clothespin to both the support and the wing wires. This will also give the wings a more natural look. Tommy also uses masking tape on the primary and secondary feathers to keep them in line and to ensure a

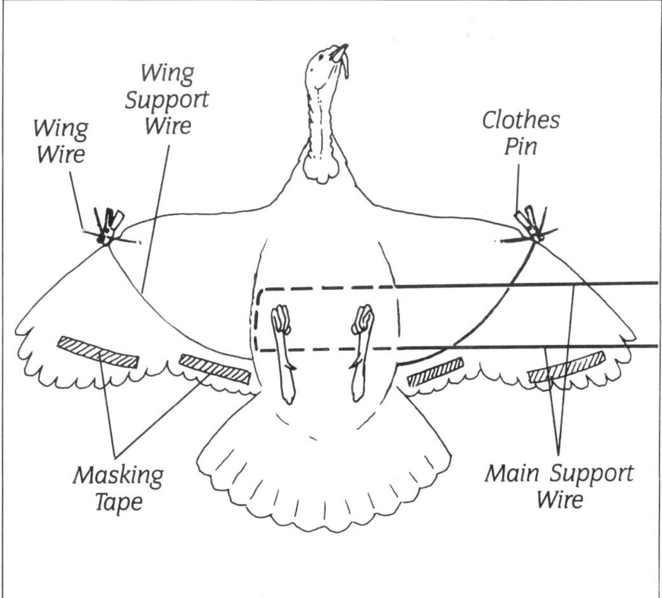

natural look. This is done by placing strips of tape on the bottom side of the feathers. When the bird has completed the drying process, the tape can easily be removed without leaving a residue on the feathers.

Attaching the Head

The next step in the mounting of the flying turkey is to attach the head to the Flex-Neck. Tommy uses freeze-dried heads which have a foam filling in them, which make them easier to attach to the Flex-Necks. Simply put a small bead of hot-melt glue on the end of the foam in the head and push it into place on the neck with the exposed wire. Hold the head in place until the glue has dried. Next, pull the neck feathers into place and with small pins lock them in place. (The skin can also be glued into place with Instant Bonding Glue). When this has been completed, inject the tarsus, toes, and ankles of the bird with a preservative (Tommy uses Formalin). NOTE: extreme caution must be used for this step.

Now the turkey is ready for the grooming process. The feathers should be naturally groomed and "taxied" into position. The turkey should not be force dried, but it should be allowed to dry naturally. It should be carefully watched and further groomed if necessary during the drying process.

Tommy likes to go over his birds every day for three or four days after the birds have been mounted to make sure that none of the feathers have moved out of place. *It is a necessity to have good reference materials handy when you groom a mount.*

After the skin has completely dried, the support wires can be pulled out and the leg and wing wires can be cut off as close to the exiting points as possible. If there is any remaining wire showing at the tip of the wings, it can be concealed by using Sallie Dahmes Finishing Wax. The exiting points on the pads of the feet can be concealed by using Sculpall and textured with a small dull instrument to recreate the pad of the bird's foot. The feet are then ready for painting. Tommy uses the same paint schedule that is in the *Breakthrough* Waterfowl and Bird Finishing Manual to finish his turkey mounts. Once the legs have been painted, the turkey is complete.

Turkey Heads

The method chosen to re-create turkey heads for taxidermy mounts varies and is the subject of some controversy. The four main methods currently in use for turkey heads are:

1. Commercially available artificial turkey heads
2. Skinned and mounted natural heads
3. Casts made from natural heads
4. Freeze dried natural heads.

Artificial heads are a less time consuming alternative to natural heads. They can be purchased commercially and exhibit no shrinkage problems. One drawback to commercial turkey heads is the lack of the tiny feather and hair follicles present on a natural head. If the taxidermist wishes the head to display these feathers, then they will have to be inserted and glued individually. Frank Newmyer uses commercially cast heads on his turkey 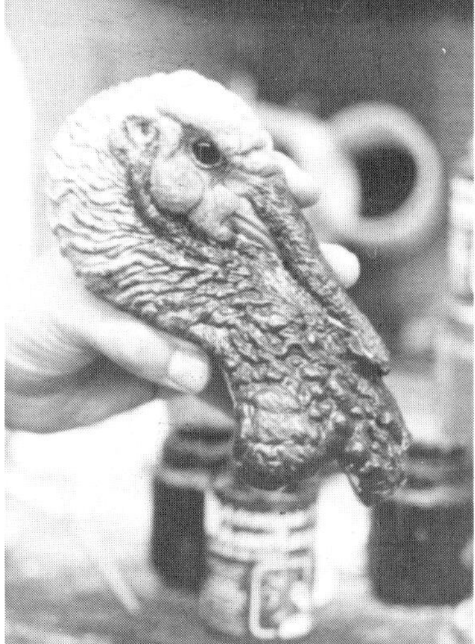 mounts as do most professional taxidermists for their commercial mounts.

A natural head which is skinned and mounted is probably the most time-consuming method. The difficulties involved with shrinkage of the fleshy head require that a great deal of time be spent fleshing and then rebuilding with epoxies and/or injecting with formaldehyde. On the plus side, the numerous hair and feather follicles present on a natural head add to the realism of the finished mount.

Some taxidermists choose to **mold and cast their own turkey heads.** This is a costly, complicated procedure, but it can produce outstanding results. A good knowledge of mold making is required to achieve realistic results and would require a complete book of its own to explain the procedures.

The fourth method in use today is **freeze dried turkey heads.** While these heads usually look very natural, there "is" the possibility that they could rehydrate, rot, and/or be attacked by insects in the future if the entire head is freeze dried. We do not recommend this method due to the large number of complaints that we have received concerning freeze dried specimens.

When it comes to freeze drying, we feel that the "jury is still out." However, many notable taxidermists employ the freeze drying process in their work and rigorously defend its use. Ed Thompson uses conventionally mounted heads "and" a freeze dryer to reduce shrinkage on most of his wild turkey mounts. It must be noted though that Ed first skins, fleshes, preserves, and mounts the head over a mannikin prior to freeze drying. In cases such as this, where the freeze dry machine is used simply as a tool to reduce shrinkage of a conventially mounted specimen, we heartily recommend it. This is an expensive luxury, but if you have access to one, as Ed does at Fernbank Science Center, then it is definitely a good way to go.

Finishing the Turkey Head

Regardless of which method is used to reproduce or mount the head of the wild turkey, the bright colors will need to be restored with paint. The following paint schedule by World Champion Frank Newmyer is designed for use on one of his artificial heads, but will work just as well for a natural one.

Painting a Sportsman Series artificial turkey head with *Polytranspar*™ airbrush paint is fast, easy, and the heads look realistic. There is absolutely no shrinkage, no bugs, and no "worries" about freeze dried heads rehydrating and beginning to decompose.

1. Prepare the head by first drilling a 3/16" hole (to accommodate a 12mm dark brown Tohickon glass eye). Drill the hole 1/4" deep (at best) on each side. (Some bird taxidermists drill a hole all of the way through. I don't like this. I prefer to drill out each eye separately.)

2. Next, I insert Sculpall, All-Game, or some type of epoxy sculpting compound in the hole to set the eye with.

3. I install and adjust a 12mm Tohickon glass eye which sets very easily into the epoxy. I rebuild the eyelid with Sculpall or All-Game and after letting it set for about 15 minutes, I use a sharp scalpel to further sculpt and press the soft looking ridges into the eyelid.

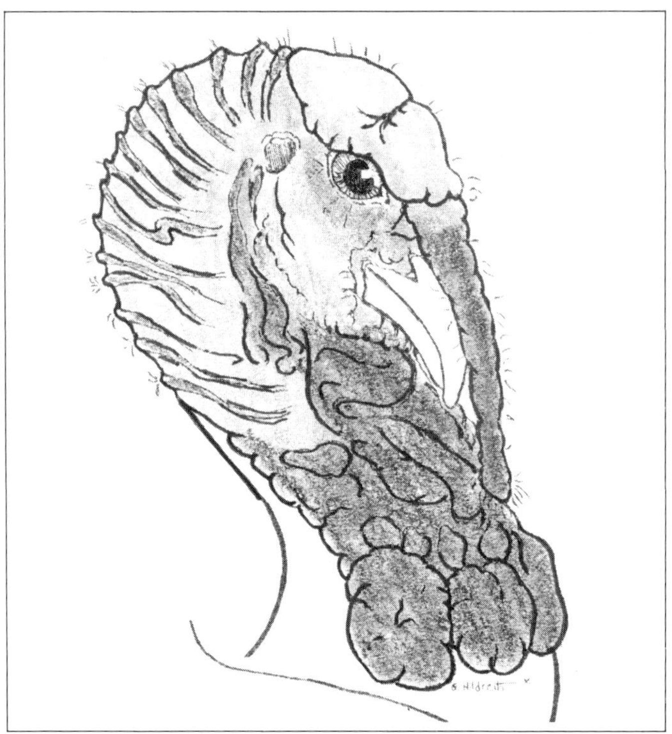

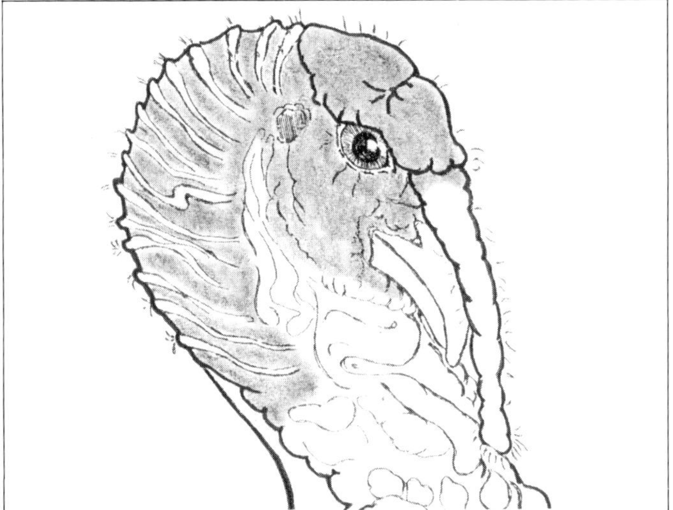

4. Once the eyes have been set and the epoxy has hardened, I begin the painting process. I begin by using *Polytranspar* FP/WA106 Widgeon Blue. This color is applied to all of the blue areas as shown in the illustration above.

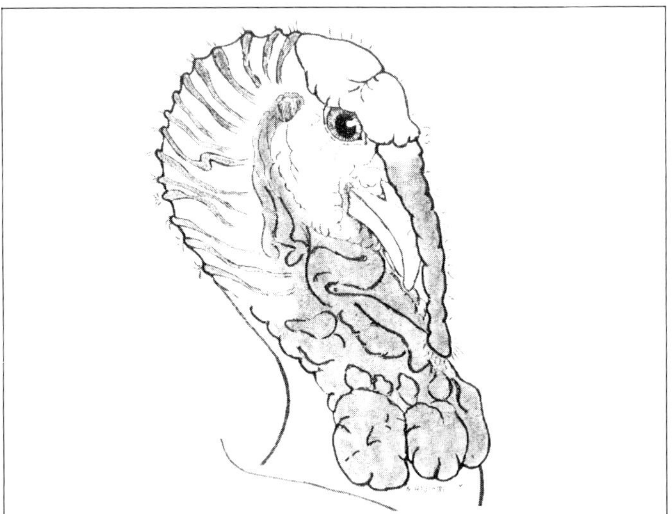

5. Next, FP/WA160 Gill Red is applied to the red areas as shown above.

6. The red area is then toned down somewhat with FP/WA105 Phthalo Blue. This gives it a "deeper" looking red. (Be sure and only "very slightly" mist this blue over the red area.) Now prepare a mixture of 80% lacquer thinner and 20% FP/WA105 Phthalo Blue and "very" lightly spray a fine mist over the entire head. This step further tones the head to give it a richer, deeper value.

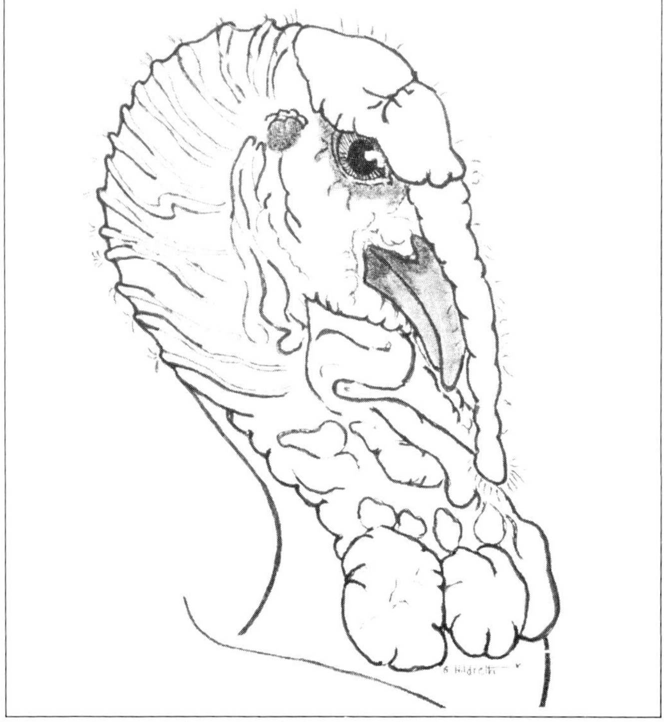

7. I always apply a very light mist of FP/WA105 Phthalo Blue lightly around the eye area. The beak is also sprayed lightly. Next FP/WA142 Teal Yellow is applied to the beak. Finally, the beak is lightly sprayed with a thin (50%) mixture of FP/WA29 Black Umber and lacquer thinner. The area around the eye first receives a base of FP/WA106 Widgeon Blue, and then is lightly misted with FP/WA142 Teal Yellow. Finally, I use a #00 artist's brush and paint the feather-like spots around the ear with FP/WA30 Black straight out of the can.

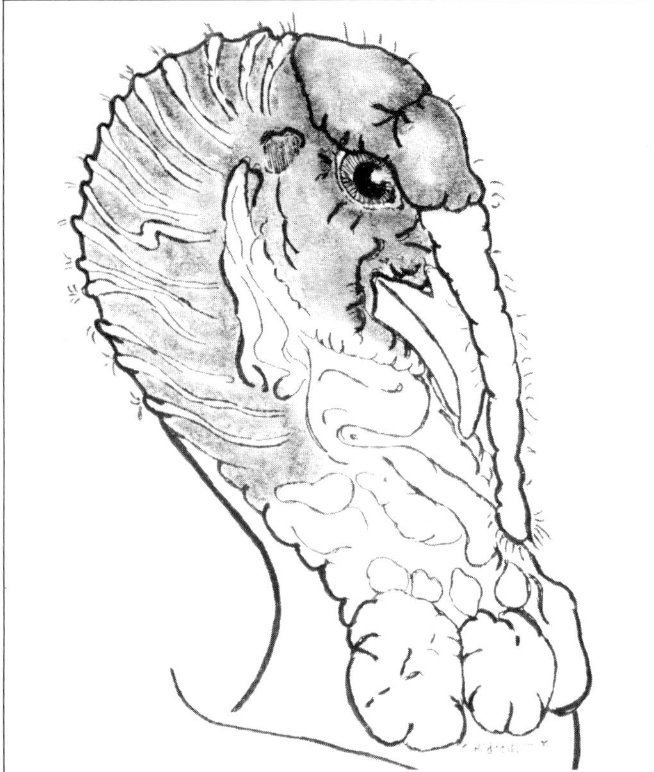

8. I then mist a "mixture" of different color values on the blue sections of the head. (Lightly spray over all blue areas.) I start misting with FP/WA32 Diver Grey. Next I mix 1 part FP/WA63 Hooker's Green to 10 parts FP/WA143 Waterfowl Base Yellow to 30 parts FP/WA106 Widgeon Blue and thin the entire mixture with 50% lacquer thinner. This mixture is then sprayed lightly on the head to create depth in the crevices.

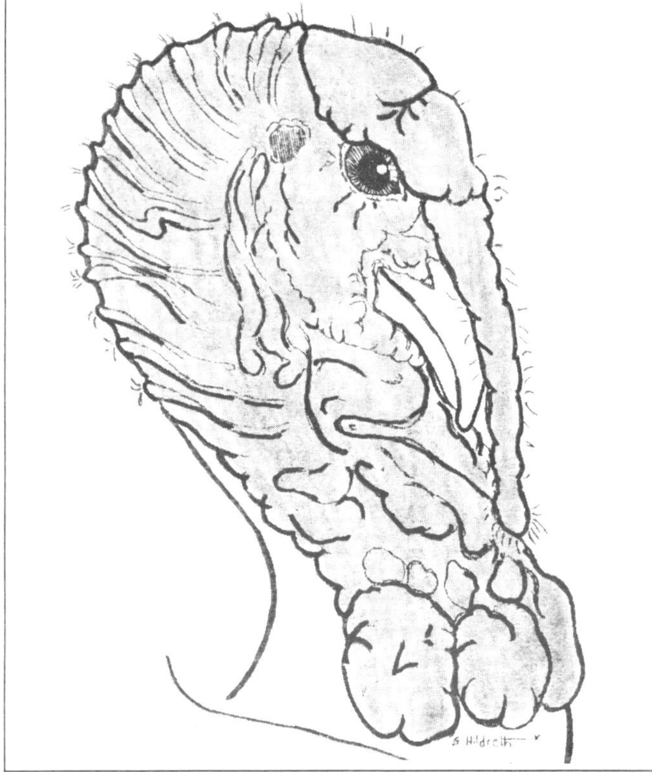

9. I "seal" the paint job with a very, very light misting of FP/WA200 Sienna. Finally, to "tone-down" the "sheen," I spray denatured alcohol over the entire head and then dry it by spraying it with plain air through the airbrush. (Any double-action airbrush will work quite well.)

10. Finally, I take a modified, stiff-bristled brush (with the bristles cut down to 1/4" long with scissors) and wash the paint off of the glass eyes with lacquer thinner and the brush. I pluck the feather follicles from the real turkey head with a pair of tweezers and I use a needle to push holes into the foam head.

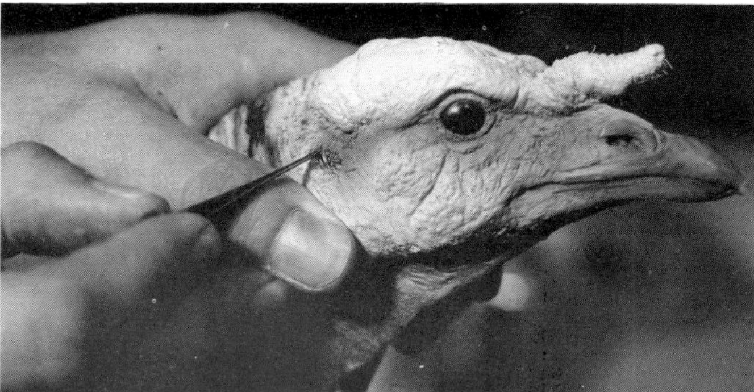

The feathers are then glued into the holes with Instant Bonding Glue. The actual feathers really "soften" the artificial head and make it look just like a natural turkey head. There are also tiny hairs on the snood that must be recreated either by painting with a paint brush to resemble these strands (or substitute whatever looks like the real thing for this step.) Use reference!

There you have it—It's fast, it's easy, and it looks realistic!

Finishing Turkey Legs

Restoring the color to the natural legs of wild turkey mounts is relatively easy. Here are the three methods that Tommy Knight uses, depending on what color the legs and feet are after the bird has dried.

Dark Legs & Feet:
Step 1. FP10 Superhide White—"Mist" over the entire leg and foot, just enough to lighten up the dark appearance.
Step 2. FP160 Cadmium Red—"Mist" lightly over the entire leg and foot to achieve a pale pink hue.

Red/Purple Legs & Feet:
Step 1. FP70 Chocolate Brown—"Mist" on just enough to tone down the natural red/purple hue on the leg and foot.
Step 2. FP167 Lavender Flesh—Apply a light coat over the entire leg and foot until a fleshy hue is detected.

Light Legs & Feet:
Step 1. FP167 Lavender Flesh—Faintly "mist" the entire leg and foot with Lavender Flesh until a fleshy hue is achieved.

CHAPTER 11

Dead Game Mounts

To the dedicated hunter, no other mount truly symbolizes a successful day afield than a "brace" of hanging birds. Because of their popularity not only with hunters but collectors of wildlife art, dead game mounts have been the subject matter of various art mediums such as decorative wood carvings and flat art renderings for many years.

Portraying dead game in a convincing manner has always been a challenge to the serious wildlife artist. Whether the medium is canvas, wood or feathers, there are many anatomical aspects that must be taken into consideration to avoid the birds having a flat, or "stiff" appearance.

The most important feeling that must be conveyed to the viewer is that the bird has recently been bagged and is indeed being suspended by one foot from a rustic nail or some type of suitable hanger. A freshly taken bird is limber, and in order to portray the bird in this manner, the head, neck, wings, legs, and tail must appear loose and supple. Carefully analyzing each part of the birds anatomy and what will take place with it when the bird expires, makes the task of convincing the viewer much easier.

To facilitate the analysis process and eventual posing of dead game mounts, a collection of reference pictures taken in the field after a successful day of hunting will prove to be a valuable aide.

"Faisandé" by David Noll (oil on canvas). Dead game has long been a popular subject for flat art renderings.

Perhaps the best way to take the guesswork out of accurately posing dead birds is to actually pose the birds that are to be mounted in the desired position before they are skinned. Simply remove the birds from the freezer and allow them to thaw in the normal manner.

Next, carefully work the wings and legs to loosen the joints. This action will give the bird a "limber" look which will enable the bird or birds to be placed in the desired position. Take reference pictures from all angles paying particular attention to the position of the head, wings, tail and feet (below left).

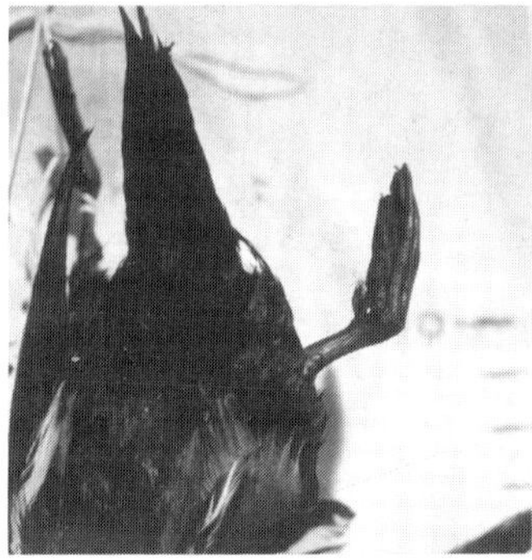

If two or more birds are to be used, note the interaction between the birds particularly if one bird is larger than the other (above right).

Another important aspect that should be captured on film, is the effect of gravity on the bird. The force of gravity will determine where the bird's body hangs in respect to the backboard and will also have an effect on all of the anatomical features of the bird.

After the bird is skinned and assembled, refer to these photographs during the posing phase. Studying reference photographs and actual dead birds suspended in the desired position is by far the most accurate form of reference material available. Studying other artist's artwork depicting dead game is enlightening and will provide insight into creating a convincing dead game mount, but should not be used by itself. Always go to the actual subject to be depicted, not someone else's interpretation of how they perceive it. Contrary to many taxidermists beliefs, posing dead game mounts accurately and convincingly is not easy—and good reference materials are definitely a necessity.

Preliminary Procedures

When the final pose and position of all of the birds is finalized, a decision must then be made as to whether to use a fabricated mannikin or a commercial mannikin. To truly "pull off" a dead game mount, certain anatomical detail must be incorporated into the mannikin whether using a commercial or fabricated mannikin. Using a commercial mannikin will definitely save considerable time and will work quite well if certain alterations are made in the area of the leg attachment points and the caudal (tail) area.

The harlequins and old squaw that were mounted for this chapter presented a particular problem because Sportsman Series mannikins were not available for these particular species of ducks. Fortunately, the Cross Reference Index in the WASCO catalog did have suitable substitutions listed for the old squaw and harlequins and a commercial mannikin could be used for two of the ducks. However, upon close examination of the two harlequins, I found a substantial difference in both the body length and circumference between the two drakes. With such a size difference, the fabricated mannikin method was the best choice to duplicate the anatomical features of the smaller, more petite, harlequin drake. (See Chapter 9 for wrapping techniques.)

A decision also had to be made as to whether to use an artificial head or the natural skull. If the bill is to remain closed, the artificial head is the best choice because of the many advantages already listed for using the Sportsman Series artificial bird head/bills. However, in order to incorporate the slightly open mouth that is characteristic on most dead game mounts, it was necessary to use the natural skulls on the old squaw and harlequins.

With the decisions finalized on the pose, type of mannikin established, and type of head to be used, the birds were ready for the skinning operation.

Skinning and Preserving

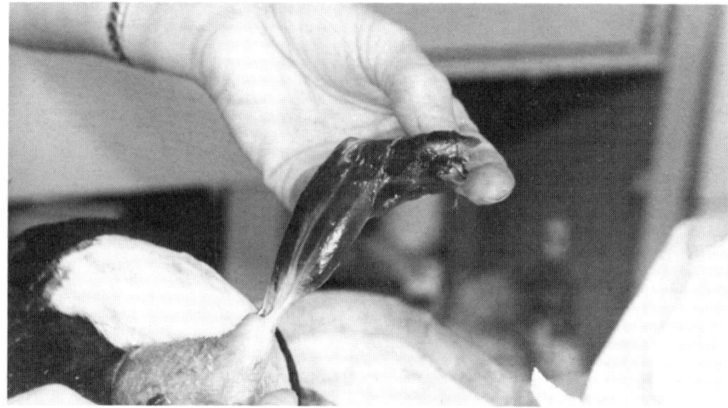

The bird(s) are skinned in the normal manner for a dead game mount with the only exception being that the femur is left attached to the tibia. Sever the leg at the head of the femur, leaving the joint intact. It is important to leave the femur attached because the "arc of travel" of the leg is changed in a dead game mount position.

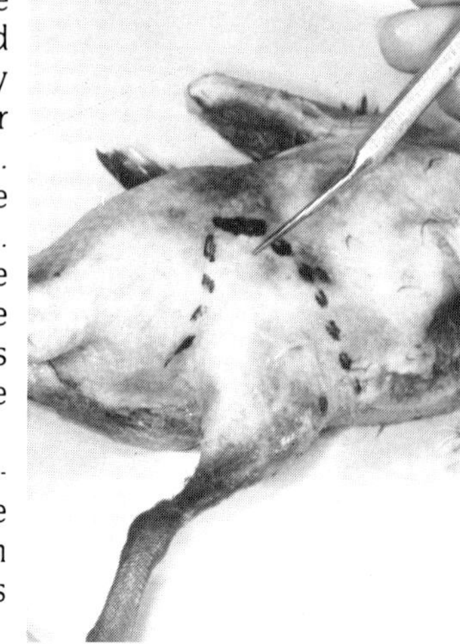

Once the bird has been completely skinned out leaving the femurs intact, then the bird skin is prepared and tanned as described in Chapter 6.

Mannikin Alteration

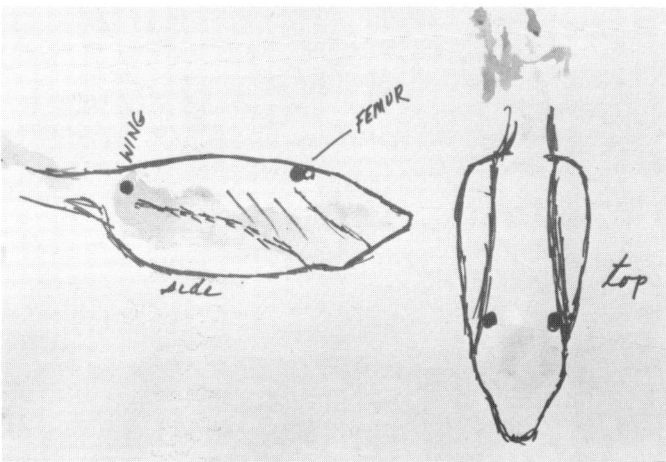

Regardless of whether a commercial mannikin or a fabricated mannikin is utilized, once the body is removed you must make carcass sketches of the body and neck. Pay particular attention to the attachment points of the wings and legs. Also make careful notations of how much bend there is in the caudal area.

If it is necessary to use a fabricated mannikin, carefully follow the steps for making it as illustrated in Chapter 9. This particular mount employed the "wrapped" mannikin method.

If a commercial mannikin is used, rasp off a sufficient amount of the side portion of the caudal area on the opposite side that the tail is leaning to. This will allow the tail feathers to fall naturally and maintain a natural curve to the body.

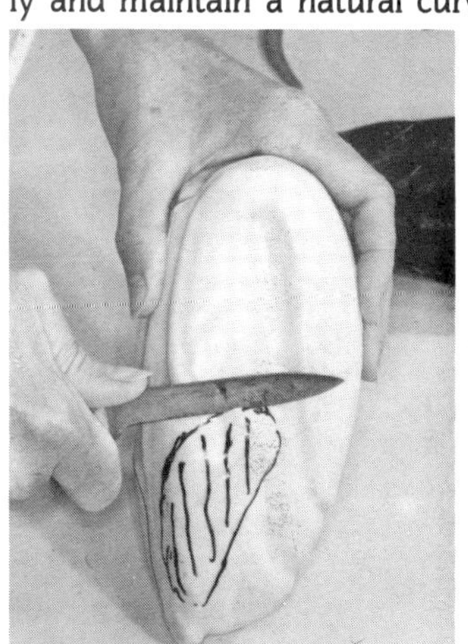
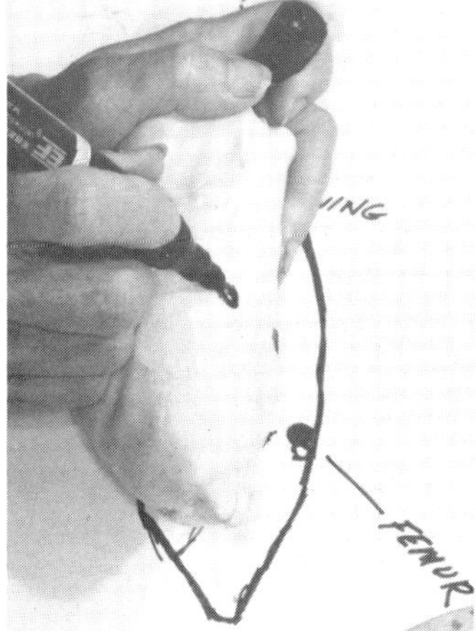

Also remove the thigh muscle that is incorporated into the mannikin. Now refer to the carcass sketch and mark the femur attachment points.

The mannikin is now ready for the neck attachment. Choose a suitable Flex-Neck of the proper dimensions for the specimen and insert the Flex-Neck into the mannikin at the insertion point.

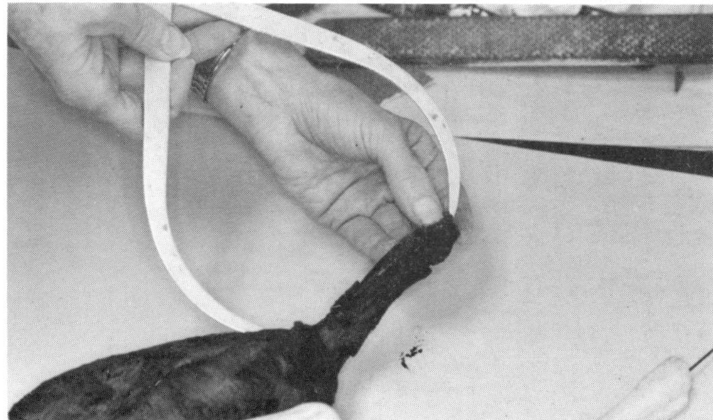

If a suitable Flex-Neck is not available for a particular bird carefully measure the length of the neck on the carcass with a pair of

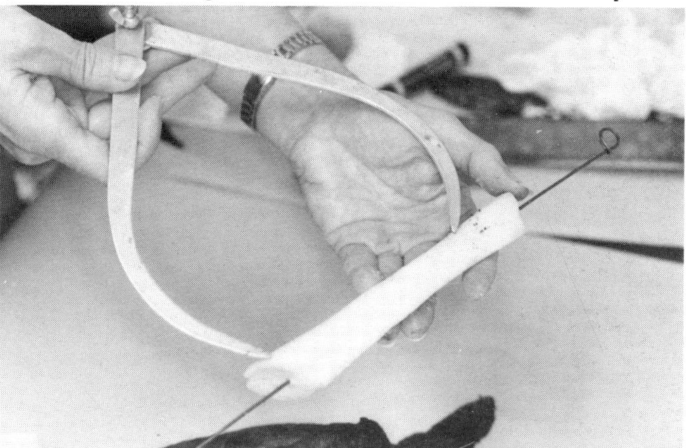

calipers. For both the harlequin and the old squaw used in this chapter, the Flex-Necks had to be shortened.

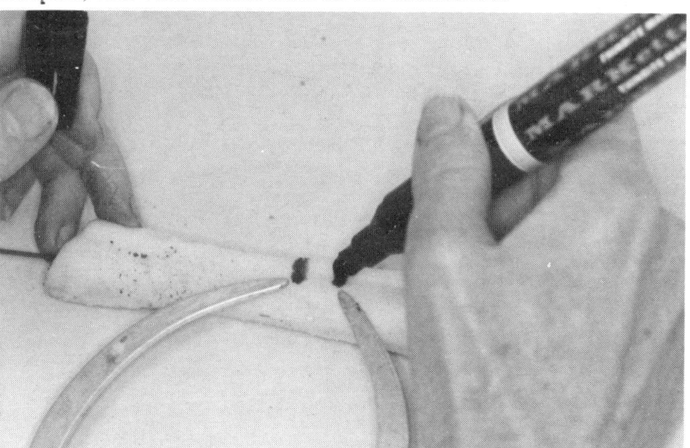

To shorten the necks, calculate the difference and mark it on the Flex-Neck.

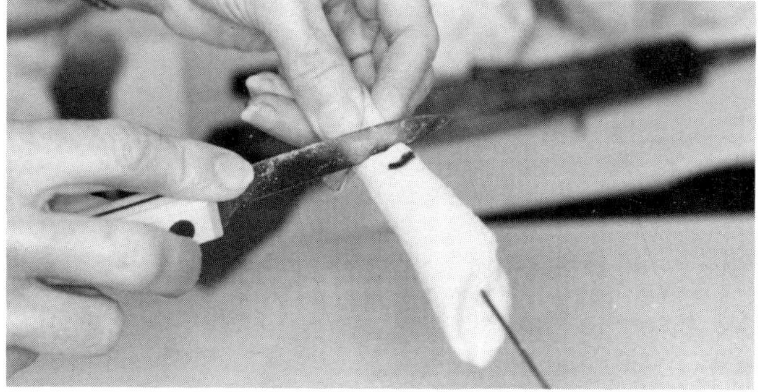

Now using the Perfect Knife, cut out the excess neck material.

Push the two halves together, securing them with hot-melt glue.

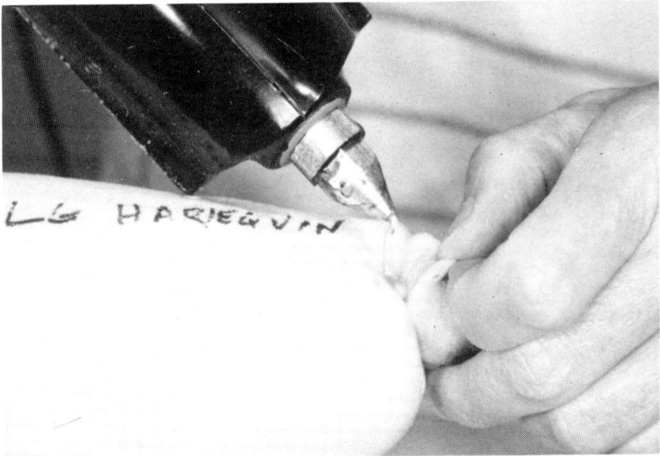

Now to secure the Flex-Neck to the mannikin, use a hot-melt glue gun to glue the neck to the mannikin.

Because of the alterations involved, it is better to use a Sportsman Series rigid mannikin because the rigid foam is much easier to carve and rasp than the Accu-Flex mannikin which is made with flexible foam.

The mounting process is essentially the same for a wrapped mannikin and a commercial mannikin. The following sequences are the step-by-step mounting procedures used for mounting both the wrapped mannikin and the altered mannikin.

Leg, Wing, Tail and Body Support Wires

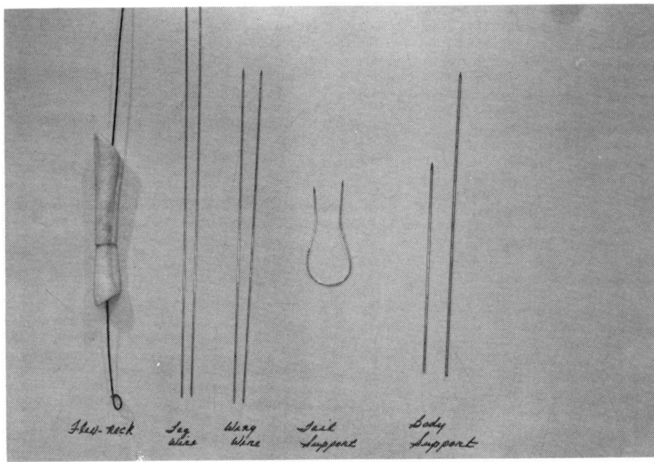

After the mannikin alterations have been completed, the next step is to cut and prepare the leg, wing, tail, and body support

wires. The leg wire should normally be cut approximately 1½ times the length of the leg bone (including the digits, tarsus, tibia, and femur).

The wire should be of a heavy enough gauge to support the leg when it is placed in the desired position. The leg wires should be sharpened at both ends with a bench grinder or file. Do not make the points too sharp, because when the wire is being run down the tarsus it will have a tendency to snag along the delicate tissue of the interior of the leg.

For the wing wires, use a heavy enough gauge wire to support the open wing without allowing them to droop. The wire should be cut twice the length of the humerus and radius-ulna. Sharpen the wing wires in the same manner as the leg wires were sharpened.

The tail support wire is cut from a medium gauge wire; it is sharpened at both ends and bent into the shape of a horseshoe. The tail support wire should be long enough to insert a sufficient amount of wire into the mannikin to support the tail feathers.

The two body support wires will be driven through the bird and into the backboard. The wires should be of a heavy enough gauge to easily "hammer" the wire into the backboard. The body support wire is only sharpened on one end and should be long enough to go completely through the backboard and into the mannikin.

Another method that works quite well is to pre-drill a hole in the backboard and insert the support wire through it. Leave approximately 3 to 4 inches of exposed wire on the show side of the backboard. Bend the remaining wire on the back side of the panel flush with the backboard and secure the wire to it with wire staples. When the bird is ready to be attached to the backboard, simply push the bird onto the preattached support wire. After the wires are cut and sharpened, the bird should be ready for assembly.

Assembling the Bird

For assembly purposes, the clean and tumbled bird skin should be placed in a tray of clean WASCO granular borax. The borax will help to keep the feathers clean during the mounting process.

Setting the Eyes Using the Natural Skull

Begin by inverting the skin and exposing the cleaned skull and brush away any sawdust clinging to the skin. Use fine jeweler's hemp to tightly wrap a small tow core which will be placed inside of the skull cavity.

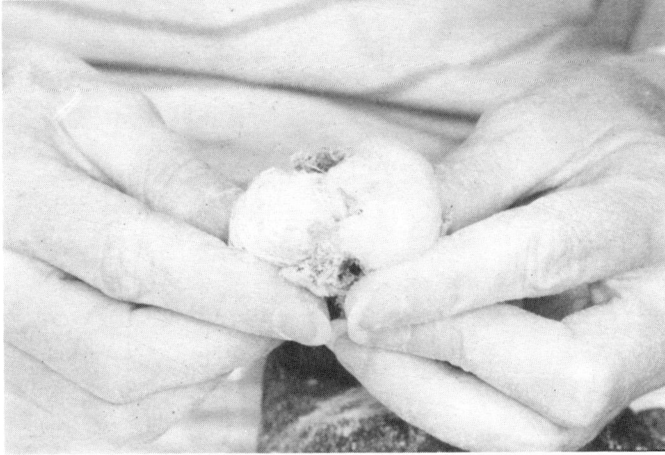

This core will help anchor the neck wire and minimize the amount of clay that will need to be used in the skull.

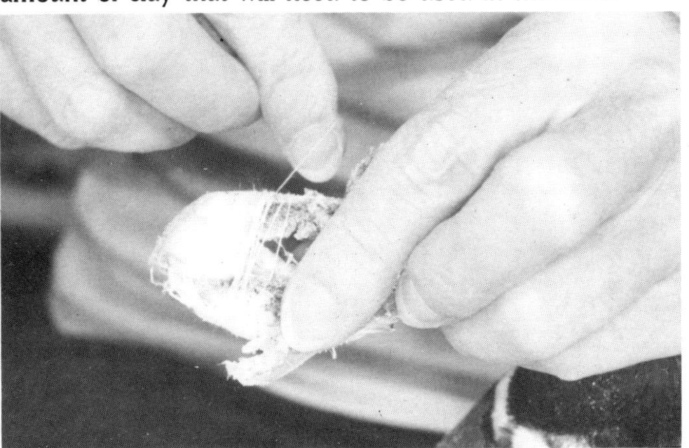

Secure the tow core in the skull cavity by making several wraps around the skull and tow with lightweight cotton thread. Be careful not to pull the thread too tight and pinch the lower mandibles.

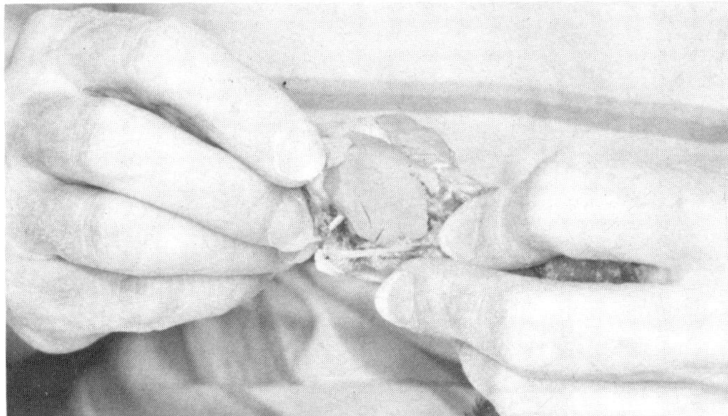

Now fill the eye sockets with WASCO Clay. The clay will hold the glass eyes in position.

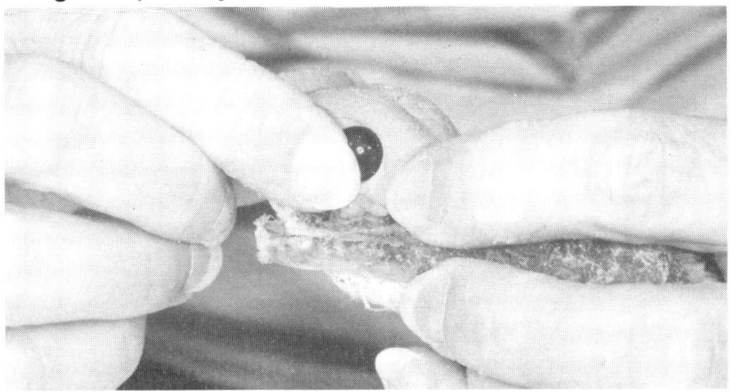

Next, carefully place each glass eye in the sockets.

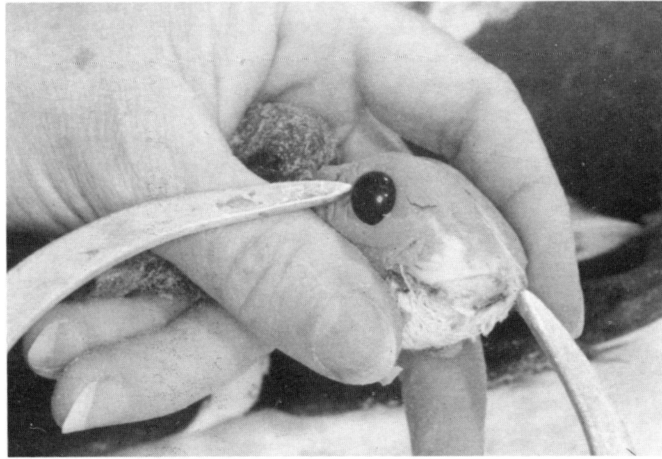

Use calipers to check the symmetry once each eye is seated in its socket. The eyes should be set with a slightly forward angle and a slightly downward cant. Symmetry is a necessity. Do not have one eye higher or further back than the other. Use calipers and view the bird from "head on" and from the top to ensure symmetry.

Seating depth is also very important in a dead game mount. Use plenty of reference when setting the eyes.

Do not set the eye too deeply as this will cause the bird to look as if it has been hanging for days instead of looking as if it were freshly bagged.

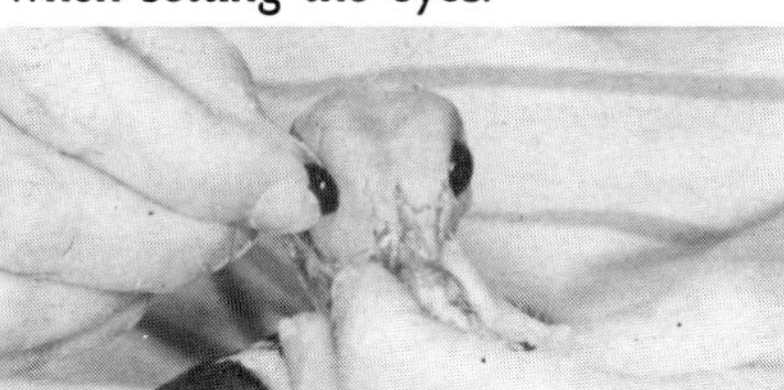

Once the eyes are set into place, finish rebuilding the muscles of the lower mandible with WASCO clay.

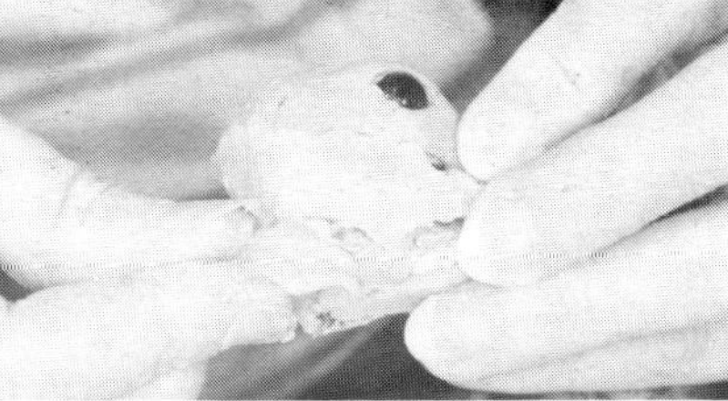

Next reinvert the skin back to the feather side out.

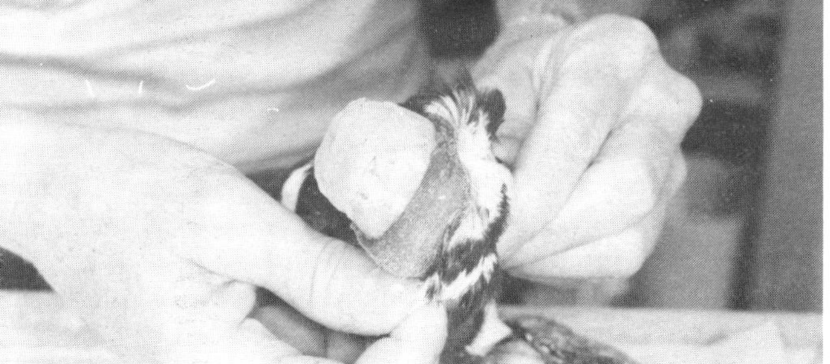

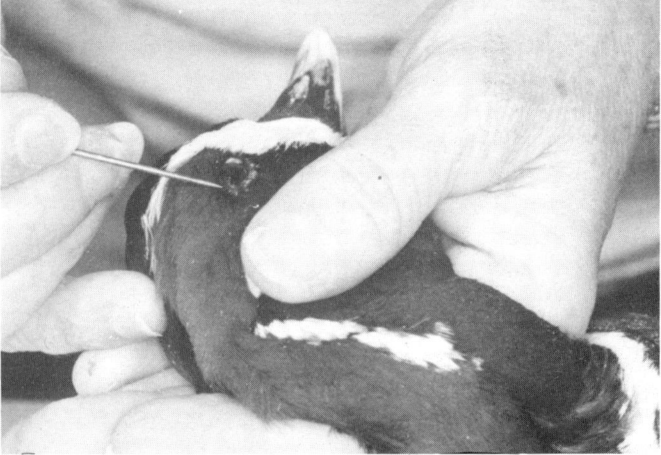

Adjust the eyelids with a small feather adjuster.

In order to achieve a convincing look, you must position the lower lid in a position that is at least halfway closed to fully closed.

When the upper and lower lids are in place, place a small insect pin in both the front and rear corners of the eye. These pins will keep the lids in place as they dry and will be removed once the mount is completely dry.

Wiring the Bird Wings

To stabilize the wings and to keep them open (if so desired), it is important to run the wing support wire into the last wing segment (metacarpal) as in a flying bird.

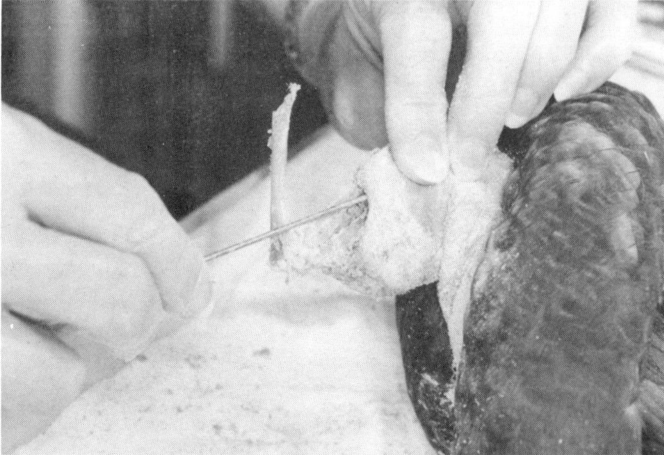

Using the sharpened medium gauge wing wire, run the wire down the humerus, radius-ulna and through the carpal (wrist) joint. The wire should lodge slightly past the carpal joint into the metacarpals.

Do not run the wire too far down the metacarpals as this will distort the primaries. The photo above shows where the wire should stop.

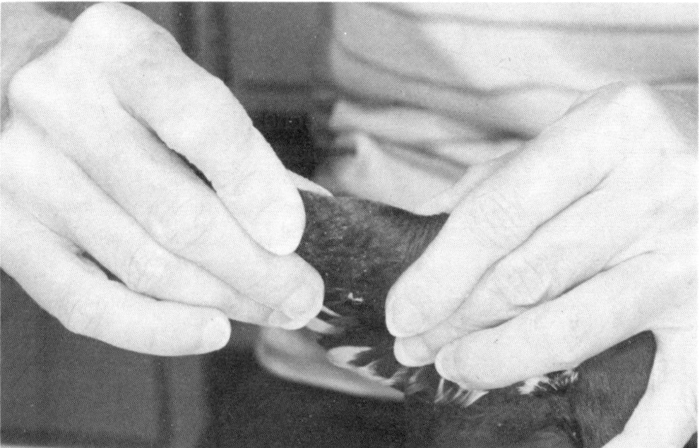

Once the wire is in place, slightly bend the carpal (wrist) joint to lock the wire in place. Now with the wire in place and properly positioned, secure it to the humerus with cotton thread. Next, use cotton batting to rebuild the muscles on the humerus. The cotton is held in place by wrapping thread around it as it is placed on the humerus. Once the muscles are rebuilt to their original dimensions, they may be wrapped with soft tissue paper or WASCO Clay.

Wiring the Bird Legs

A wire is run through the legs in order to support and hold them securely in position. On dead game mounts it is important to run the leg wire to the end of the digit, but *do not exit the foot with the wire.*

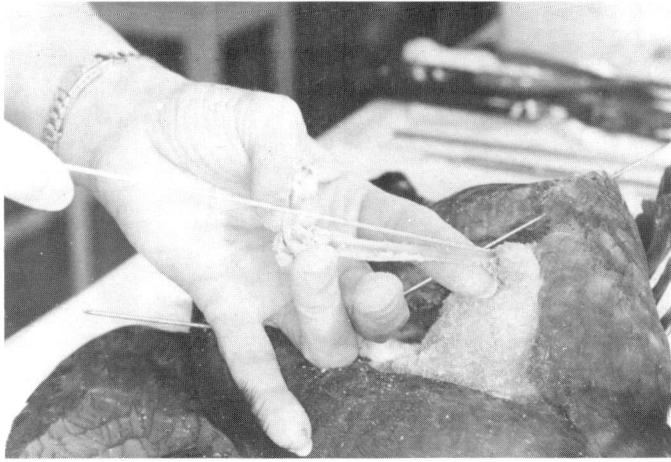

Hold the tarsus in one hand and expose the tibia/femur. Place the wire on the back (posterior) side of the ankle joint and gently push the wire down the back of the tarsus.

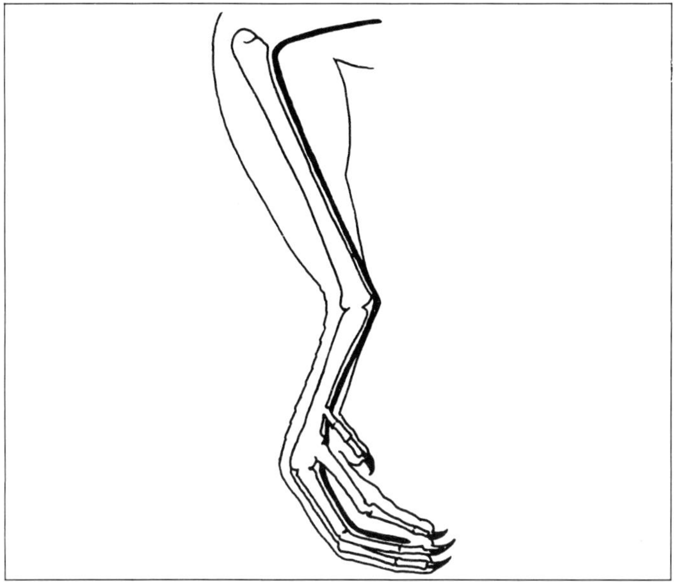

When the joint between the tarsus and digits is reached, straighten the joint to allow the wire to continue past the joint. Once the wire is past the joint, continue pushing the wire to the end of the third digit. Do not exit the wire.

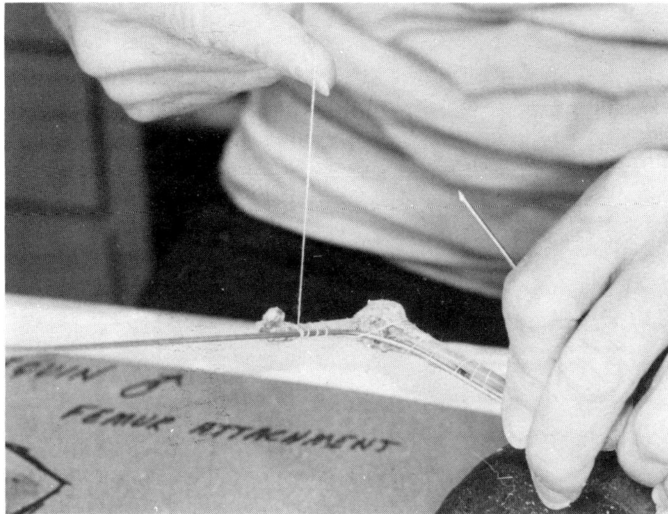

Secure the wire to the tibia/femur by wrapping with cotton thread. Remember to refer to the carcass sketches when rebuilding the tibia and femur.

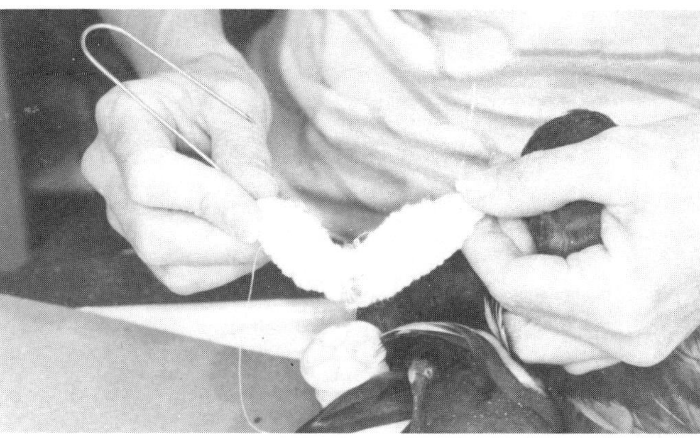

The muscles are rebuilt to the original dimensions with cotton batting which is again held in place with cotton thread. Once the muscles are rebuilt to the final dimensions, a fine layer of tissue paper or WASCO Clay can be added to give it a "softer" touch.

Now with both wings and legs wired, the skin should be ready to accept either the Sportsman Series mannikin or wrapped mannikin.

Installing the Mannikin

Make a small loop on the tip end of the neck wire. This loop will prevent the wire from snagging on the skin as the wire is pushed through the neck skin towards the skull.

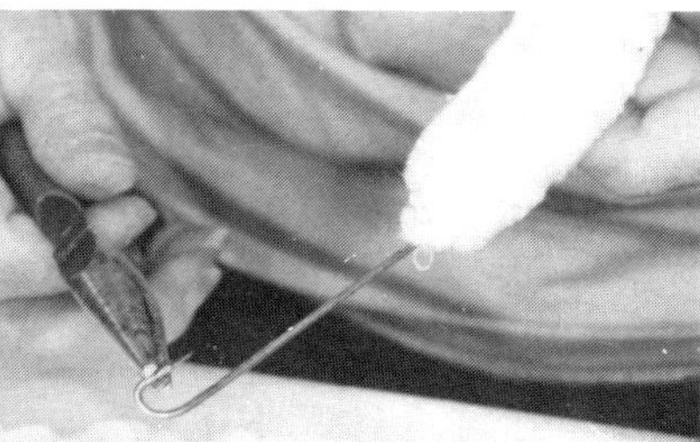

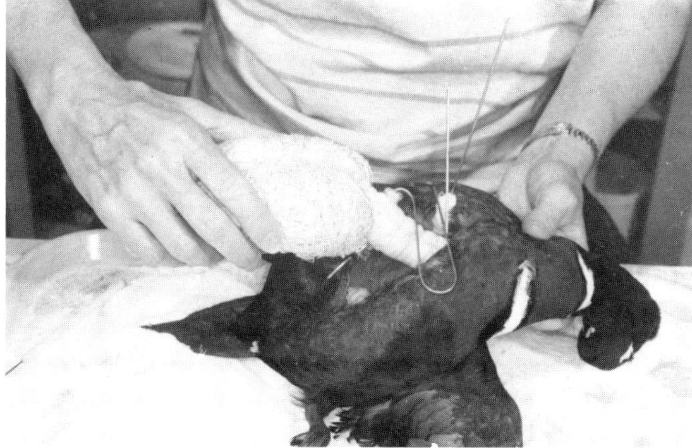

Now insert the mannikin through the breast incision and work the skin down over the neck.

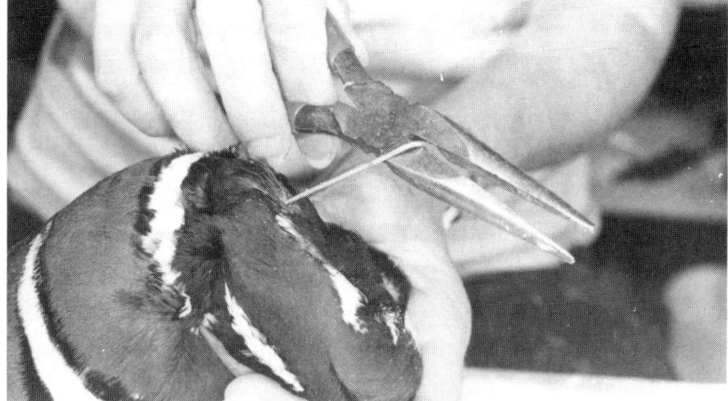

When the loop reaches the skull, snip it off at an angle with wire cutters so that it can easily be pushed through the fill material that is in the skull cavity.

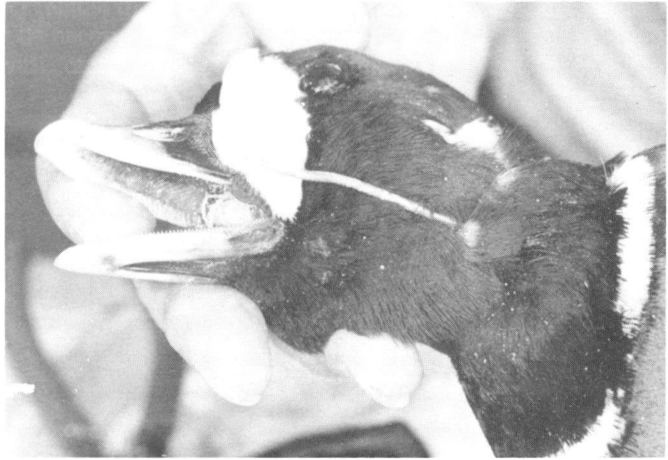

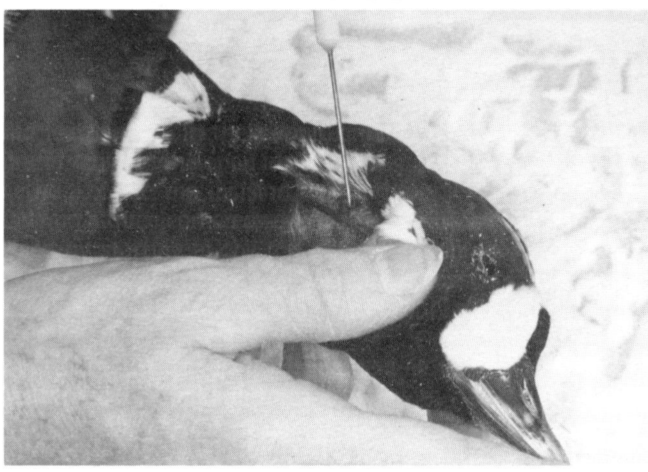

Position the head at such an angle so that the wire can be pushed through the back of the skull and imbedded into the rear portion of the opening of the mouth as the photos above and illustration below indicate.

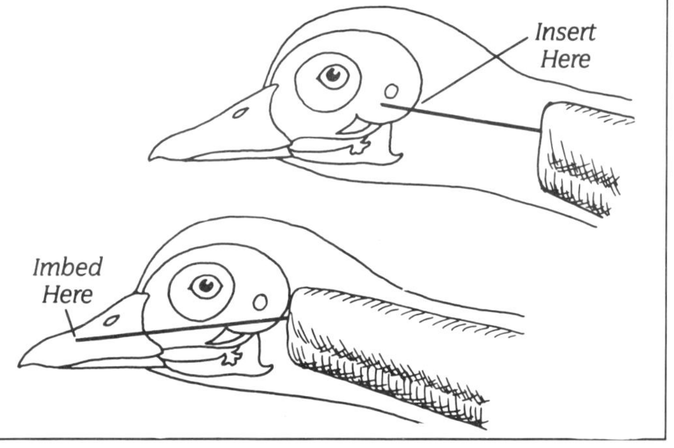

The skull should then be firmly pushed down and anchored solidly against the neck. Do not try to make any further adjustments to the neck until the wing wires are attached to the body.

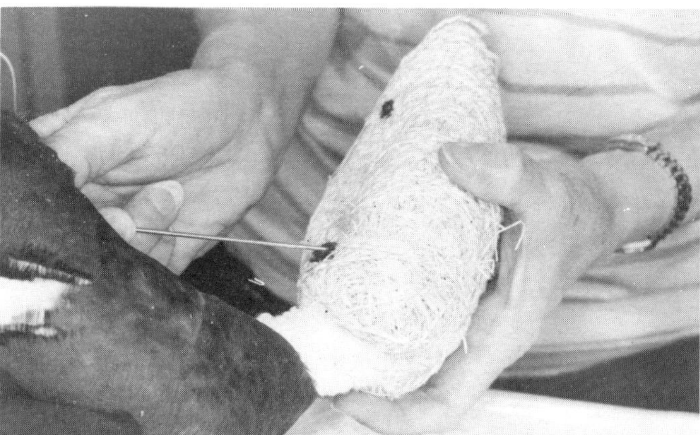

Insert the wing wires at the insertion points marked on the mannikin. Angle the wires so that they exit the body on the opposite sides of the lower quadrant of the breast. Pull each wing bone (humerus) snugly against the body.

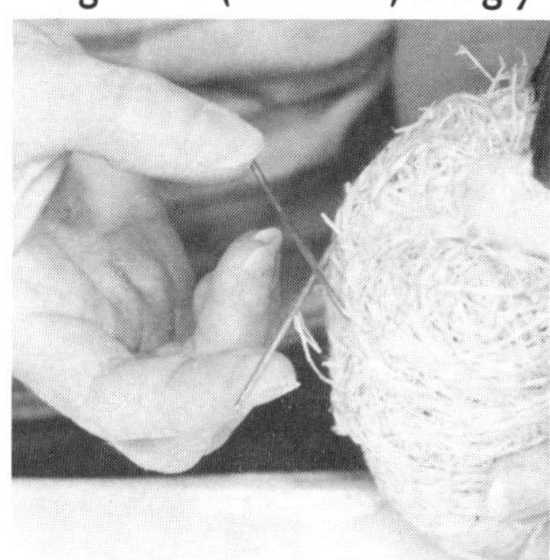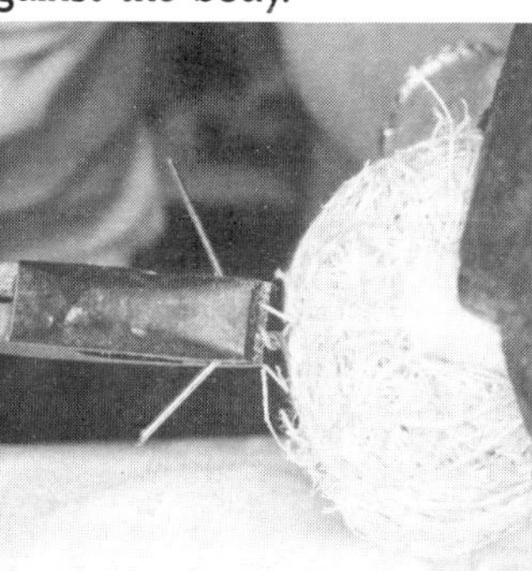

Make "certain" both wings are securely in place and snug against the body. Now cross the exposed wires and, using a pair of lineman's pliers, make several twists on the wires, locking them together.

Finally, bend and "clinch" the remaining wires into the mannikin.

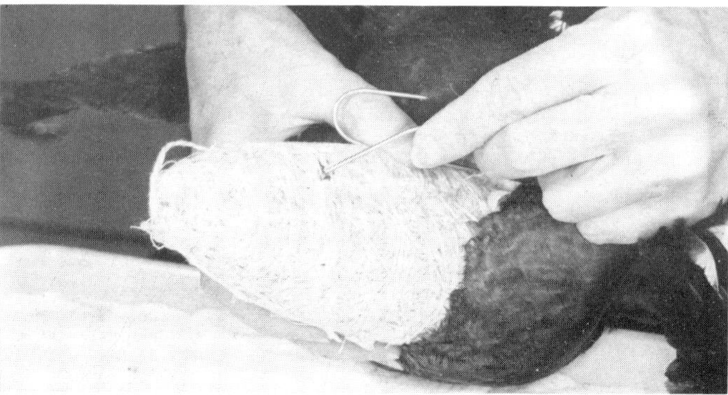

The leg wires are now inserted at the premarked insertion points.

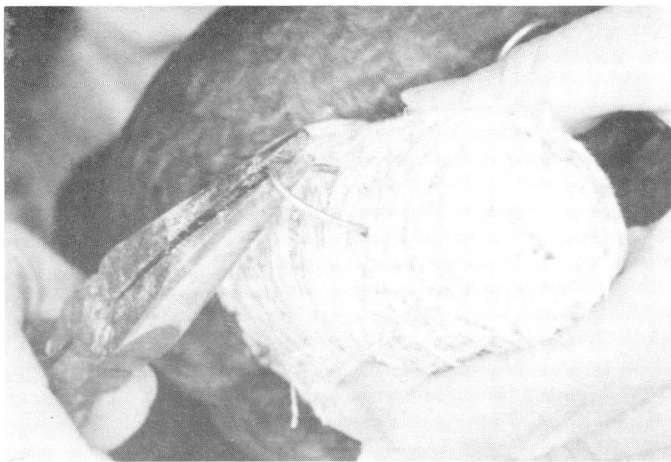

Instead of securing the two wires by twisting and clinching into the body, simply run each wire through the body, make a horseshoe bend on the end of the wire with pliers, and push the bent end back into the body to anchor the wire.

Make sure that the femur is fitted snugly against the mannikin. Be careful when attaching the femur to the mannikin. It is a common mistake to attach the femur to the mannikin "backwards," which could mean that you will have to "rewire" the leg support wires. Remember before attaching the wires, to make certain that the femur and tibia are positioned at the correct angles with respect to the mannikin.

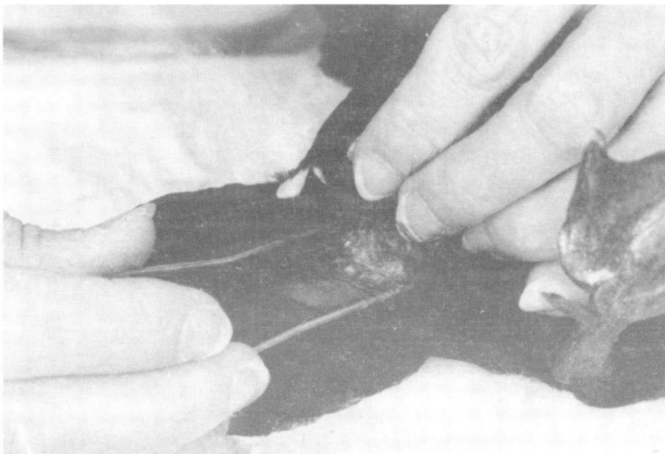

With the wings and legs now secured, gently pull the skin around the rear section of the body and carefully adjust the tail feathers. Separate the under tail covert feathers from the tail feathers. Now insert the horseshoe-shaped tail support wire between the under tail coverts and the tail feathers. The tail support wire should *not* be pushed snugly against the body, but should be positioned with enough distance between the mannikin and the tail support wire to allow you to further align and adjust the feathers naturally in the rear section of the bird. Depending on how the skin fits, it may be necessary to fill slightly on either side of the tail support wire with cotton batting. If so, now is the time to do so.

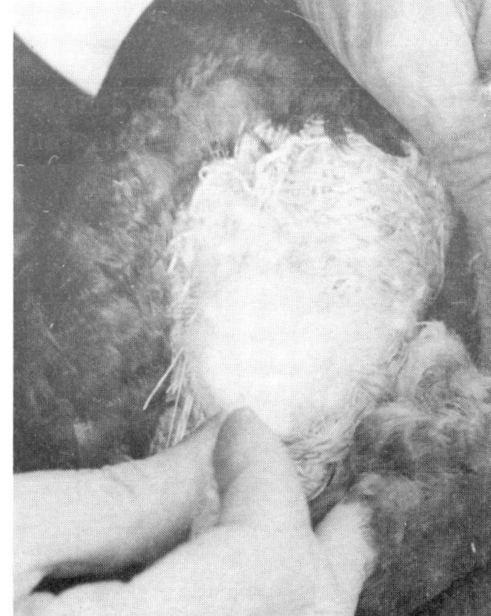

Starting at the neck, adjust the skin and position the feathers. At this time it may be necessary to fill such areas as the side pockets and crop with cotton batting to achieve a full looking mount.

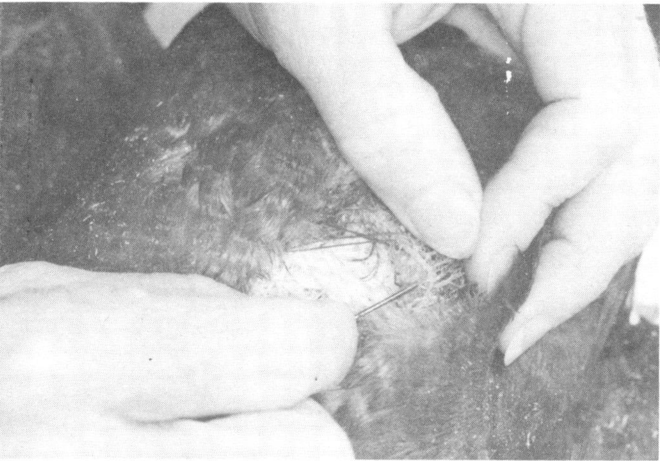

When the skin and feathers are in their approximate positions (sprinkle dry preservative in the open incision if using the dry preservative method), close the incision using a double strand of nylon thread and a three cornered sewing needle. Now sew the incision closed by starting at the top of the incision and sewing towards the tail section using a baseball stitch. Make small neat stitches, being careful not to get the feathers caught between the stitches.

After the incision is sewn closed, adjust the feathers along the breast and flank areas. Finish fluffing the feathers with a hair dryer set on a cool setting or a vacuum cleaner set on a "reverse" setting. The bird is completely dry when all of the down feathers next to the skin have a fluffed, "full" appearance.

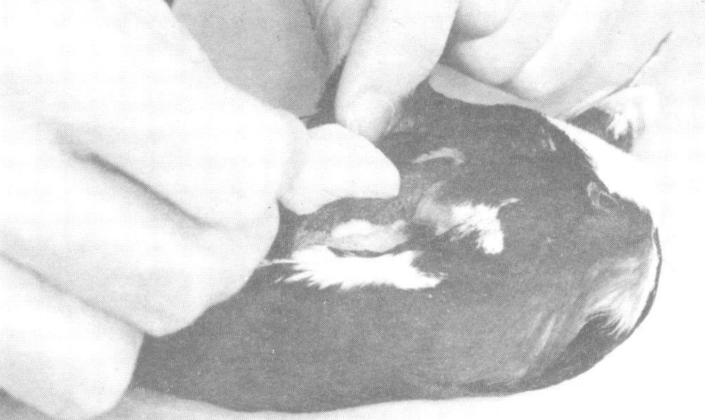

Pose the body in the basic position that the bird is to remain in, and then finish adjusting the head and neck. It may be necessary to add additional WASCO clay or cotton to the lower mandible or to the area where the neck joins the skull.

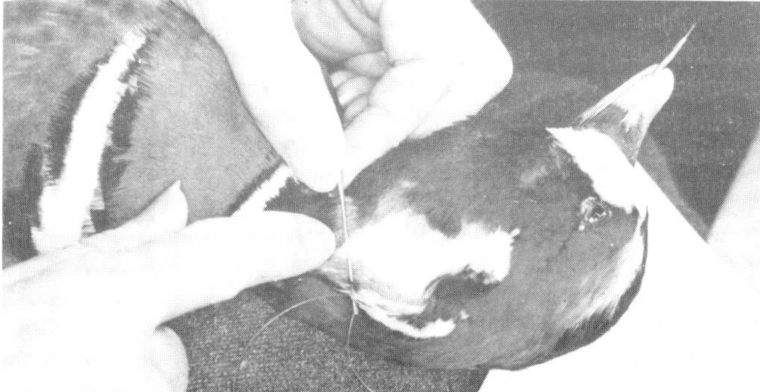

Carefully adjust the feathers in the head area and along the incision. Close the incision with a single strand of nylon thread and a three cornered sewing needle. Start at the bottom of the incision and sew toward the ear opening using the baseball stitch. Make small neat stitches, again being careful that no feathers get caught between the stitches. Once the head incision is closed, use the feather adjuster to smooth the feathers.

Attaching the Bird to a Backboard & Posing

Refer to reference photos that were taken of the birds before they were skinned. Pose the head, wings, and legs, in their approximate positions and then attach the birds to a rustic backboard. Each bird is held to the backboard by driving a sharpened heavy gauge wire through the mannikin into the backboard or as already discussed by presetting a wire in the backboard.

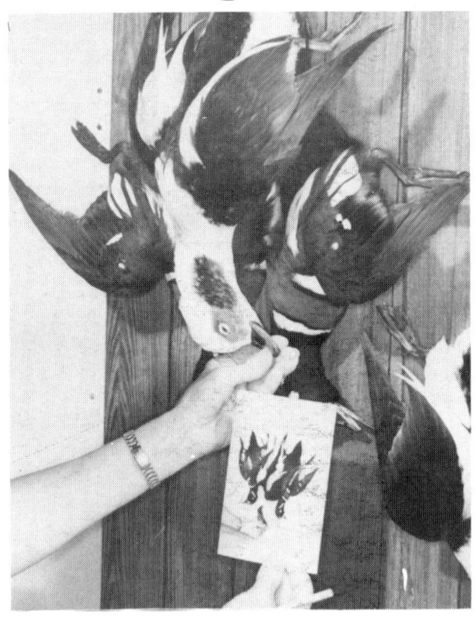

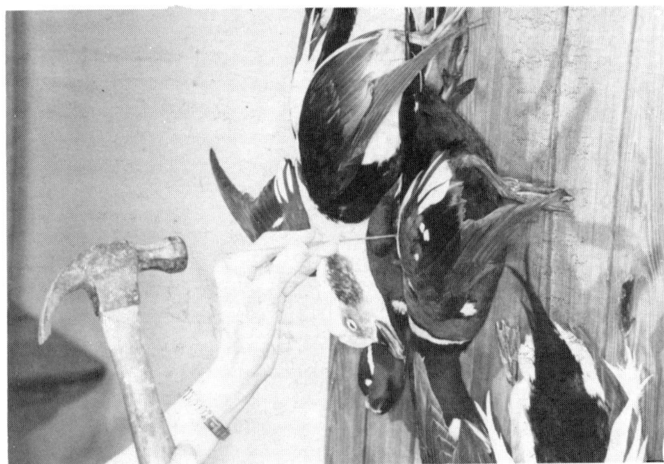

It is always advisable to use two wires to secure the bird. The second wire will stabilize the bird and prevent it from turning on the backboard. The second wire should be placed in the bird after the neck and body are positioned with respect to the foot that is suspending the bird from the nail or other rustic hanger. When securing the bird to the backboard, be careful not to attach the bird "flat" against the backboard. Gravity "does" affect the body, wings, legs, tail and head; however, it "does not" force the bird against the backboard. Some mounts are depicted in such a manner as if the entire mount was tied to a jet plane during take off. This is a common mistake, so be careful and check your references. Carefully refer to reference photos and begin final adjustments and positioning of the bird.

A convincing dead game mount should incorporate subtle anatomical characteristics such as: the bill being slightly open, eyelid half closed, feathers on the head and neck fluffed, wings hanging loosely

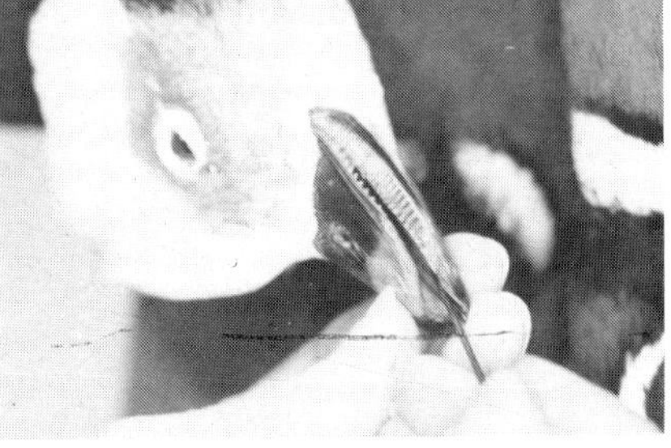

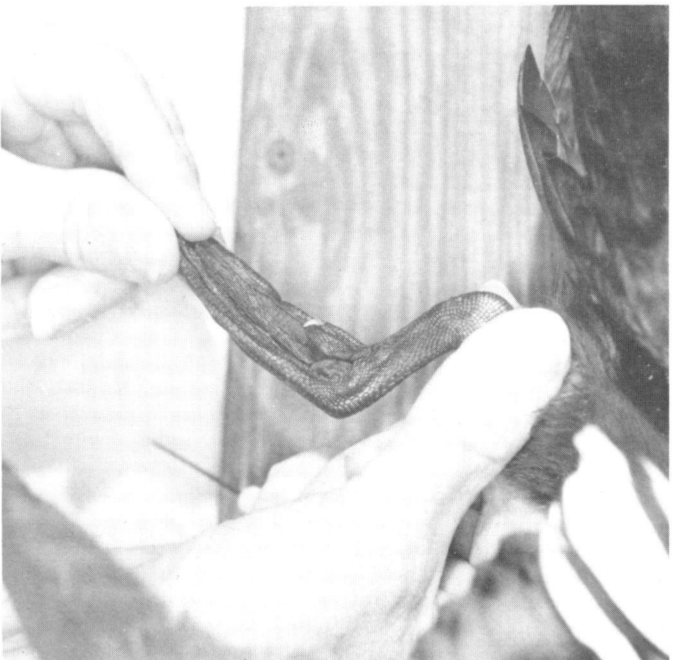

from the body, the unattached leg limply dangling from the body and perhaps the tail leaning off to one side.

Once the bird is in the desired position, begin grooming and aligning the feathers and placing them in their proper positions. A convincing dead game mount does *not* mean that the bird must look like it was carried in a game bag for two days with fifteen other birds. On the other hand, it should not look as if the bird has just spent two days grooming itself either. Obviously a happy medium must be met.

To keep the primaries and secondaries in place, it may be necessary to card these feathers with medium weight poster paper. The carding paper is held to the primaries and secondaries by placing the feathers between two pieces of carding. A jumbo paper clip is then attached to the end of the carding and the first primary feather. T-pins are then placed every inch to apply enough pressure to hold the remaining feathers in place. The carding should only help keep the feathers aligned, not *flattened.*

Setting the Feet

In order to keep the feet full and plump looking, it is necessary to inject them. Most taxidermists use Formalin for this purpose. When using Formalin (formaldehyde) it is imperative to wear protective goggeles and surgical gloves and to have a good ventilation system. Formalin has a devastating effect on the eyes, skin, and lungs; therefore, it is best to inject the feet at the end of the work day to allow the fumes to dissipate overnight. When injecting the feet with Formalin, use a small, insulin type needle. This small needle will prevent the Formalin from leaking out of the puncture site. To ensure that the foot will remain full and plump looking, apply another injection of Formalin on the second day if necessary.

When using the natural skull it is impossible to totally eliminate the shrinkage that occurs in the bill during the drying period. However, it is possible to control some of the shrinkage by packing the nostril openings with cotton. Force the cotton down into the nasal passage with a probe or feather adjuster. Once the nasal cavity has been filled, carefully inject the nostrils with Formalin. The combination of saturating the cotton with Formalin and the pressure of the cotton against the interior of the bill will help control some of the shrinkage problem. When the mount is dry, all of the cotton packing should be removed with a pair of tweezers or a feather adjuster.

Detailing and Finishing the Dead Game Mount

Refer to "The Breakthrough Waterfowl and Bird Finishing Manual" to accurately finish the bill and feet. Freshly taken birds still retain the natural color of their feet and bill; it is important that the color is restored to these areas. After the bill and feet are painted, attach a leather strap from the support leg to the rustic hanger or nail. If new leather is used, a mixture (50/50) of turpentine and WN Burnt Umber applied with a artists' brush will give the leather a worn looking effect.

The "glassy" look of the eye is achieved by applying Wet Look Gloss or Competition Wet Look Gloss to the eye with a fine tipped artists' brush.

Depending on the wishes of the customer, a hint of blood may be applied to the inside of the nostril. This is best accomplished by adding a very small amount of WN Alizarin Crimson to five-minute epoxy, and mixing thoroughly with a tongue depressor. This mixture is then applied to the inside of the nostril.

There are any number of personal memorabilia that a customer may wish to add to a dead game mount to truly make the mount a "special" trophy. An old duck call, decoy, dog leash, shotgun, hunting bag or hunting vest incorporated into the mount will add pleasant memories of a special day afield for many years to come.

CHAPTER 12

Swimming and Other Artificial Water Poses

Water, water everywhere and all the ducks did swim, dive, feed, drink, stand and splash. It seems that the incorporation of artificial water and habitat scenes with standard bird mounts has provided the taxidermist with an endless source of possible combinations to test their creative imaginations.

Not only has the use of artificial water challenged the taxidermist's creative nature, but the addition of elaborate water scenes to standard taxidermy mounts has been largely responsible for both the changing attitude toward taxidermy as an art form and the improving image of taxidermists among sportsmen and collectors of wildlife art. (Not to mention the increased revenues to the taxidermist for the creation of these scenes.)

The availability and use of casting resins and epoxies has finally provided the taxidermist with the medium to portray waterfowl in their natural environment—*water*. With the demand being so keen for new and innovative ideas each competition season, mastering the art of "artificial water" is of paramount importance.

Water can be simulated by using many different materials and techniques. The most popular materials used by wildlife artists for representing water in habitat scenes are **1.** clear polyester resins, **2.** acrylic polymers and epoxies, and **3.** plexi-

glass sheets. Often, one habitat scene may contain water effects produced by using a combination of all three materials.

Clear polyester resins (such as *Polytranspar*™ Artificial Water) are two-component systems that include a liquid resin and a catalyst. The catalyst must be carefully measured and thoroughly mixed with the resin. When the molecules of the catalyst react with the resin, the "liquid" mixture changes into a "solid" through a process called "polymerization." If the resin is correctly manipulated before it hardens, it can quite accurately produce the effect of simulated water.

Resins are best suited for creating splashes, ripples, water 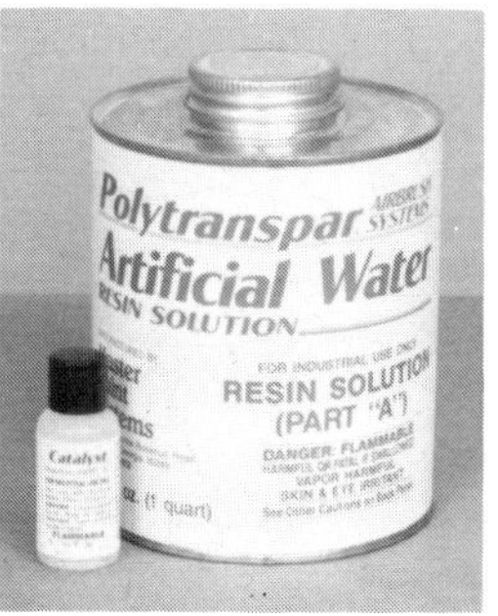 droplets, waterfalls, and irregular water surfaces. They may be poured over plexiglass, molded, manipulated while in a gel state, or shaped in any number of ways to produce acceptable results. Resins may also be colored to create various effects; however, most artists prefer to color the surface "underneath" the artificial water to achieve the desired effect, rather than try to color the water itself.

Acrylic polymers (such as Envirotex) and epoxy compounds are generally composed of two components which are mixed at a 50/50 ratio. The gel time usually cannot be controlled by varying the amount of the components as the curing time is "set" within the formulation of the chemicals; however, the "set" time can be hastened by adding heat to it. Acrylic polymers are best suited for repre-

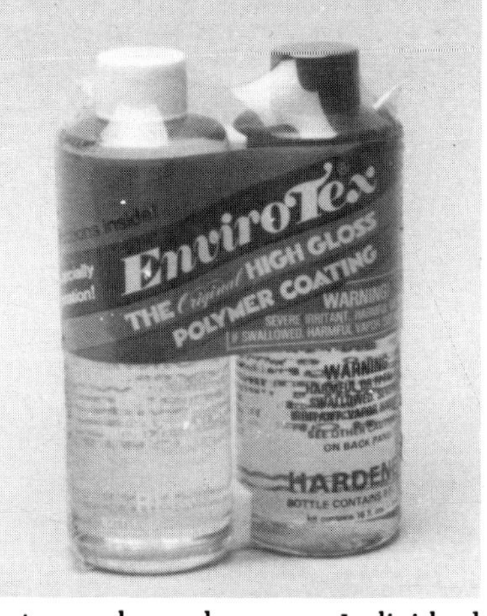

senting shallow pools, still water, and marsh scenes. Individual pours should not exceed ½" in depth. To achieve greater thicknesses will require multiple pours with time allowed for curing in between.

Plexiglass is also used when recreating water surfaces. It is usually used as a base with a layer of clear resin poured over the top of it to provide surface detail. When plexiglass is cut, carved, or ground to produce ripple effects, the resulting "frosted" appearance can easily be restored to a crystal-clear look by simply painting a layer of catalyzed resin over its surface.

Pertinent Facts About Resin & Epoxies

As with any type of chemical, the materials used to create artificial water are potentially dangerous to use. Thoroughly heed all safety precautions and directions on the containers of materials used and *observe all* of the instructions. Children should not use these materials without adult supervision. Adequate ventilation is a must and proper storage and fire precautions must be followed to the letter. Many of the catalyzing agents can cause chemical burns or tissue damage to exposed skin. Always wear protective clothing (such as a heavy canvas apron and latex gloves) whenever working with any two-component mixtures.

Experimentation and creativity can lead to unique new methods for producing water scenes. However, the artist should remember to follow safety precautions and use sound judgment and common sense when experimenting with new techniques.

Measuring

Epoxies, polyesters, casting resins, urethanes, and silicone RTV's, all contain two components (resin and catalyst/hardner). In order for the components to react properly, exact measurements must be made. Carefully follow and thoroughly understand the directions before you attempt to mix the components. "The Breakthrough Habitat and Exhibit Manual" has a complete chart on mixing ratios for polyester resins. I would suggest purchasing this manual in order to thoroughly understand the complexities and techniques involved with using polyester resins and epoxies.

Mixing

In order for any two-component epoxy or polyester resin to harden properly, the components must be *completely* and thoroughly mixed together. Simply inserting a stirring tool into a container and giving it a couple of quick swirls *will not* properly mix the components. A good habit to form when mixing components is to pass the stirring tool on the sides of the wall of the container, stir and pass the stirring tool on the bottom of the container and vigorously stir the center. This method will ensure a thorough mixing of the components. Remember the mixture cannot harden by simply "air-drying;" each component must come in contact with the other in order to properly cure.

Curing Time

Epoxies and polyester resins have always been viewed by many taxidermists as a "witch's brew," because there never seems to be any rhyme or reason as to when or how these chemicals are going to cure. However, if you understand and take into consideration the factors that affect the curing time, the process is not so mystifying.

The amount of time between the addition of the catalyst and the point where the resin begins to "firm up" is known as the "gel time". Many factors will affect the gel time of a resin mixture, among them; the room temperature, the humidity, the amount of catalyst used, the temperature of the resin, the total volume of the resin, and the age of the components. Learning to control these factors is an important step in the creation of successful simulated water scenes.

All these factors should be analyzed every time the components are mixed in order to have complete control and predictability over the final outcome of the project. Even artists extremely familiar with resins will always make test-batches of resin/catalyst ratios before adding them to their habitat scenes. There are simply too many variables to these chemicals to risk pouring them into a scene without first testing a small batch.

Problems

"Troubleshooting Artificial Water Problems" is excerpted from "The Breakthrough Habitat and Exhibit Manual" and should help in clarifying some of the problems that occur when using artificial water.

Troubleshooting Artificial Water Problems

Problem	Probable Cause	Corrective Measures *(NOTE: More than one method often can be used to correct the "same" problem.)*
FINGERPRINTS	Handling before resin has completely cured	A. Spray with *Polytranspar*™ Competition Wet Look Gloss FP241 B. Spray with Surface Coat Spray C. Paint thin coat of catalyzed Artificial Water on the area
CLOUDINESS	Moisture	A. Allow to sit in sun B. Apply heat lamp C. Apply Artificial Water with heavy catalyst ratio
WARPAGE	Excessive Catalyst Excessive room temperature	A. Reheat with heat lamp, hair dryer, or hot water and reshape B. Prevent by pouring resin in 70–77° room temperature or reduce catalyst ratio.
YELLOWING	Excessive Catalyst Inferior Resin Contamination	A. Cut back on catalyst ratio B. Use *Polytranspar*™ Artificial Water C. Replace lids securely in place after usage.

Mounting a Bird for a Swimming Pose

One of the most well-accepted simulated water scenes that a taxidermist can offer a customer is a swimming pose. However, the taxidermist must remember that when the water scene is utilized and the bird is placed in any number of the various swimming positions, there are anatomical changes that occur and the mannikin must be altered to successfully "pull off" this particular type of pose. If a wrapped or carved mannikin is used, the anatomical changes must be taken into consideration when fabricating the body.

References are also a "must" and should be utilized not only for duplicating the anatomical detail of the bird but for the accurate representation of the special artificial water effects and habitat composition. Additionally, the choice of underwater habitat is critical in creating a convincing scene and should be researched prior to producing the scene.

The pose and action of the bird must be "in harmony" with the habitat. You would not want to portray a duck casually swimming along with a giant "tidal wave" trailing behind.

ANATOMICAL CHANGES AND MANNIKIN ALTERATIONS

Breast

When a duck assumes a swimming position, the breast drops forward and flattens as the bird pushes forward. The flattening and buoyant effect of the breast is essentially created when the mass of the body meets the mass of the water. This flattening action is somewhat similar to a person sitting in a chair; the posterior portion of the body actually flattens and widens when we sit.

To achieve the flattening effect of water pushing against the breast, mark the area of the breast from the point of attachment of the humerus to the midsection of the body. With a bench grinder or farrier's rasp, grind or rasp no more than 3/16 of an inch off of the mannikin.

Remember, don't go overboard and flatten the breast too much. This would have a reverse effect and cause the bird to look as if it is being pushed backward by the water.

Neck

Ducks swimming in water assume any number of distinct attitudes and expressions which are achieved by the length and curvature put into the neck. If the head and neck are lowered in a typical swimming position, the neck drops into the recessed area of the clavicle.

To maintain the proper profile of the swimming position and to execute the curvature of the union of the breast and neck, the clavicle must be deepened on the standard standing mannikin. This is accomplished by cutting a "V" shaped wedge from the front portion of the mannikin (under the neck on Accu-Flex birds). This void will then allow the neck to drop easily between the breast.

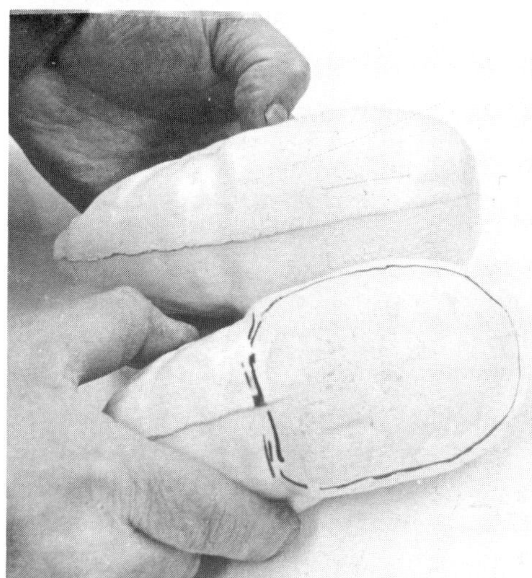

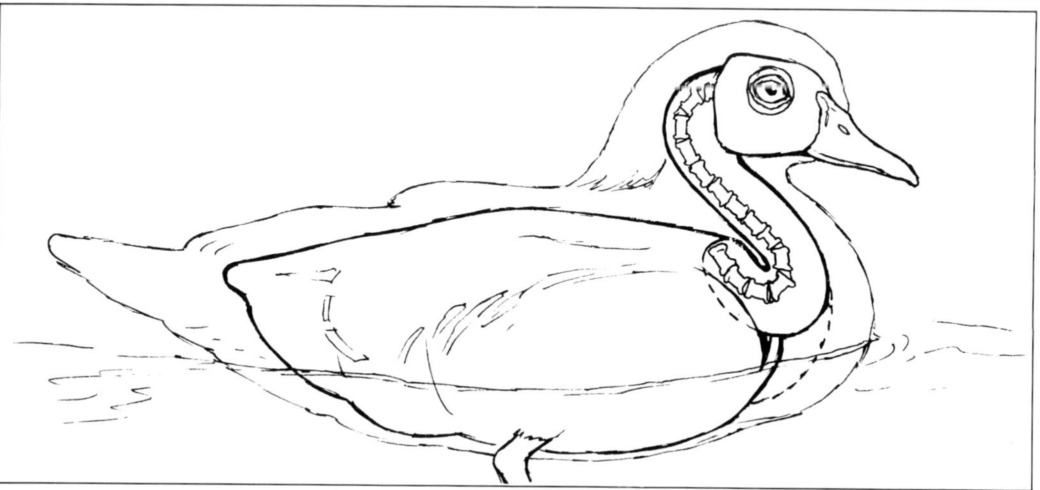

Legs and Feet

A thorough understanding of the arc of travel of the femur and tibia is mandatory. Proper placement and correct angles of the femur, tibia, tarsus and digits with respect to the body are the key to a convincing swimming pose.

The position of the tibia will determine whether the mannikin should be altered. If the tibia is extended, it changes the position of the femur. The thigh muscle should be rasped off of the form. It is important to remember that if the tibia is extended and the thigh muscle removed from the mannikin, then the femur should remain  attached to the tibia during the skinning process. The thigh muscle is then rebuilt when the bird is wired.

Once the necessary alterations have been made to the mannikin, the bird can be assembled as described in previous chapters.

Swimming Habitat

There are probably as many ways to present a swimming bird in a habitat base as there are wildlife artists. Some of the more creative of these scenes are pictured in this chapter. Water scenes can include vegetation, roots, stream banks, ice and snow, waterfalls, rocks, marshes, islands, and underwater scenes. Some artists frequently add a "porthole" to the base of executive cases in order for the viewer to witness the underwater habitat first hand.

For the following explanation, a generic version of a standard swimming mount is presented. Readers are encouraged to use this example as a "starting point" only, and to creatively expand upon this concept.

A "standard" swimming duck exhibit case might be constructed as in the diagram below. A hardwood exhibit base is used as the basic structure. In the bottom of this base, an underwater habitat is constructed from urethane foam, artificial rocks, artificial lily pads and stems, moss, twigs, and assorted materials. The water surface is represented by a sheet of 1/8" plexiglass, which is supported by the ledge around the interior of the case. The ripples upon the surface of the water have been simulated in polyester resin. The duck mount is positioned inside a carefully-made hole in the water surface and has been "sealed" with an additional pour of resin. As a final touch, a protective glass cover guards the scene from dust and debris. The result is a handsome executive case which can easily bring a taxidermist four to five times the price of a duck mount alone.

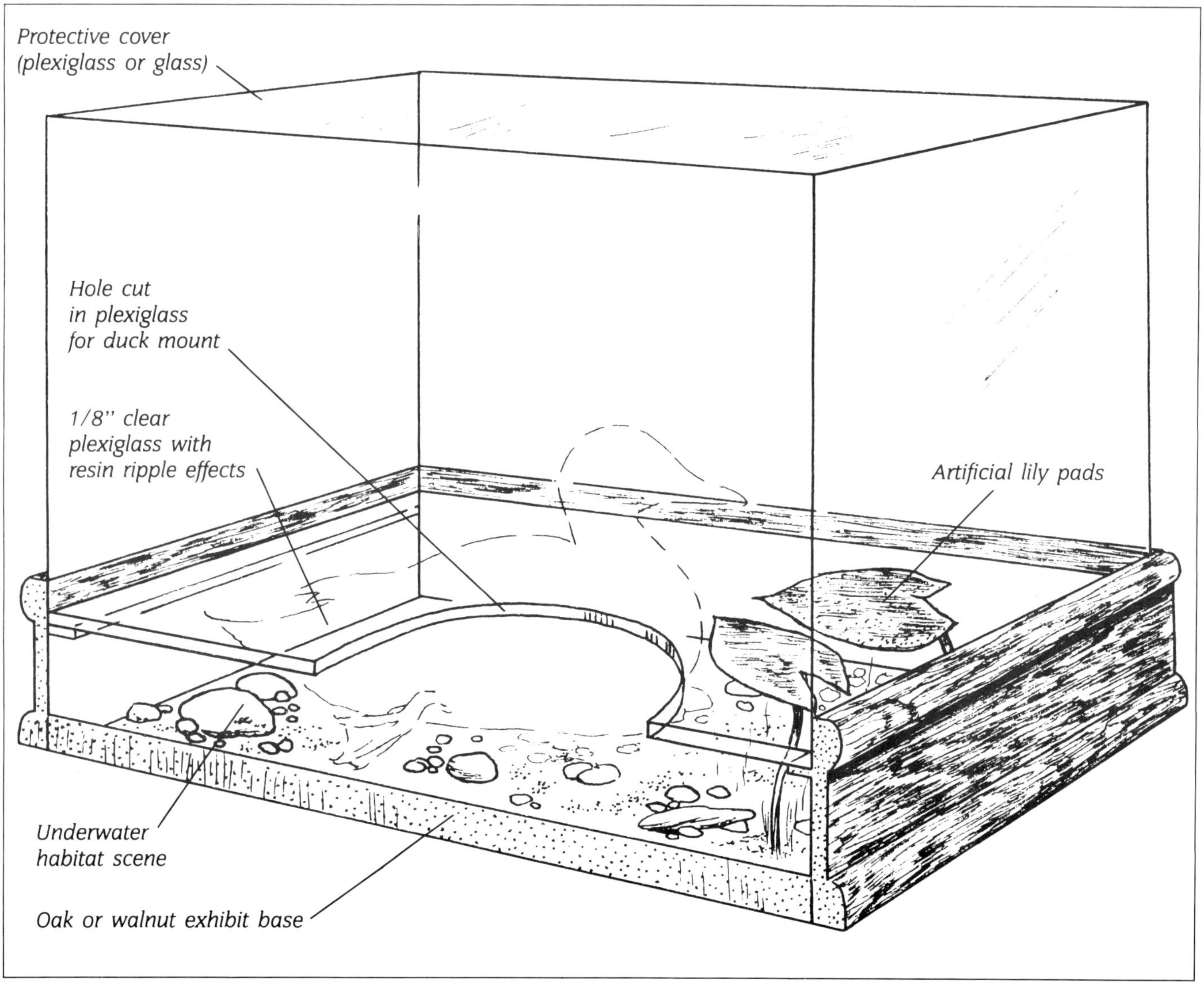

To produce a swimming duck habitat, the first step is to completely finish the bird. The bill and feet should be painted and the feathers groomed before the mount is placed into the scene.

The next step is to produce the underwater habitat. The more extreme the ripples on the water surface are, the less detailed the underwater habitat must be. This is due to the optical distortion of light passing through a rippled surface. In some cases, good results will be obtained by simply painting the base interior a dark color and not using any habitat materials; however, in most cases the artist will want to incorporate a variety of realistic materials underneath the water's surface.

Once the underwater habitat scene has been completed, a template should be made of the portion of the duck's body that will be recessed into the plexiglass. Carefully check your references to determine the "water line" of the bird. When the position of the water line has been established, use calipers to measure the length of the template between the waterline at the front of the breast and at the flank.

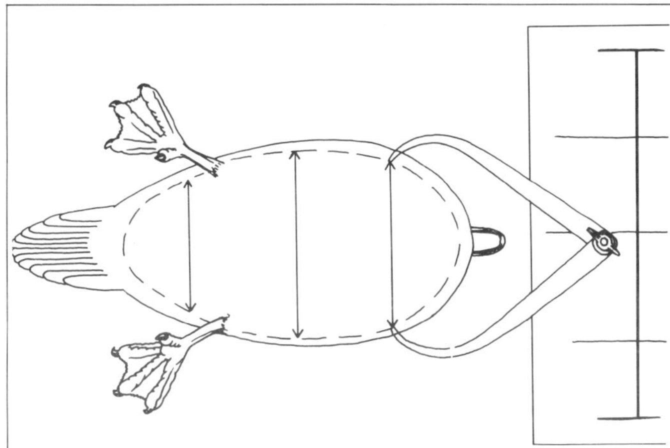

Transfer this measurement to a piece of cardboard and connect the two points with a straight line to serve as a centerline. Take several measurements of the width of the body at the centerline and transfer them to the cardboard.

Now, connect the measured points and cut the shape out of the cardboard. Test fit the mounted bird in the hole in the cardboard and make any additional adjustments as necessary. When satisfied that the cardboard template provides an accurate, "snug" fit around the waterline of the

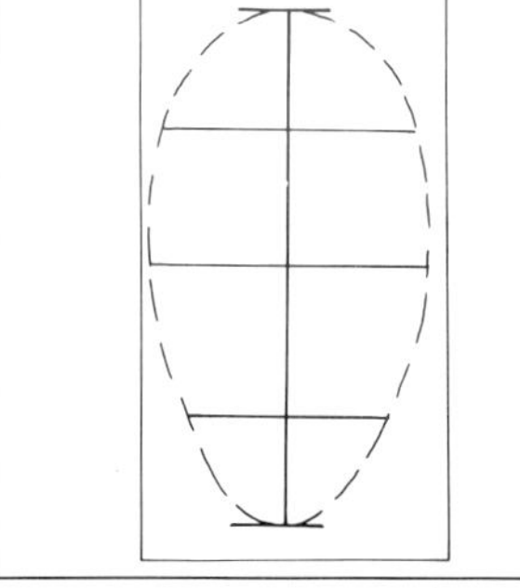

bird, position the template on the plexiglass and carefully trace the pattern onto the plexiglass. The hole for the bird may now either be cut out with a saber saw with a plexiglass cutting blade or a Foredom tool with proper cutter. Cut the plexiglass at an angle that will help to support the bird.

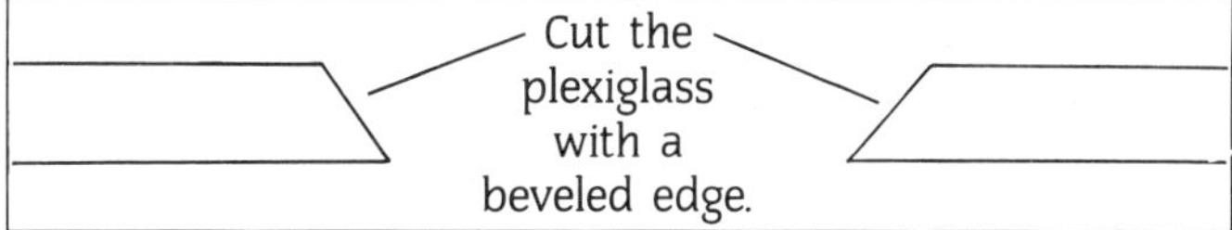

Now test fit the mount in the plexiglass to determine the final position.

Creating Ripples

Next, the mount is removed from the plexiglass and the water effects of the water's surface can be created. There are many different methods of simulating ripples on a water surface such as this. Three popular methods are outlined below.

1. Carving Ripples. Some artists choose to "carve" the ripple shapes directly on the surface of the plexiglass using a Foredom tool. Once the basic shapes are roughed-in, the surface can be further refined with sandpaper. Grinding and sanding the plexiglass will result in an opaque, frosted appearance to the surface. When satisfied with the final arrangement of the ripples, the artist merely has to brush on a thin layer of catalyzed resin to restore the clear, transparent look to the surface. The carved indentations will recreate the ripples quite accurately once the resin is applied.

2. Molding Ripples. Several artists choose to sculpt ripples from wax or clay and make molds of these sculptures into which catalyzed resin is poured. When the resin sets up, the individual ripple com-

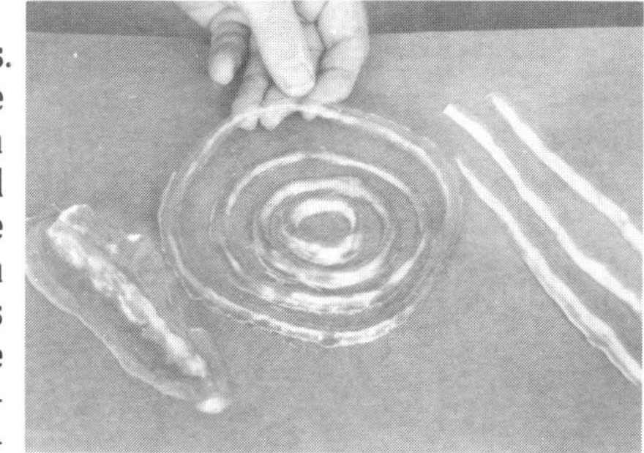

ponents are removed from the molds and positioned individually on the plexiglass surface. A final layer of catalyzed resin is then poured over them, effectively "locking in" all the elements in place while making the assembly points invisible (blending the separate ripples into the surface smoothly—with no discernable transition point).

3. Gelling Ripples. The third method of creating ripples involves physically manipulating the catalyzed resin as it approaches the gelling stage. For this method, the resin is poured over the surface of the plexiglass and ripples are defined by the use of an artists' brush, a spatula, a tongue depressor, or even compressed air. When the resin begins to set up, the manipulation is stopped, and the ripples remain.

When the surface of the water complete with ripples has been prepared to the artist's satisfaction, it may be placed into the base and allowed to cure overnight. A cardboard box placed over the scene will prevent dust from ruining the appearance of the resin surface.

Inserting the Bird

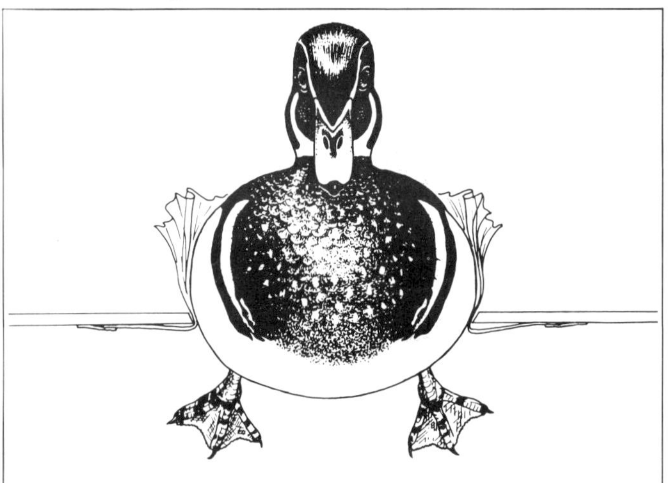

The bird is then placed into the hole in the plexiglass. If necessary, it may be held in position with long Jumbo Head Pins (the pins will be removed once the artificial water starts to set). If there is a space between the plexiglass and the duck, cut a plastic sheet to fit the curves of the duck's body. Place the plastic next to the duck and tape it to the underside of the plexiglass. This will prevent the resin from running through (and onto the duck's belly) when it is being sealed. After the resin has cured, the plastic sheet will be removed.

To seal the bird in place, catalyze a sufficient amount of resin and stir it until it "just" begins to gel. The gelling resin should be placed around the circumference of the bird to create a dam (see illustration). It should be brushed to fill any gaps. By using resin that has already begun to gel and creating a dam will prevent any saturation of the feathers by the resin, and a "wet look" to the feathers will be avoided. Once the "dam" has "set," a second pour will be necessary to form a smooth transition from the outer edges of the duck's body to the main body of water. The "dam" will prevent this coat from ruining the feathers.

Droplets

A nice finishing touch to any artificial water scene is the application of artificial water droplets to the duck. They are quite easy to create and add an interesting bit of realism to the rendering.

One excellent way to apply the droplets is with a transfer pipet. Thoroughly mix the artificial water and catalyst in small quantities in a 3 oz. paper cup. Immediately before the resin begins to gel, squeeze the bulb of the pipet to draw the catalyzed mixture into it. With the mixture in the pipet, simply place the drops on the bird.

Creating a Standing Pose

The depth of the artificial water surrounding the bird will determine the "plan of attack" to be used in creating this type of scene. A deep water pool will be treated much the same as the swimming pose, while a shallow water pool may be accomplished much easier with a solid resin pour.

Deep Water Pool

If the bird is to be standing in deep water (mid-tarsus) with an elaborate base, the use of plexiglass as the foundation for water surface is the best choice. Problems can occur if the artificial water is poured to depths greater than 1/2" thicknesses.

The plexiglass should first be fitted to the support structures of the base. This is important as the plexiglass with the duck installed should fit easily so that there is no unnecessary handling of the bird.

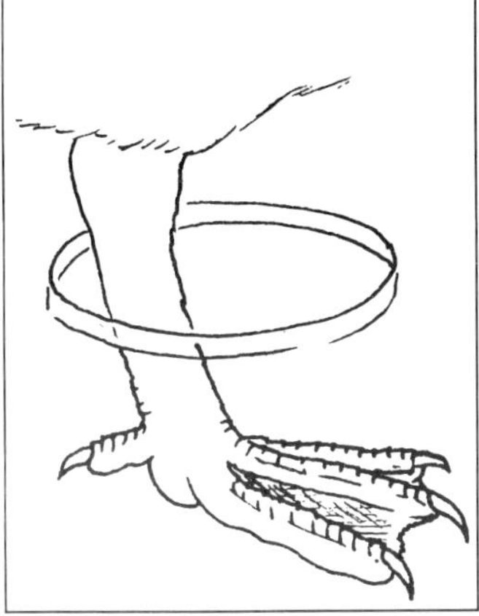

A hole will have to be cut to position the bird in the plexiglass, much in the same manner as inserting a swimming bird. Unfortunately, because of the webs on the feet, the opening will have to be cut rather large in order to accommodate slipping the entire foot through the plexiglass. Once the bird is in position, careful measurements of the distance of the base surface to the bottom of the plexiglass must be taken to situate the plexiglass at the proper height.

When the proper depth has been determined, the hole must be "plugged." To make the "plug," cut a plastic sheet (or the plexiglass that was cut from the hole) to fit snugly around the legs and the opening of the plexiglass. The plastic (or plexiglass) should be hot-melt glued in the plexiglass and a coat of resin poured over it. When the "plug" is tightly secured to the plexiglass and the birds legs, mix enough resin and catalyst to seal the opening made for the legs. Several pours may be needed to have a smooth, even look to the surface of the plexiglass.

Next, set the plexiglass and bird aside to cure, again using a cardboard box to cover it. Immediately *before* the plexiglass is placed back on the support structure, five minute epoxy may be mixed and put on the bottom of the bird's feet to help secure and stabilize the mount to the bottom of the base. Now place the plexiglass back in place and finish the water surface to your satisfaction with additional pours of artificial water.

Shallow Water Pools

Creating the illusion that the bird is standing in shallow water presents less of a problem because the bird is attached to the base and the resin is poured onto the scene to the desired depth. These scenes are easily created by using Envirotex. Simply pour a quantity over the scene to create a pool and then let it harden.

However, if other special artificial water effects are incorporated into the scene, one method to consider is attaching the bird to the base *after* the "special effects" are completed. This will faciliate designing and constructing the scene without the bird being in the way and hampering the construction of the special effects.

The position of the bird's feet should be marked and dammed off with an oil-based clay. The resin is then poured and the special effects placed into the scene.

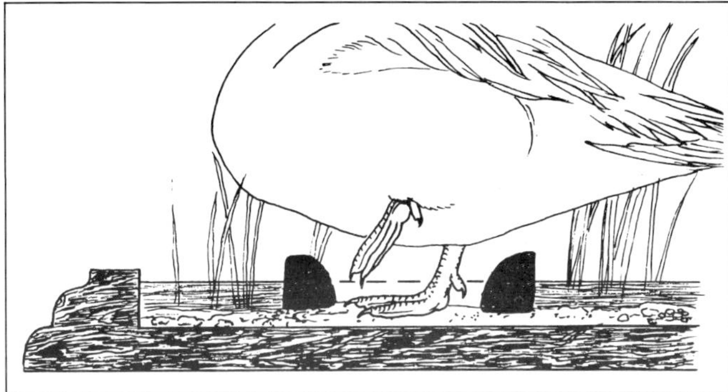

Once the scene is completed, the clay dam is removed and the hardened resin surrounding the dam is cleaned of any clay residue with lacquer thinner. Next, the bird is attached to the base in the pre-marked spots. Five minute epoxy works extremely well for securing the bird to the base. When the epoxy has cured, the immediate area surrounding the bird is poured with catalyzed resin. Depending on the desired thickness of the artificial water or Envirotex, several individual pours may be required until a smooth transition is achieved.

Flushing Duck

Another excellent artificial water scene is the flushing duck taking off from the surface of a pond. This can be accomplished by mounting the bird over a clear acrylic rod which is then attached to a sheet of plexiglass representing the surface of the water. The support rod is disguised by splash components made from catalyzed polyester resin. When properly designed and executed, these types of mounts can reach the epitome of the bird taxidermist's art.

Artificial water has provided taxidermists with another dimension to test their creative skills. And like every phase of taxidermy, mastering the use of epoxies and casting resins will require time, experimentation and a lot of patience, but the end result will certainly be worth the effort—both in the financial rewards from satisfied customers and in the personal satisfaction of accomplishing a difficult task.

Chapter 13

Finishing Techniques

Throughout this entire book, we have stressed the importance of certain procedures that would dramatically affect the end results and longevity of the mount; however, the final finishing procedure is what will have an *"immediate"* impact on your customer's reaction to their mount.

It is important that every taxidermist understand that finishing techniques incorporate more than just painting the bird's feet and bill. Finish work encompasses a final inspection and attention given to the *whole* mount.

The quality of your finish work projects more about yourself and your shop than any other aspect of your taxidermy work.

The complete removal of all paint from the eyes is a perfect example of paying close attention to detail. It does not matter whether an artist is a professional, full-time taxidermist or a part-time amateur, it requires no special talent or ability to clean the glass eyes on a completed mount. But even in competitions, a great percentage of entries lose points because paint was not thoroughly removed from the glass eyes. Something as "seemingly insignificant" as cleaning the eyes on a finished mount, shows that the taxidermist did pay close attention to *detail.*

Spending the necessary time to make sure that the eyes are cleaned, all pins removed, colored parts restored to their natural color, and the finished mount is dusted and clean is what finish work is all about.

The most efficient way to finish waterfowl and upland game is to develop a routine and a check list system and follow this system *every time* a mount is to be completed.

Check List

1. Remove all pins, wires and carding
2. Clean the eyes
3. Clean the nostrils
4. Rebuild the shrunken areas
5. Paint the feet, bill and eyelids
6. Check the feather arrangement
7. Clean and dust the base/panel
8. Clean the eyes again!
9. Give the entire mount a final "once over"

1. Remove all pins, wires and carding

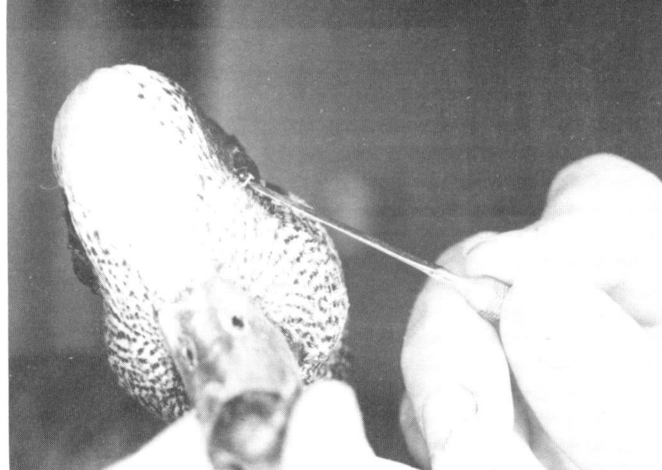

Small insect pins are used to keep the eyelids in place during the drying process. As a rule, these pins are easily removed by placing a fingernail under the head of the pin and gently lifting up on the head of the pin. However, sometimes the pins stick and are difficult to remove. If this happens, the end of a thin instrument will give enough leverage to remove the stubborn pin. Be careful when prying the pin loose, as the skin is very delicate in this area of the eye.

Pins are also used to hold certain feather tracks in place. These pins should only be removed if they are visible when viewing the bird.

Wires such as tail supports and supporting neck wires should remain in the bird. However, a careful inspection should be made to assure that no pins or wires are visible when viewing the finished mount.

2. Clean the eyes

This seems like a simple task, but is one that often causes problems to the taxidermist.

The eyes should be cleaned thoroughly before the eyelids are rebuilt and painted. It is much easier to clean the eyes *then*, before any work is done on the lids, simply because there is less chance of damaging the filler or removing any paint on the lid.

Remember to clean the eye thoroughly with Windex, paying particular attention to the front corner of the eye and any spaces where the lid has pulled away from the eye.

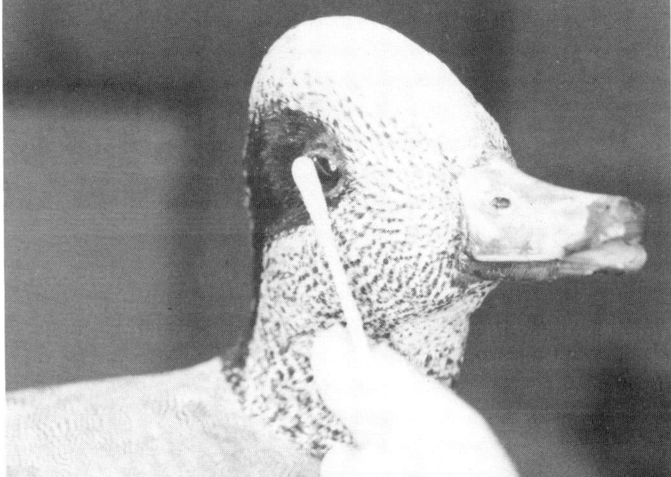

There are several ways to approach the problem of cleaning the eyes. Cotton-tipped swabs saturated with Windex work well on the main surface of the eye.

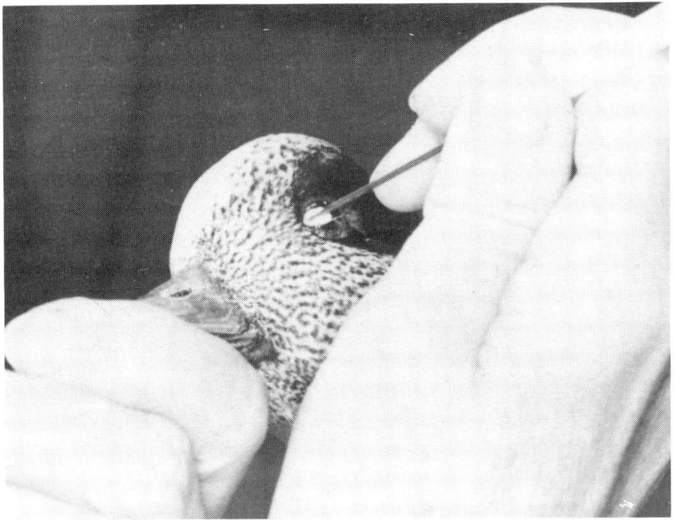

To get into the corner of the eye and right next to, or underneath the lid (if the lid has pulled), take a tiny piece of cotton saturated with Windex and place this on the tip end of a long pin. Gently follow the contour of the eyelid with the wet cotton. This method lets you clean right next to the eyelids.

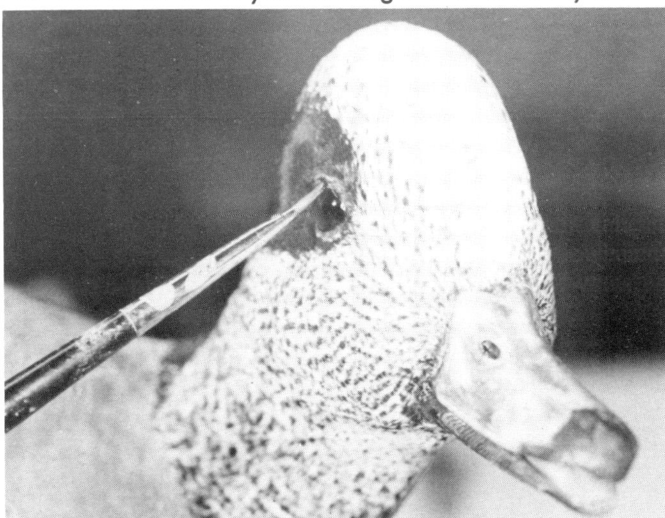

Windex may also be applied to a pointed artists' brush to clean the eye after the bulk of any residue has been removed.

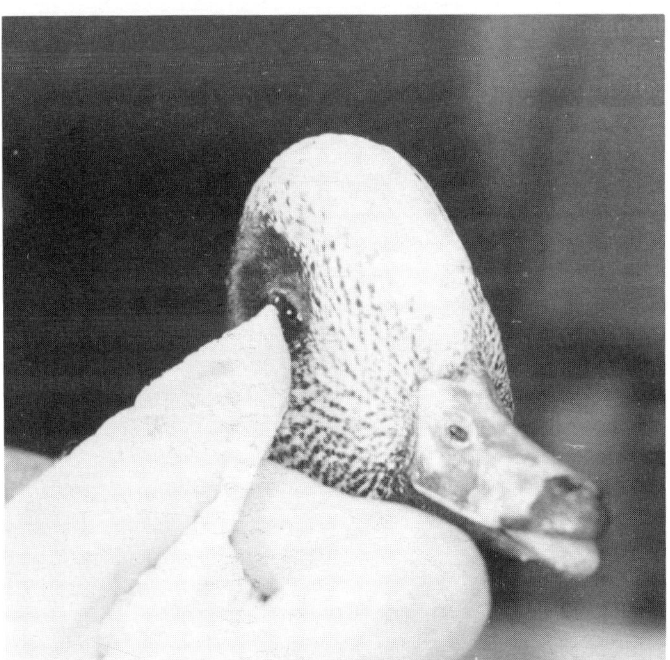

My favorite way of cleaning eyes is to place a small part of a lint-free rag, moistened with Windex, over the end of my thumbnail. Needless to say, this is always a quick and handy way to clean glass eyes.

The main objective is to clean the eye of any film or residue left from the mounting process. Remember, this will be the last opportunity to aggressively clean the eye before the finish work is begun on the eyelids.

3. Clean the nostrils

This may seem like an insignificant step; but, how many times have you completed the last finishing touches on a bill to find that when you turned your back, several grains of borax or dry preservative have fallen from the inside of the nostril and have adhered stubbornly to the paint on the bill? Many times this will ruin the paint job and you will have to start the project over again.

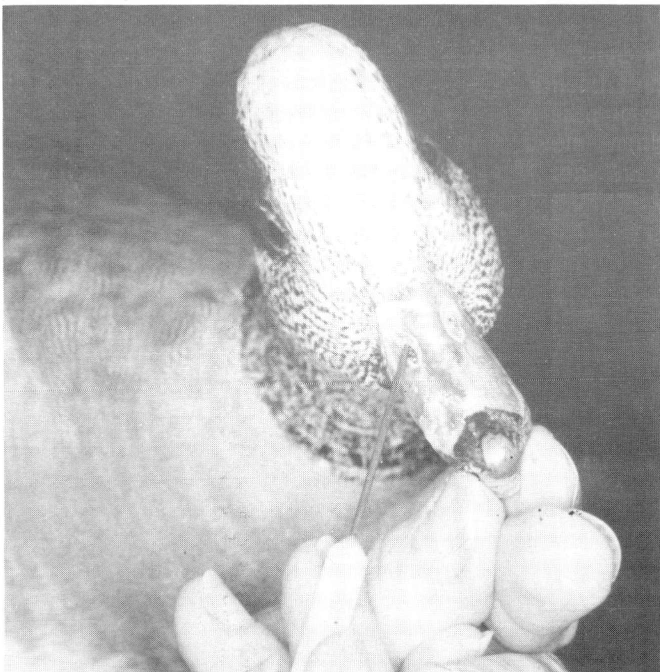

To eliminate this problem, take a feather adjuster or similar probe and insert the tip into the nostril opening, freeing any granules of borax or dry preservative that remain.

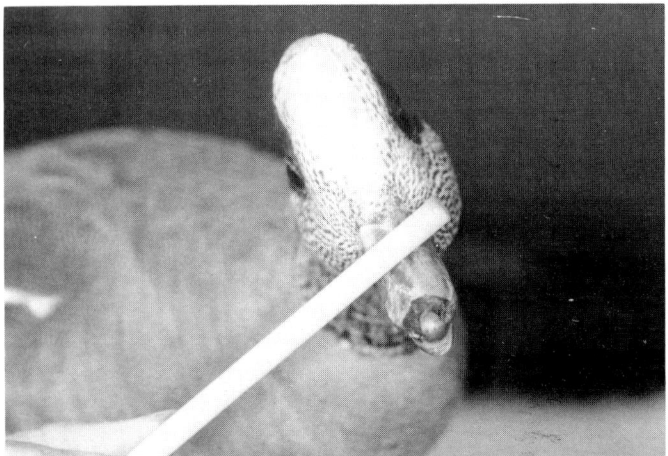

Three or four light taps on the tip of the bill will usually free any stubborn particles clinging to the inside of the nostril.

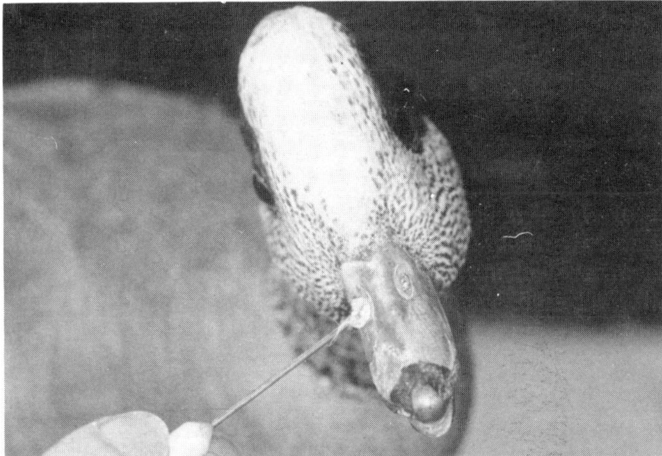

Once the particles or any other foreign objects have been removed, place a damp piece of cotton on the end of the probe and wipe out the inside of the nostril cavity. This will remove any borax powder or sawdust that is left from the mounting process.

After this, the inside of the nostril will be clean and free from residue that may hamper the final finishing process or inhibit the paint from adhering to the surface.

4. Rebuild shrunken areas

Eyelids—Eyelids vary from specie to specie so it is best to have reference pictures available and close at hand when re-creating them. Rebuilding the eyelid is best accomplished using Sculpall or All Game.

These putties are somewhat aggravating to use because of their tackiness but their redeeming features are their adhesive

qualities and the amount of detail that they will accept.

There are several techniques that will make using these putties somewhat easier. First, mix the two components together and allow this mixture to set for several minutes before attempting to sculpt or mold the detail.

Secondly, any tools that you may use to mold the putties should be dipped in water before touching the putty. This eliminates the epoxy from sticking to the tool and pulling off the sculpted eyelid when you are trying to put in the detail. This method also facilitates the process of placing the sculpted lid onto the eye. Many times the sculpted lid absolutely refuses to come off of the tip of the modeling tool and becomes totally distorted in this operation if the tool is not first dipped in water.

A moistened artists' brush also works well to shape and texture the eyelid once it is in place.

Certain species of waterfowl (such as wood ducks, geese and swans) have a very pronounced ring around their eye; this "eyelid" is much thicker and more fleshy than found on most species of waterfowl. In order to portray the bird accurately, this "eye-ring" must be rebuilt. Sculpall or All Game epoxies work well for this operation.

Ed Thompson uses a unique method which gives more "body" to the Sculpall and also permits easier handling when rebuilding this fleshy layer.

First, thoroughly mix a small amount of Sculpall and allow this mixture to set for several minutes.

The *Breakthrough* Bird Taxidermy Manual

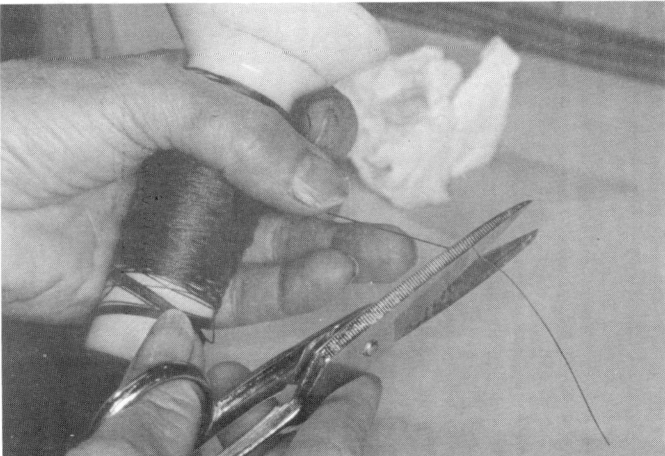

Then cut a small section of thread (enough to go completely around the eye opening). When the Sculpall has set up enough to permit easy handling, flatten the mixture into a thin strip approximately the same length as the thread.

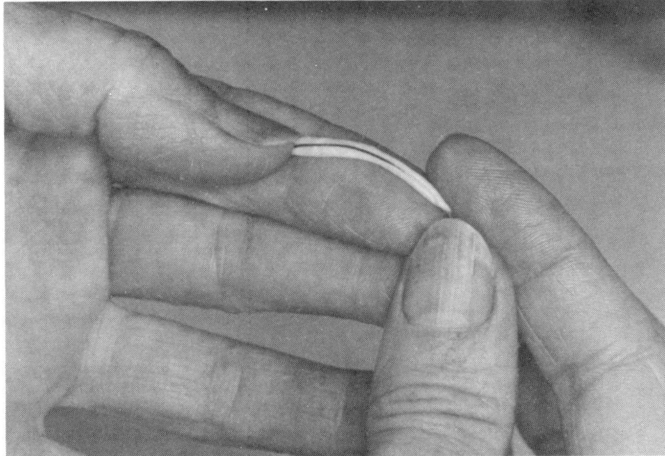

Place the thread in the center of the flattened Sculpall and proceed to wrap it around the thread.

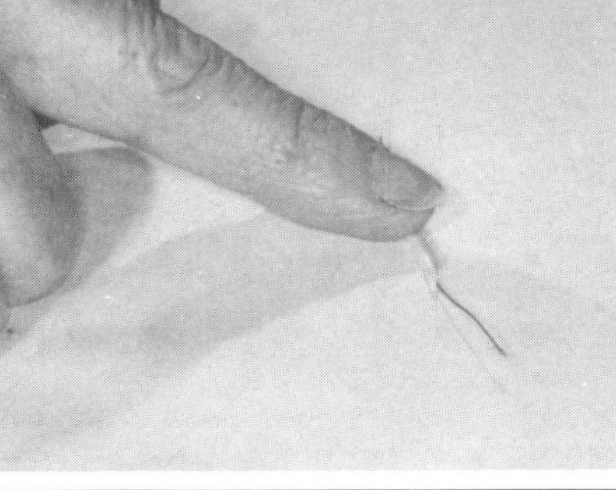

Once the thread is completely enclosed, lightly roll the cylinder with your fingers over a hard, flat surface. "Rolling" in this manner will give the Sculpall a smooth, even appearance.

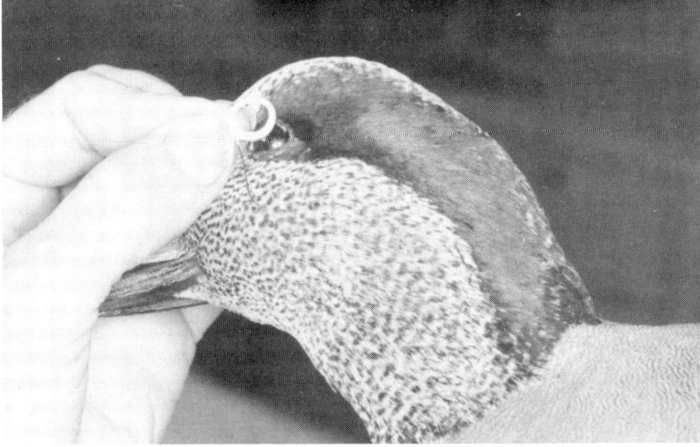

When satisfied with the appearance of the Sculpall, make a loop the same size as the area to be rebuilt and carefully place this on the edge of the natural shrunken "eye-ring." Apply

enough pressure to the Sculpall loop to securely affix it to the bird; however, be careful not to distort the shape of the sculpted loop.

When the loop is securely in place, use a combination of a feather adjuster and moist artists' brush to shape and texture the loop. Remember to keep modeling tools and artists' brushes moist with water. Once the shaping and texturing is completed, allow the Sculpall to "set-up" before painting.

Bill—Unless an artificial bill is used, a certain amount of shrinkage will occur to the natural bill during the drying process. In order for the bill to look full and natural, shrunken areas around the base, rim, and nostrils should be rebuilt with Sculpall or All Game. Rebuilding these areas is an absolute must for competition mounts.

Mix the two components together and allow them to set for several minutes before applying to the bill.

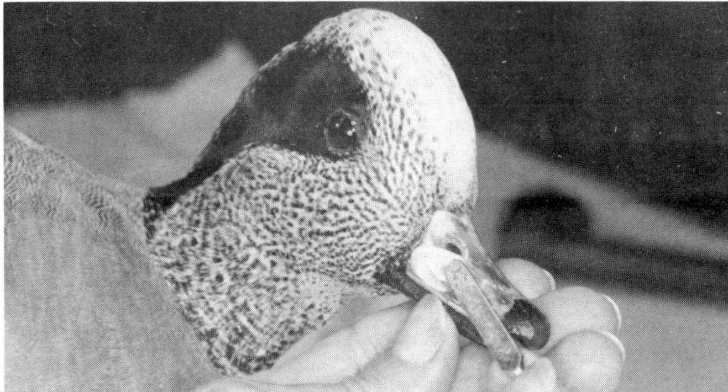

Apply the epoxy with a modeling tool or small spatula.

Work and smooth the mixture with your fingers. Remember to keep your fingers wet with water. To give a smooth look to the bill or to lightly texture the bill, use a moist sable artists' brush. Be sure to clean the brush thoroughly when the operation has been completed.

The epoxy may also be used to repair any shot holes in the bill. These shot holes should be repaired before the actual rebuilding of the bill takes place.

Feet—Even with the use of formaldehyde, the feet will experience a certain amount of shrinkage; however, it is not feasible to rebuild the entire foot because of a slight amount of shrinkage.

Shot holes, web damage, pin holes, etc., should all be repaired with Sculpall or All-Game.

After the shrunken areas of the eyelids, bill and feet are rebuilt, the bird should be set aside to allow the epoxy to set up and harden. This will usually take several hours.

5. Paint the feet, bill and eyelids

Whether airbrush paints, oil paints or a combination of both methods are used, the end result that the artist is trying to achieve is the same—a soft, natural-looking paint job.

Before the paint can be applied, the feathers on the head and the area surrounding the legs and base will require some type of protection from overspray from the airbrush or from the unsteady hand of the artist.

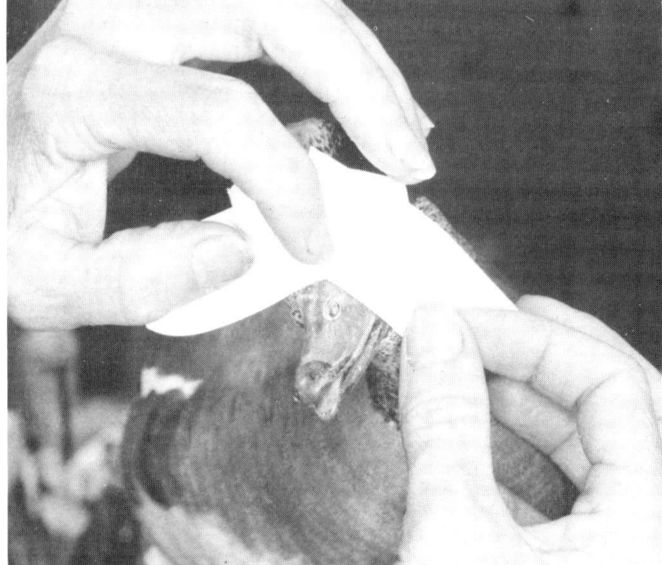

Lightly applying masking tape around the bill will protect the feathers in this area. Be careful not to apply too much pressure to the tape as this will make it difficult to remove.

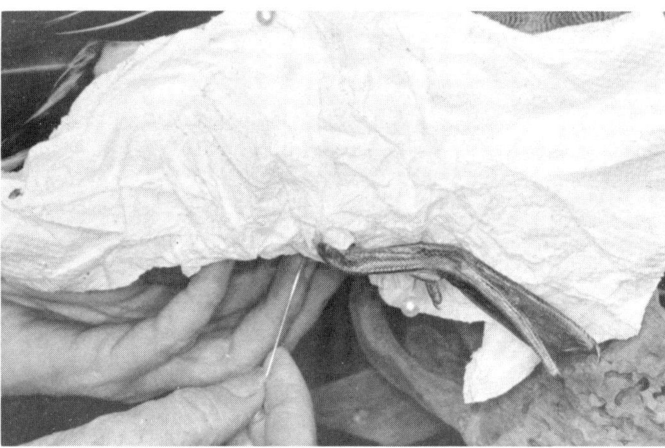

Waxed paper or paper towels work well under both the belly and the feet to protect the feathers and the base.

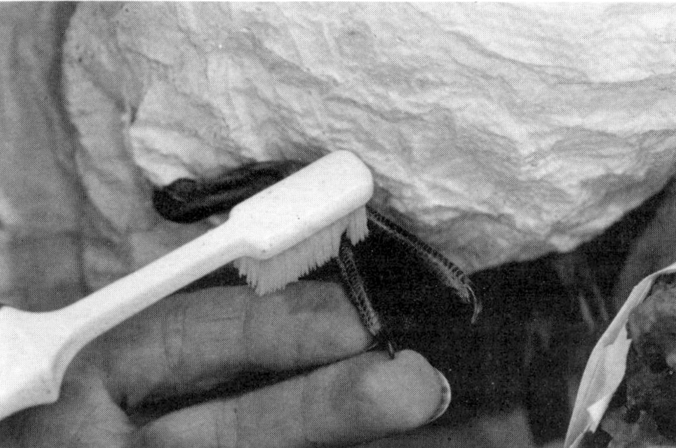

Once the feathers and base are protected, the bill and feet should be cleaned to make sure there is no residue remaining from the mounting process. An old toothbrush works well for this process.

After the surface area is free from sawdust, preservative, etc., wipe the feet and bill down with lacquer thinner to remove any grease or moisture.

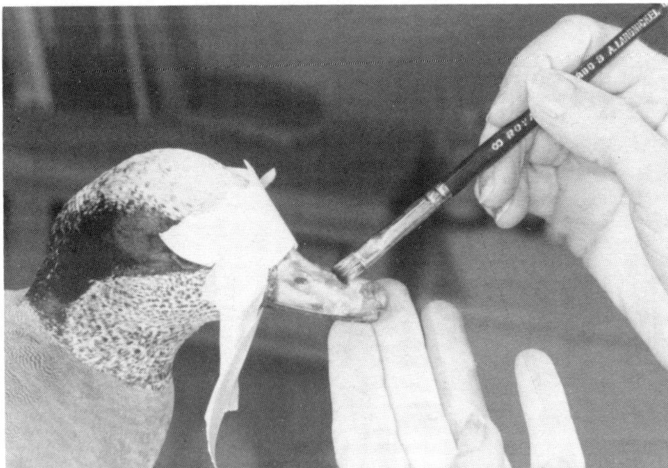

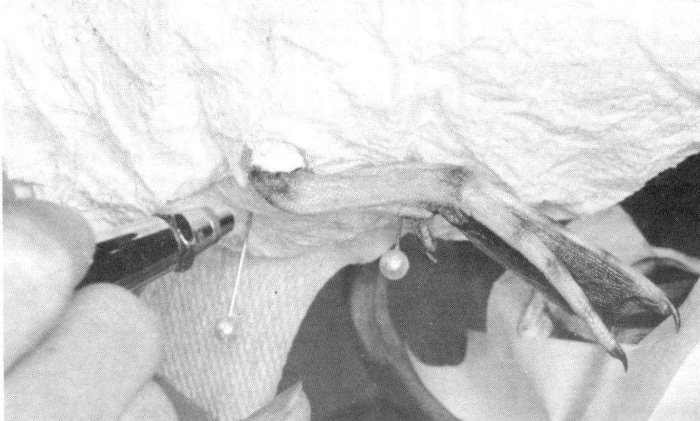

Whether *Polytranspar* lacquer-based or water-based airbrush paints are used, the appropriate Fungicidal Sealer should be applied to all surfaces. The sealer may be applied with either an airbrush or artists' brush. Fungicidal Sealer should also be applied before painting with oils or acrylic paints.

The bird is now ready to have the natural color restored to the eyelids, bill and feet.

of the color of the base coat will drastically affect the final outcome.

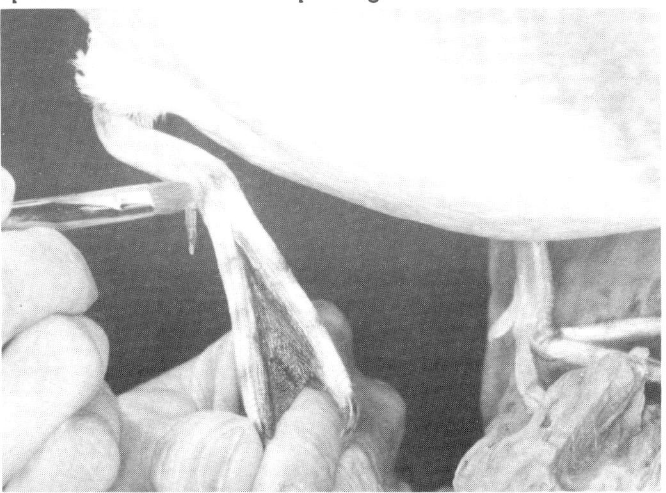

Before actually applying the paint, it is imperative that every available reference photo is within easy viewing. No matter how familiar or how many times you have painted a particular species of bird, use your references and use "The Breakthrough Waterfowl and Bird Finishing Manual."

The best results are obtained when the artist utilizes a combination of *Polytranspar* airbrush paints and either acrylic or oil washes. The most important thing to remember when applying either airbrush paint or acrylic/oil bases is that *less is better*. It is always easier to add a little more paint to a foot or bill than to remove the paint.

Only in rare instances will one color be adequate on the bill or feet of a particular bird. Achieving a natural, life-like color will require a series of different color applications.

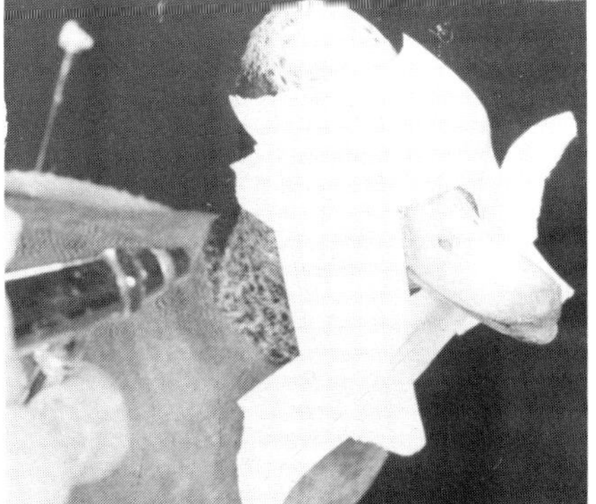

The first application of paint applied to the foot will be the base coat. The base coat is important because it will help you to achieve the desired color with less paint. Light will penetrate through colors painted over the top of the base coat and reflect off of it. The intensity

Refer to "The Breakthrough Waterfowl and Bird Finishing Manual" to expand on the techniques used for the various species of waterfowl and upland game.

When the proper color has been achieved on the feet and bill, a "wash coat" will be utilized to enhance the illusion of depth by highlighting and shading and to lend a more natural look. A by-product of the "wash coat" process is a softer overall finish on the legs, feet and bill which is so important in finishing waterfowl.

Airbrushing with *Polytranspar* lacquer paints and using a wash coat of acrylic water-based tube or Winsor & Newton artists' oils for inlay work is compatible and will not affect the *Polytranspar* work. Remember, whatever the medium that you choose to paint with, make sure that it is compatible with the base coat.

When following the waterfowl finishing instructions in "The Breakthrough Waterfowl and Bird Finishing Manual," the finishing sheen is incorporated into the paint schedule. The mixture of Satin Sheen and Refracted Frost will ensure a soft, natural look to the feet and bill. If you choose another method to finish the specimen, remember to check your references. Bills and feet are *not* glossy, so use an appropriate finish coat to maintain the look of a live bird.

The important thing to remember is that paint can be removed; so don't be afraid to experiment and try different techniques that are described in the *Breakthrough* Finishing Schedules.

6. Check Feather Arrangement

When removing pins, carding material and protective material from the bird, some of the feathers may get out of line. Use either a small feather adjuster or tweezers to put the feathers

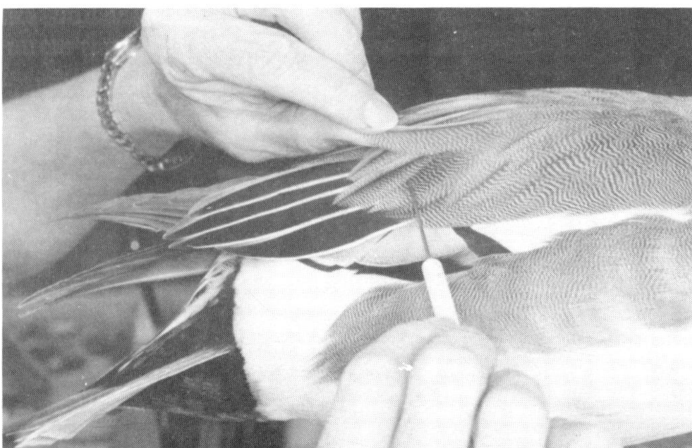

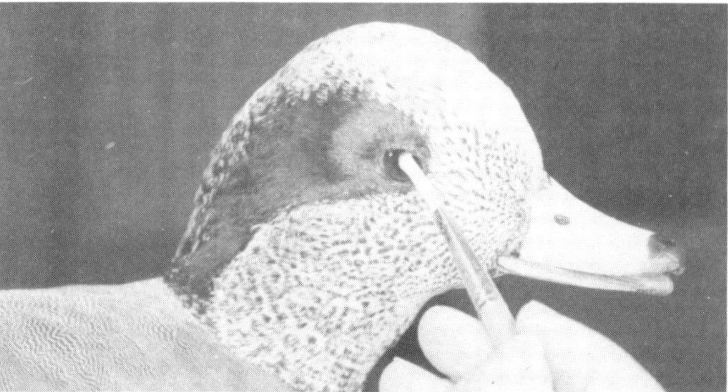

back in place. A gentle lifting motion with the feather adjuster or a slight pull with the tweezers will get the feathers back on track.

Check the feathers for any paint overspray and remove lacquer paint with lacquer thinner, water acrylic paint with water, and oil paints with Silicoil. Use a cotton-tipped swab for this operation.

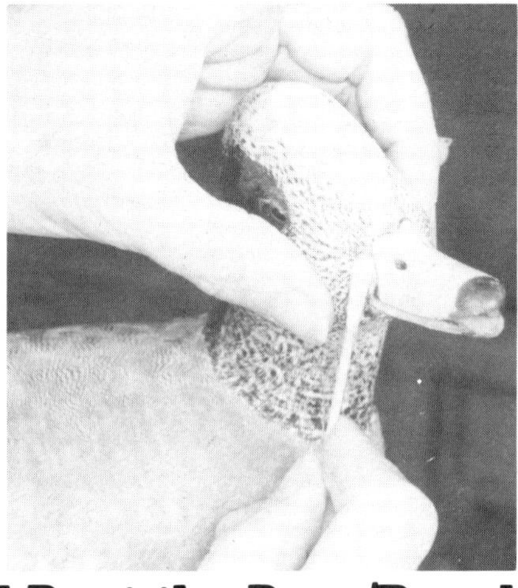

If lacquer paint was used on the eyelid, dip an artists' brush in the lacquer thinner, remove excess lacquer thinner from the artists' brush with a paper towel and paint over the glass eye with the brush. The thinner will remove any paint remaining on the glass eye. Using a pointed tip artists' brush will enable you to remove the paint next to the eyelid. Once the paint film is removed, use an artists' brush with Windex to finish cleaning the eye. If water acrylic paint or artists' oils were used, the same procedure is used substituting either water or Silicoil.

The majority of the paint is removed during the actual process of painting the eyelid. The eyes are much easier to clean if the paint is not allowed to thoroughly dry.

9. Final Once Over

It is very easy to overlook a pin, piece of thread or a feather out of place, especially if you are intensely involved in another project on the bird.

7. Clean and Dust the Base/Panel

Depending on the habitat material used in the base, use an air compressor or feather duster to remove any dust or debris

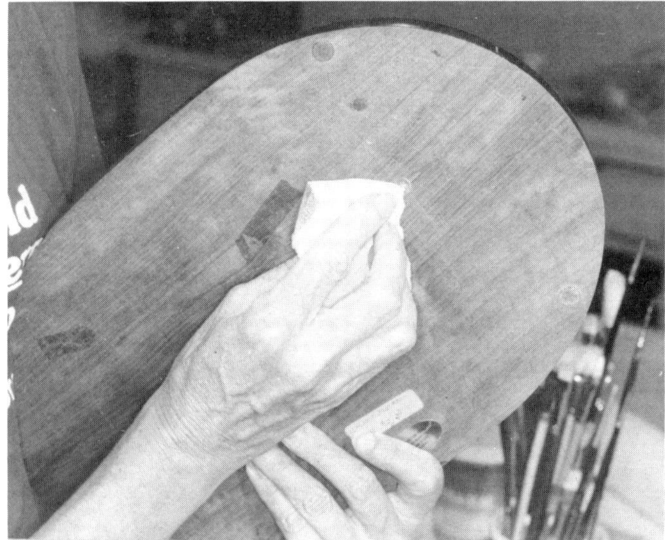

For the final once over, stand back and objectively study the bird as if you are looking at the mount for the first time. If you are pleased with the results, and feel that you would display the mount in your den, then give the bird one last pass with the feather duster and put it in your showroom for the customer to pick up.

from the base. Check the base for mechanical soundness and make sure that there are no sharp wires or screws protruding from the bottom. Nothing would distress a customer more than to have a base scratch a good coffee or end table.

If a wooden panel is used to display a flying bird on the wall, wipe the panel down with Liquid Gold. Check the hanger and support wire, making sure both are secure. Make certain there is nothing on the back of the panel that will scratch the wall.

8. Clean the Eyes AGAIN!

At this point the only residue that should be left on the eye is a slight paint film.

CHAPTER 14

Marketing Techniques

The old adage, it takes money to make money, applies to taxidermy as much as any other business. Taxidermists must spend a portion of their "money" on advertising and other business oriented necessities. Likewise, they must spend a portion of their "time" to develop a good professional advertising program and a decent showroom in order to make customers feel confident about leaving their prize trophies. Finally, it almost goes without saying (but for many it needs to be said anyway), attention "must" be given to your own "personal" appearance as well as to the neatness and cleanliness of the studio surroundings.

An attractive building not only attracts potential customers, but also projects your company image giving clients the confidence to entrust you with their hard-earned throphies. The shop above, owned by World Champion Bob Elzner, is a perfect example.

Advertising

The "heart" of any business is its advertising program. No living being can exist without its heart nor can a business. If your business is to remain healthy and thrive, you must take care of your advertising program. To begin an advertising program, you first need to select your own personalized "logo" or trademark. This will identify "you" and your company to potential customers. Put some thought into your logo; it is very important. It should appear on everything that pertains to your business. Having a professionally designed logo for your business is an investment well worth the money spent. A local commercial artist or advertising agency should be able to design a professional quality personal logo for your business that is completely "camera ready" for a price of $300 to $800. Sound like a lot of money? Maybe; but it may mean the difference in a customer choosing your ad out of the phone book over someone else's. A professional quality, "eye-catching," logo designed into your building sign, business cards, price lists, yellow page ad, etc. will attract many a passing sportsman's attention. These "attention getters" recruit life-long customers. One good customer will pay for a professional quality logo design many times over.

Personalized price lists, business cards, letterheads, and envelopes are all very important in creating a first class, professional and "lasting" impression. Consider your own mail for instance. When you are going through a stack of envelopes, which do you read carefully, and which ones go into the

wastebasket? Chances are good that the mail that is dull and unimpressive with plain type will go straight into "file thirteen." Well designed, professional looking literature will deserve a closer look.

How do you go about buying this literature? One way is through WASCO. An entire line of custom printed, professional quality price lists, letterheads, envelopes, business cards, and hunting/fishing license "water resistant" envelopes have been especially designed for the wildlife artist. WASCO has a selection of beige, linen, and white bond stock that can be printed with your own custom logo or one of WASCO's standard logo designs. All are reasonably priced and guaranteed to help you project a professional image for your studio. This eliminates the need and expense of dealing with commercial artists, typesetters, and printers. If you already have easy access to such professionals then by all means take advantage of them. (Perhaps you can even swap out work with them.)

The "important" thing is that you do it!

Business cards are probably the best and least expensive form of advertising for a wildlife artist. They are inexpensive enough to put out by the thousands. You "need" advertising like this. Once you obtain quality business cards you should immediately put forth some effort into getting them into your customer's hands. Consider where the bulk of your potential clientele congregate: bait shops, marinas, sporting good stores, barber shops, etc. You should display your cards at each of these sportsman stops.

A simple way to project a professional appearance, which will help convince store owners to allow you to leave your cards in their place of business, is to put them in an attractive business card display holder. This display is always preferred by customers and store owners over business cards stacked in disarray on their countertops.

Another item that can be displayed in these same holders are hunting and fishing license water resistant envelopes. These can be left at any location that sells hunting and fishing licenses. Usually store owners welcome taxidermist placing these handy envelopes in their shop as they can "give" them away whenever they sell a license. It is good advertisement for you because the moment the hunter or fisherman pulls out their license or big game tags to properly tag their trophy, your studio name will be "jumping out" in front of them.

A successful advertising program also encompasses the advertising aides to further the recognition of your business. Business cards, license holders, etc., are obviously essential to this phase of marketing your business; however, a highly visible building sign, displaying your logo is another very good way to attract walk-in customers. If your sign is poorly placed or not large enough to be read from the road, it is simply a waste of hard-earned money. Always incorporate your logo and business name into such a sign. You want the customer to readily identify "you" with your logo, just as the golden arches are considered synonymous with McDonald's. Put a sign on your truck, your bass boat, your friends truck, or anywhere else that it can be seen. Design it the same as the sign on your shop. Let it be seen "anywhere and everywhere" that

A professionally painted sign projects a quality image.

sportsmen are likely to congregate. Don't have an amatuerish hand scribbled sign. It projects a poor image of you. Even if you must have a small sign, have it professionally painted.

While we are on the subject of signs, did you ever consider the possibilities of incorporating a walking sign into your bag of advertising tricks? Consider custom printed T-shirts, hats, and aprons with your company name on them for additional exposure. Put a few of these on several of your best customers and watch the results and profits roll in. Name recognition! It is important. Put hats and T-shirts in your shop for sale. Sell them to marginal customers and give them away to good customers or customers that will wear them and provide advertising for you. The customer will sure appreciate a free gift and its walking advertising for you. Remember you need to spend some money on advertising.

Being a taxidermist, I'm sure that you have had a customer who did not know anything about field care. As you know, there are all too many sportsmen driving all over town with their trophy deer getting ruined on the hood of their trucks or wringing the neck of every duck they bring to you. Consider the value of teaching basic field care techniques to your customers. It will make your job much easier, and you can advertise at the same time. Basic field care tips printed on one side of a hunting and fishing license envelope with your company name printed on the other side is just the ticket. Don't forget that along with the field care tips is your company's name telling them where to take their trophy after they properly tag and administer field care techniques. You should always give a free hunting and fishing license envelope to your best customers and definitely staple a business card to the backboard of every mount. These cards and envelopes will keep on advertising for you for years to come.

Key points to remember: Your name and company reputation are your most valuable assets, and you should do everything within your power to protect and promote it. Why not visit a few shops in your area and see if your business rivals have a competitive edge. Notice everything about their business in comparison to yours (especially their advertising strategy) and objectively ask yourself the question, "Who, as an uneducated customer, would you rather have mount your trophy?" Always keep in mind: Without customers (old and new) any business is out of business.

When a customer enters your shop, they should get the impression that you are the most innovative taxidermist in the area. Hang completed customer mounts in your display area until they are picked up. This allows potential clients to see the high quality of the mounts and feel confident about handing their hard-earned "once-in-a-lifetime" trophies over to you. Change your showroom around occasionally and display the latest poses either with mounts or photos of mounts. Make it interesting for even frequent visitors. Take advantage of fads and innovative ways of presenting mounts. Take advantage of impulse buying by the customer. Have items on hand for sale to walk-in customers. Here are three excellent examples:

Chicken in a Basket

Why should a successful wildlife taxidermist even consider mounting a domestic bird such as a chicken? Does $120 to $155 profit for about three hours work sound like a pretty good reason? Your showroom, local art and craft shows, and retail stores can become the market for some potential big, easy bucks selling "chickens in a basket."

In this article, World Champion Frank Newmyer demonstrates his step-by-step techniques for mounting a chicken. Frank used an artificial head and Accu-Flex chicken mannikin to mount the Rhode Island Red chicken pictured above. He covers "attitude" of the hen, placement of the cloth, placement of the chicken, and attaching the bird to the basket.

Listed below are the tools and supplies used for this project:

Liner cloth
Basket
Chicken (obtained live from a local chicken farm)
FN25A Accu-flex chicken mannikin (includes neck)
FNB220 artificial chicken head
Plastic syringe w/needle
Reference photos
Tohickon BZ10 glass eyes (pheasant eye is almost identical to the chicken eye)
Heat gun/hair dryer
Polytranspar Airbrush Paint
Drill
Hot-melt glue gun
Instant Bonding Glue
WASCO Clay

Sculpall	Scalpel
Annealed wire	Perfect Knife
Wire cutters	Ultimate Scissors

The most noticeable difference between a chicken and most other commonly mounted birds is the head. This difference becomes very obvious during the skinning process.

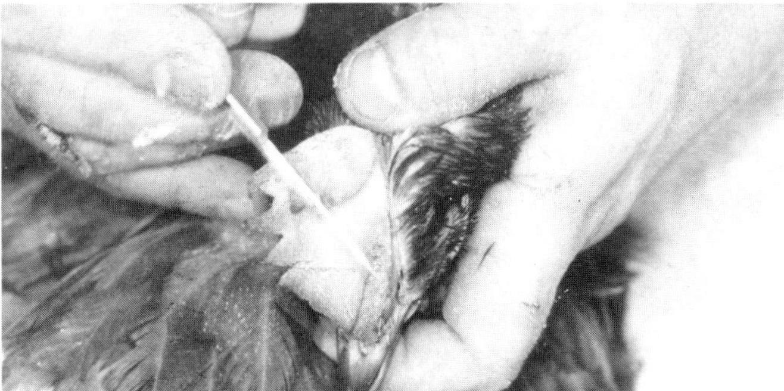

by making a smooth, even incision as close to these areas as possible.

It is a good idea to use the artificial head (FNB220) as reference for making these incisions.

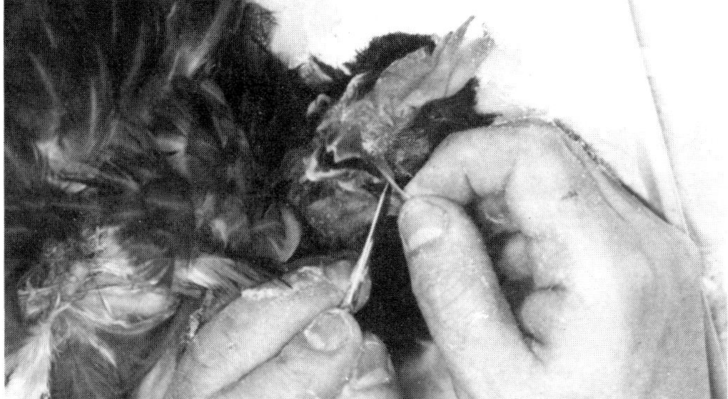

First, the skin is removed from the beak, wattles, and comb

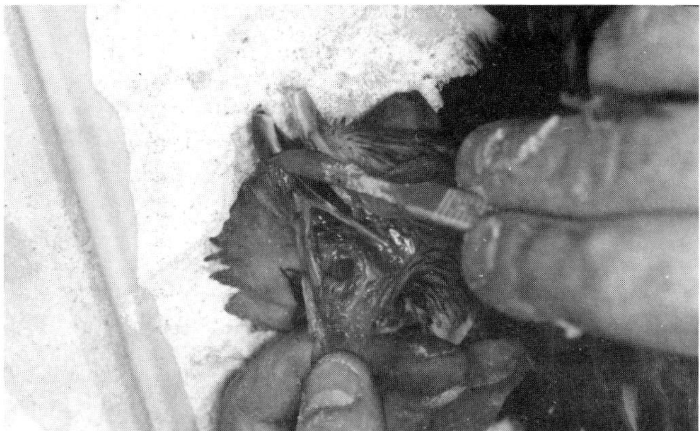

Remove the skin from the front of the head, rolling it back

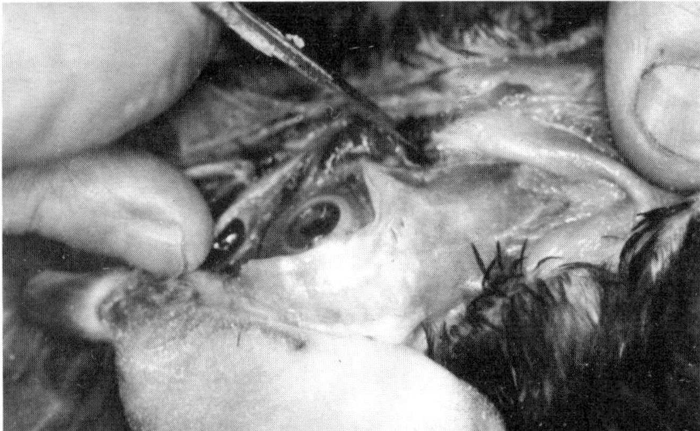

as the incisions around the eyes and ear canals are made,

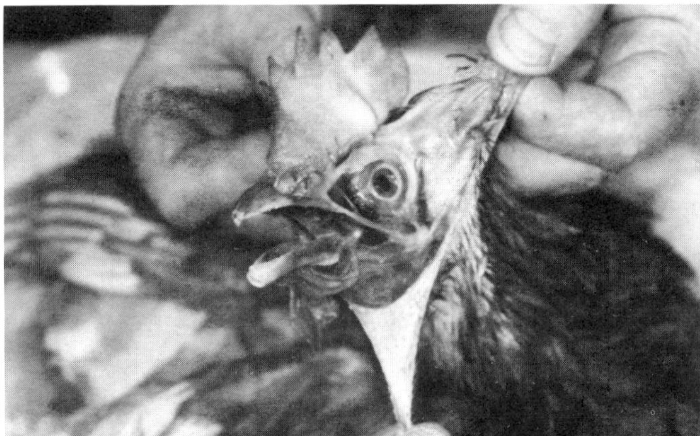

stopping at the base of the skull.

At this point, the bird is skinned the rest of the way in the conventional manner using the ventral incision. The feet of the chicken are totally concealed on the finished mount and do not require much attention. They can either be removed entirely during the skinning procedure or conventionally wired, preserved and mounted in position under the bird.

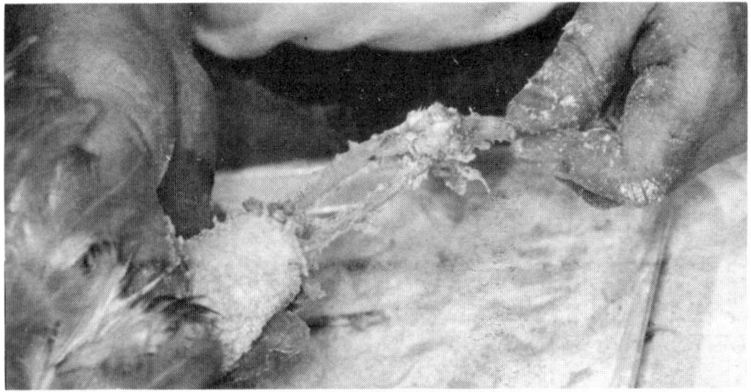

Both wings are skinned and inverted and all of the tissue is removed leaving all of the wing bones (ulna, radius, and humerus) intact. Once skinned, the bird is fleshed, degreased, and preserved with WASCO dry preservative or tanned with the *True-Tan* Bird Tan System.

This Rhode Island Red was then mounted on a Sportsman Series FN25A Accu-Flex mannikin with a wire inserted vertically through the back and out of the chest of the body to 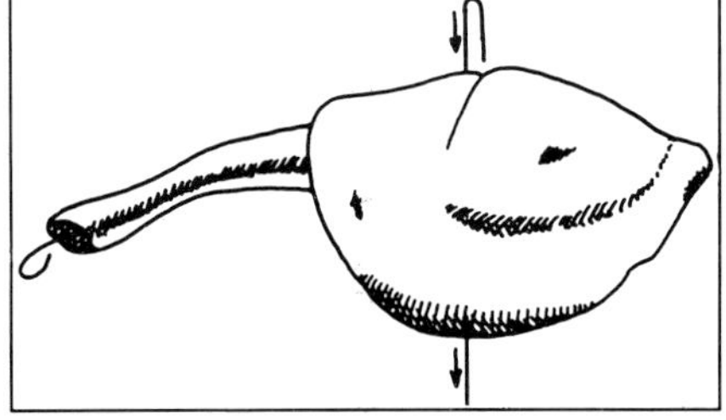 hold the completed chicken securely in the basket. Frank then wrapped the humerus with cotton batting and cotton thread (the wing was not wired or attached at this point).

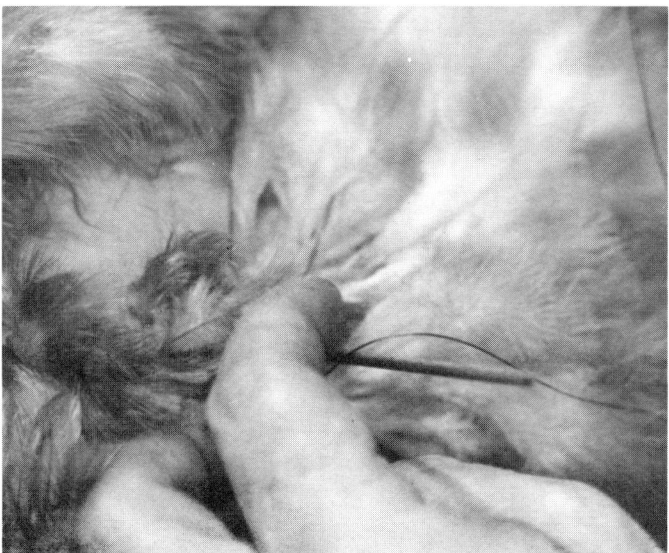

Next, the ventral incision is closed. This can be done by either sewing, pinning or stapling since the incision will be hidden in the basket.

Before attaching the head, Frank works out a few fine details. The comb and wattles on the artificial head are easily manipulated and positioned once they are heated.

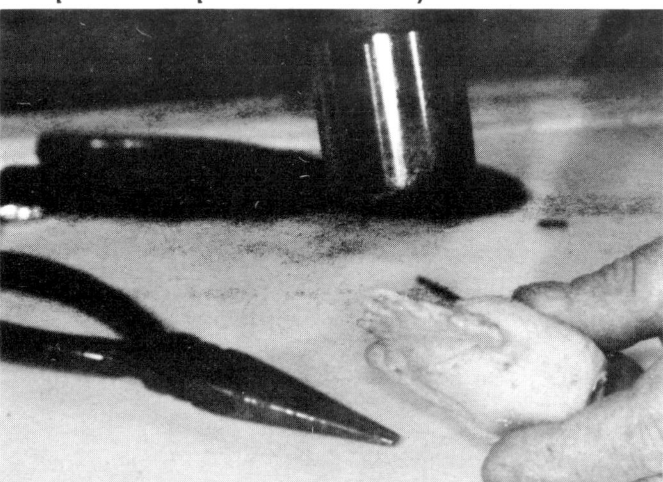

A hot air gun (paint stripping gun) works best for heating the artificial comb and wattles, due to the greater amount of heat that this tool generates. (A hair dryer or microwave oven, although less effective, can also be used if a hot air gun is not available.)

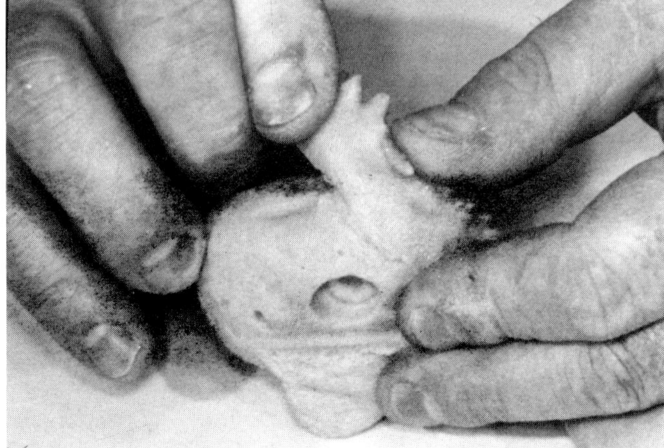

Once heated, the comb is bent into the desired position and held in place until set. Each wattle is adjusted in the same manner.

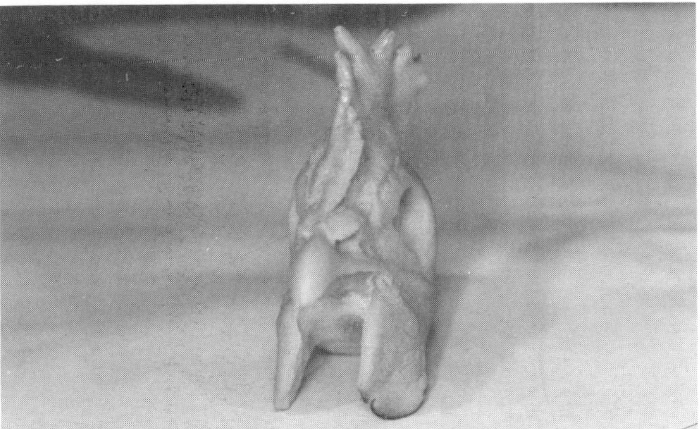

By doing this, Frank gives each individual bird its own distinctive look and attitude.

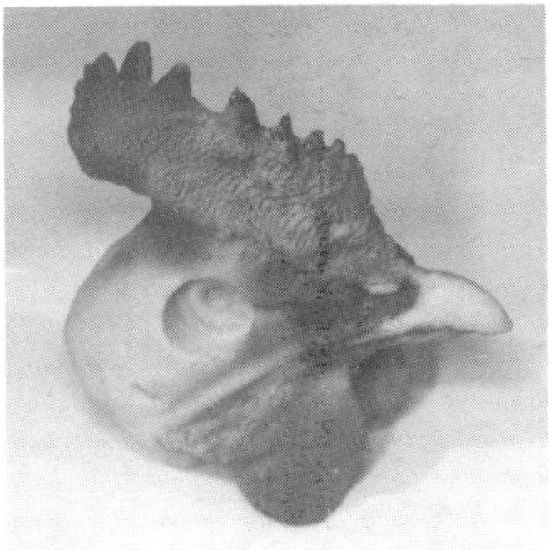 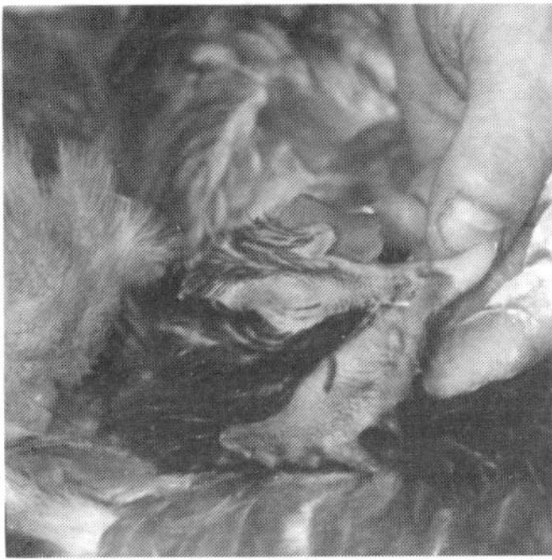

When the comb and wattles are in position, they are painted a pinkish-red to match the facial skin as closely as possible. A mixture of *Polytranspar* Superhide White (FP10), Teal Yellow (FP142), and Gill Red (FP160) is used as a a base coat (the final touch-up is done after the mounting is completed).

The skin from the face and head is then pushed down on the neck as far as possible to allow the artificial head to be glued into position.

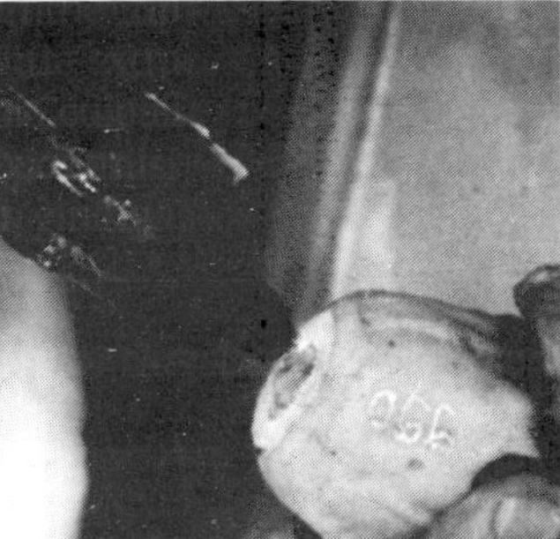 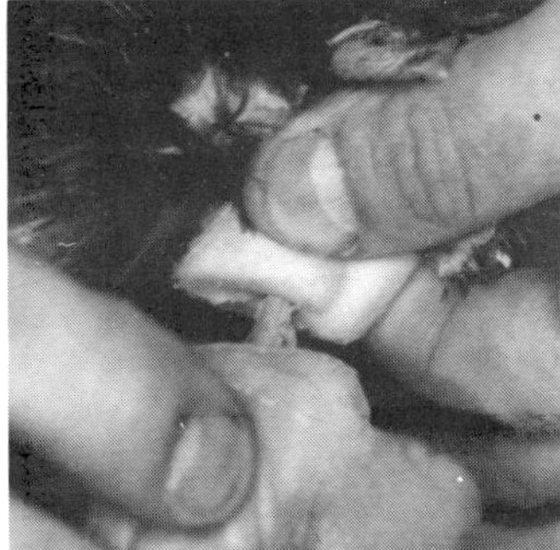

A hole is drilled in the back of the head to accept the neck wire from the Accu-Flex mannikin. Once this hole is drilled, it is checked for the proper fit.

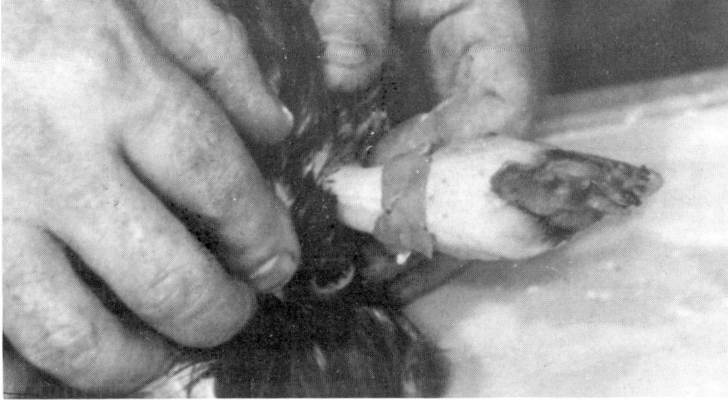

Next, the neck wire is cut and the end is formed into a loop. The neck and head are then hot-melt glued into place.

Using WASCO clay, the eye sockets are filled along with the area where the head joins the neck.

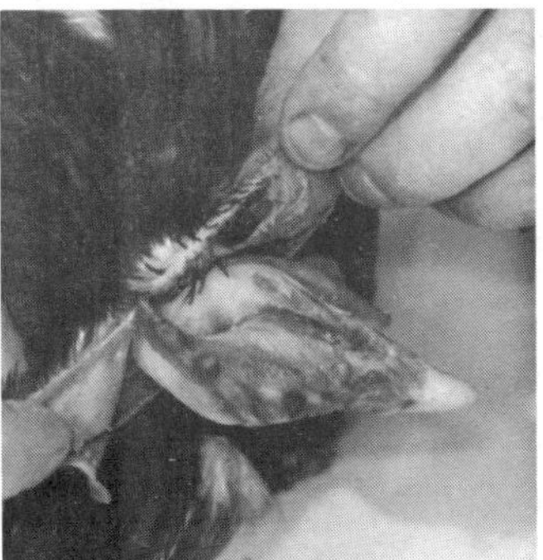 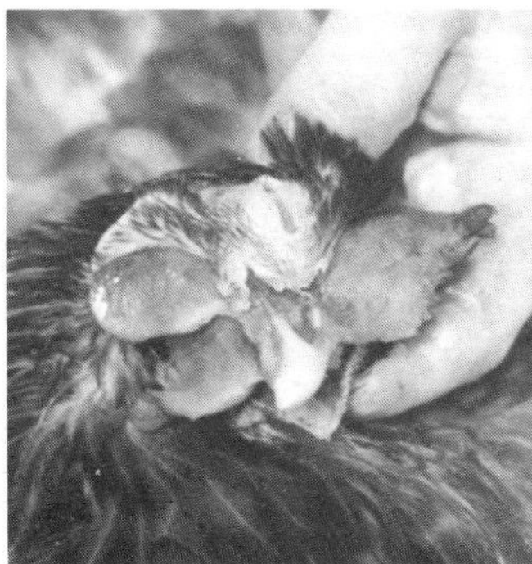

The next step is to carefully pull the skin up over the head and "taxi" it into position.

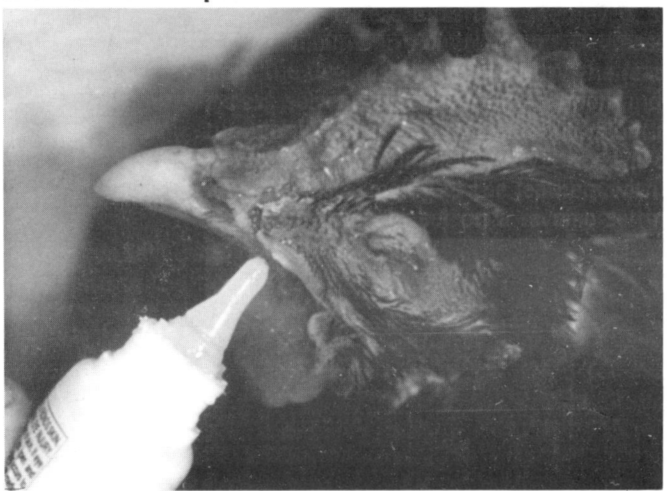

At this point, Instant Bonding Glue is used to attach the delicate head skin to the wattles, beak, and comb. For a convincing "true-to-life" looking mount, be sure to get a good union and tight fit in these critical areas. For those areas on the top of the beak, and in the corner of the mouth where a little extra fullness may be required, a small cigar-shaped piece of Sculpall may be rolled and placed in them. This allows the taxidermist the opportunity to sculpt in fine "detail" even after gluing.

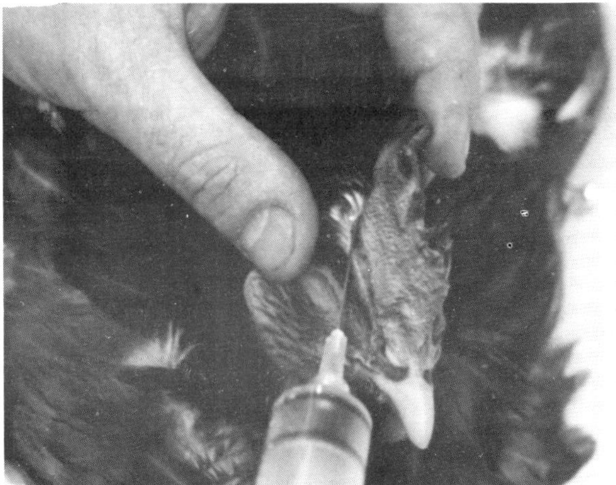

When gluing areas that require less glue than the tip of the bottle will allow, a plastic hypodermic syringe and needle is used to place "minute" amounts of glue exactly in these tight spots. (The syringe and needle should be cleaned immediately after use with acetone or fingernail polish remover).

At this point, the entire facial area is securely in position. Carefully check over the head area for any low spots that might need to be built up. If any of these areas exist, insert cotton or Sculpall through the eye opening and work it into position.

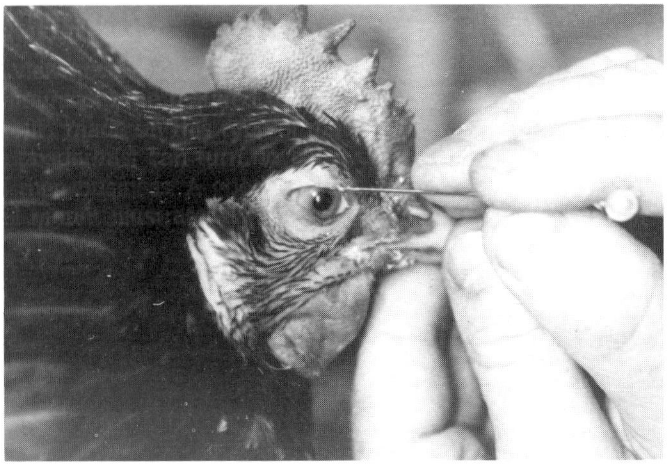

To set the eyes, place a small amount of Sculpall beneath the skin, above and below the eye socket, and insert the glass eye while holding the skin away from the head with a pin. Correctly position the eyes (use reference), and tuck the lids into position by manipulating the Sculpall underneath.

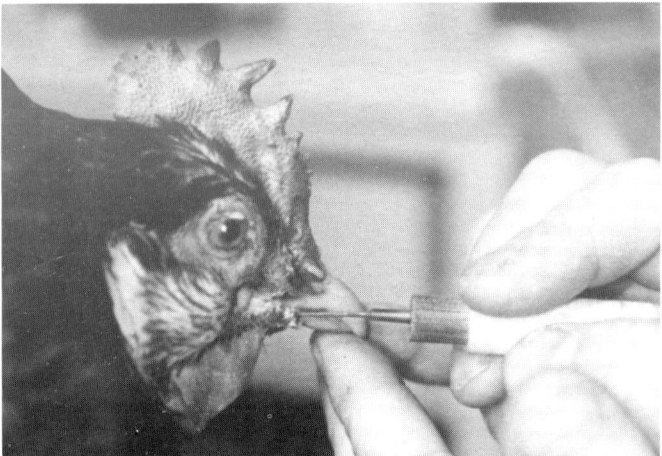

If using Sculpall, any touch-up work needed on the head and face should be taken care of at this time, because it will "set" fairly rapidly.

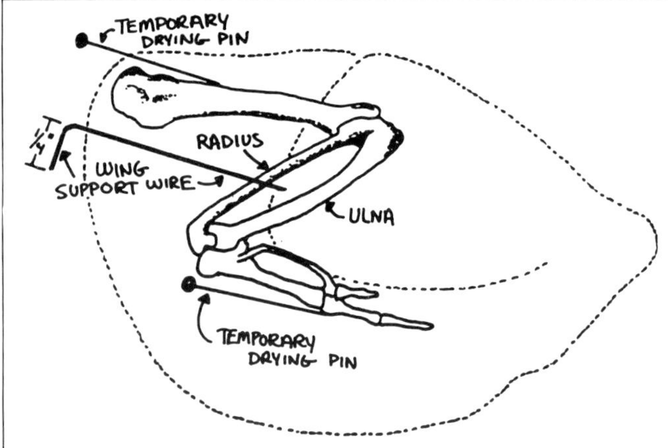

The bird is now ready for posing and positioning. Place the attachment wire in a vice or similar instrument with the bird in a natural upright position. Adjust the wings into the desired position. Next, using a 14 gauge pinning wire, with 1/4" at the end of the wire bent into a 90 degree angle, insert the wire between the ulna and radius, and into the foam of the Accu-Flex mannikin (see illustration). With this method, only a couple of extra pins will be needed to hold the primaries in position until dry. The 14 gauge wire will be hidden by feathers and left in the bird. With the wings now in position, arrange and straighten the feather groups. Using reference photos, try the head in different positions. Once satisfied with the "attitude" of the bird, it is ready to be placed in the basket.

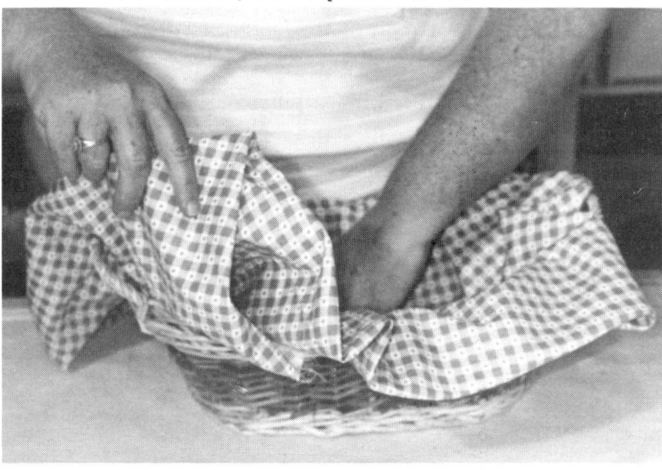

For this Rhode Island Red, a blue "country print" cloth is placed inside the basket. (A cloth material is probably the best filler for the basket because natural fillers such as hay or straw are very messy and may contain insects.)

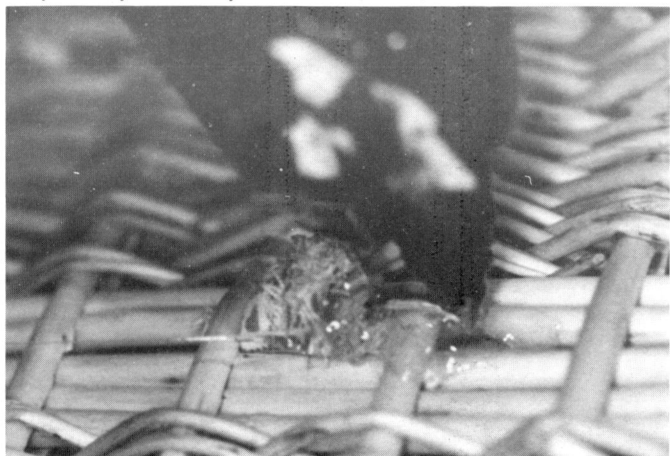

When the arrangement of the cloth is complete, place the chicken in the basket by inserting the sharpened wire through the cloth and the bottom of the basket. Bend the wire flush with the underside of the basket and cut off any excess. Apply an ample portion of hot-melt glue to the wire and bottom of the basket and cover with a small scrap of cloth or felt to prevent the wire from scratching table tops, countertops, etc.

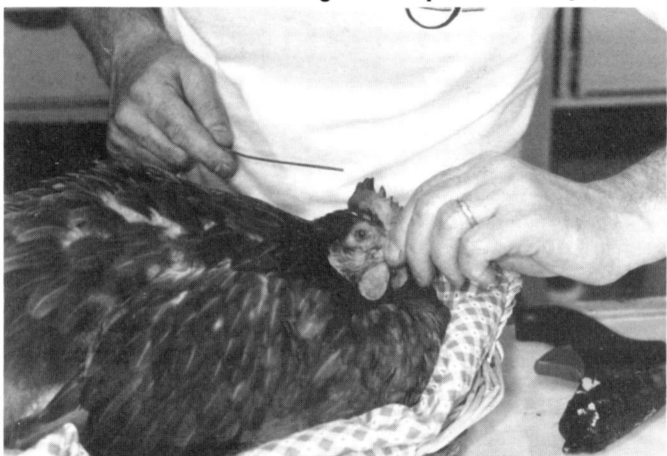

With the chicken now resting comfortably in her basket, final adjustments are made to "bring her to life." Check the head positioning once again (the way the chicken "settled" into the basket might justify altering the original pose).

The Breakthrough Bird Taxidermy Manual

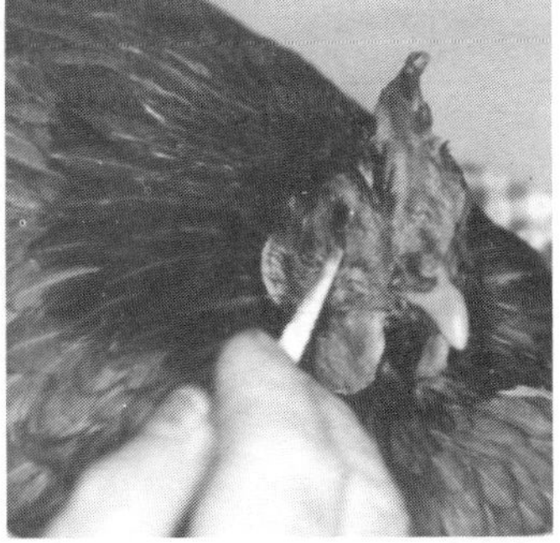

Once positioned, "touch-up" the painted areas by blending the color of the wattles and comb into the skin.

Clean the facial hairs and feathers with lacquer thinner and touch-up with *Polytranspar* Chocolate Brown (FP70) and a small artists' brush.

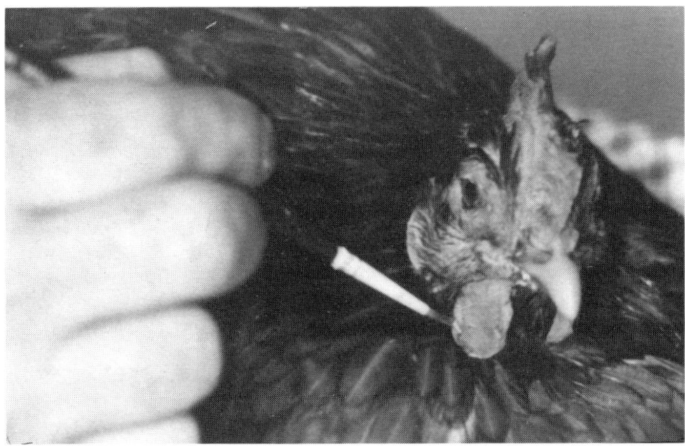

Lightly brush the comb and wattles with the brown paint, touching only the raised detail to add depth to these areas.

For a final "touch," a matching cloth bow can be added to top off this very eye-pleasing creation. This piece would appeal to practically anyone with a taste for country living.

This entire rendering cost approximately $20 to produce and took about 3 hours to complete. These creations are currently selling for anywhere from $75 to $250 and the people capitalizing on this idea at the moment can not keep enough made.

Why not take it a step further? Scour old barn yards, keeping in mind that a chicken will nest almost anywhere; look for milk cans, butter churn bottoms, old crates and other such artifacts. They can easily be located and increase your profit margin even more. Even the most successful taxidermists should be able to squeeze a few "high profit" chickens into their busy schedule every once in a while. As a side benefit if you do enough of them, you can regularly treat your family to some fine eating, too!

Turkey Tail Mount

Many taxidermists view novelty mounts as a real "pain" to do because of the amount of time required to complete one versus the actual "profit" made on this type of mount. With a little ingenuity, the taxidermist can turn novelty work into a profitable phase of their business. A good example of this will be the turkey tail mount illustrated in this section.

To complete this mount all that will be needed is a WASCO antler mounting kit, hot-melt glue and glue gun, Ultra Lite Filler, Skin Prep and *Polytranspar*™ Degreaser.

First, remove all fat and meat from the base of the tail and between the tail quills using both the wire wheel and Ultimate Scissors.

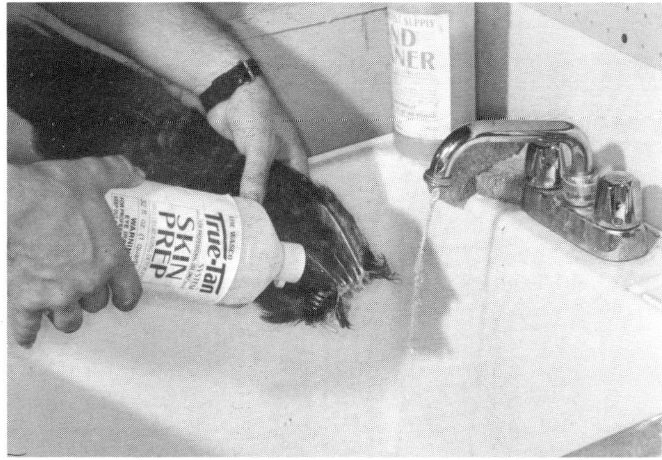

The defatted tail should then be thoroughly washed in Skin Prep and patted dry with a towel.

Once the excess moisture is removed, soak the turkey tail for 20 minutes in *Polytranspar*™ Degreaser. Remove the tail from the degreaser and towel dry the feathers to absorb the excess solvent. Fan the feathers open and place the tail in a tumbler with Fur Dresser's Sawdust. Continue to tumble until the feathers are completely dry. Once the tail is dry, blow out any sawdust clinging to the feathers with an air compressor or vacuum cleaner set on a reverse setting.

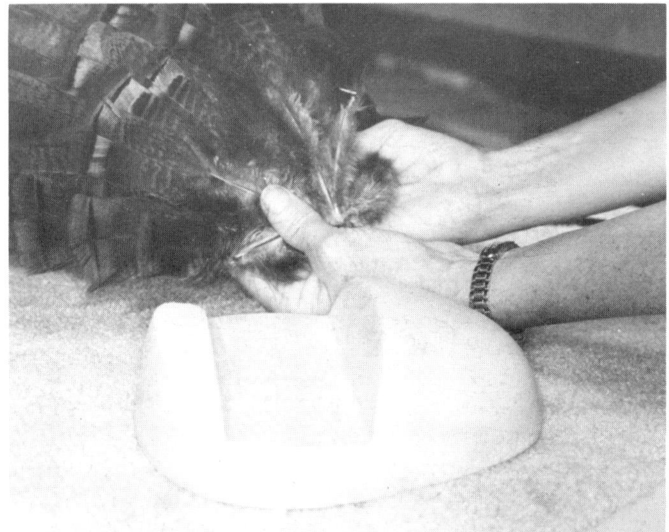

Fan the tail open and test fit it into the slot on the antler mannikin.

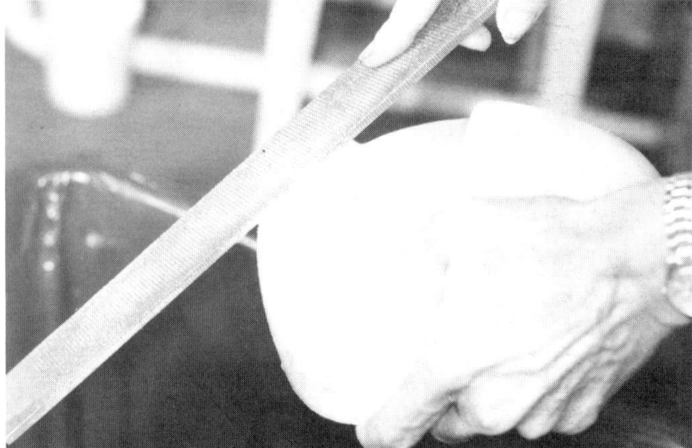

Depending on the size of the "fan," the back portion of the antler form may need to be rasped down. It is important when rasping down the form that the curve and shape of the form is maintained. The shape of the form will give the turkey fan a realistic look of a bird actually fanning his tail rather than the look of the standard "flat" turkey tail mount.

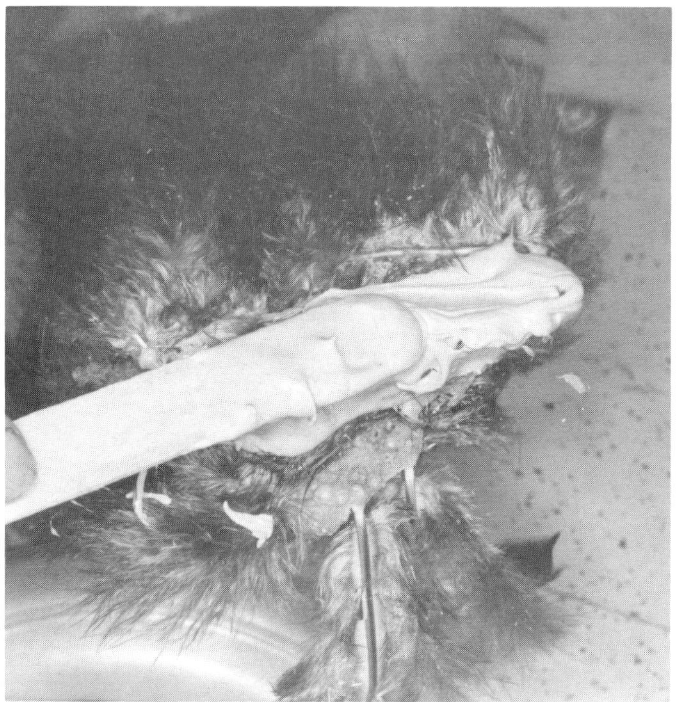

When satisfied with the fit, mix a "medium set" batch of Ultra Lite Filler and place the filler on the base of the tail quills.

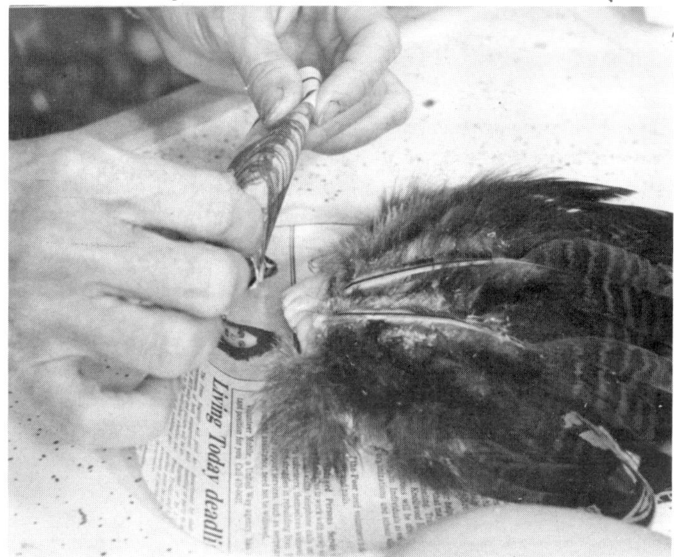

To avoid getting filler on the tail feathers, sandwich the base of the tail between newspaper.

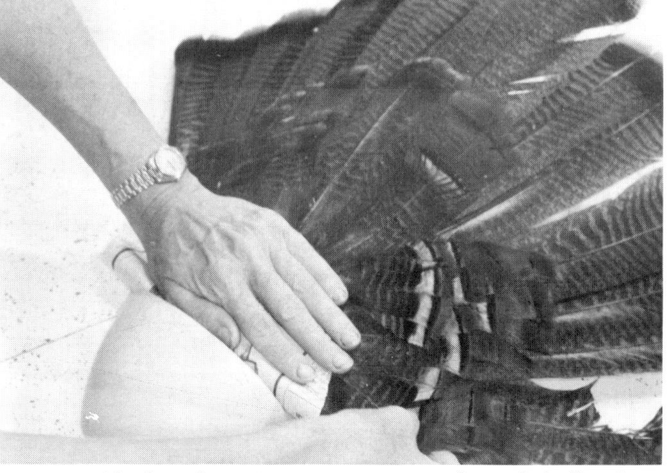

Now, with the Ultra Lite in position on the tail quills, place the turkey tail into the slot on the antler mannikin.

The Breakthrough Bird Taxidermy Manual

When satisfied with the fit, fan the tail open and pin the feathers into place. Use Jumbo Head Pins to hold the tail feathers in the proper position.

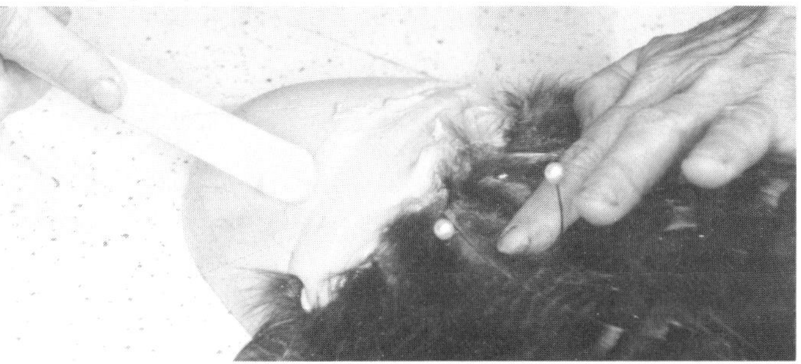

Then mix a "hot" batch of Ultra Lite and fill the antler slot. This will hold the tail in position. Try to maintain the shape and curve of the mannikin when applying the Ultra Lite filler to the slot in the antler form.

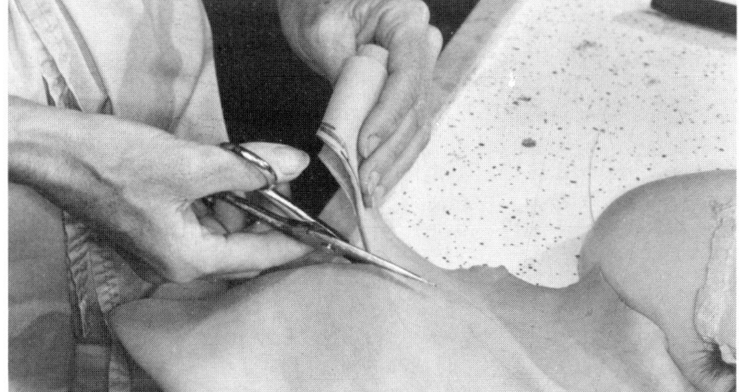

Once the filler has set, carefully measure and cut the appropriate amount of buckskin needed to cover the base of the tail and the antler mannikin.

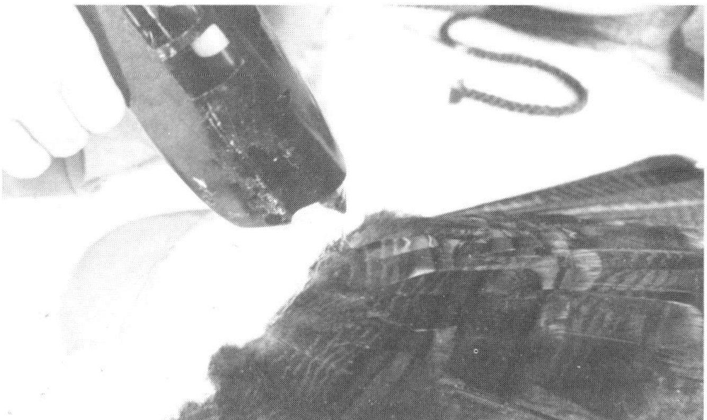

Before attaching the buckskin, hot glue cotton batting over the Ultra Lite Filler. This batting gives a smooth appearance

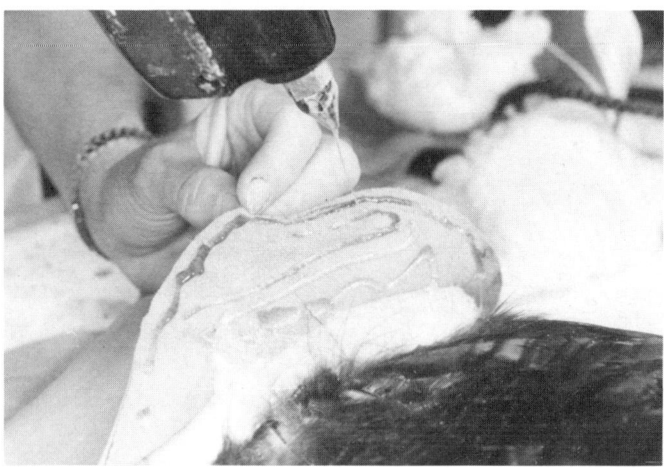

to the buckskin when it is placed over the form. When satisfied with the fit of the buckskin, apply hot-melt glue to the underside of the buckskin and firmly press the buckskin down on the base of the turkey tail.

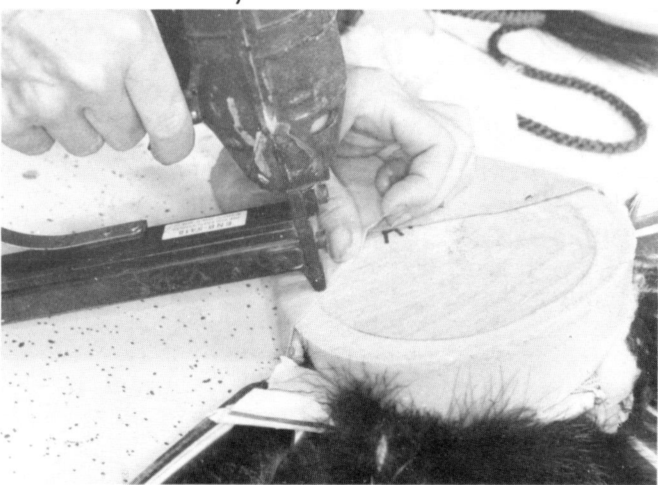

When the hot glue has set, staple the remaining loose buckskin to the back of the antler form. Remember to adjust and pull out any wrinkles in the buckskin as you staple around the form. When the buckskin is secured to the antler form, test fit the decorative braiding (included in the antler mount kit) and cut the braid to the appropriate length. Remember when cutting the braided cord to tape both ends to prevent the braid from unraveling when cut.

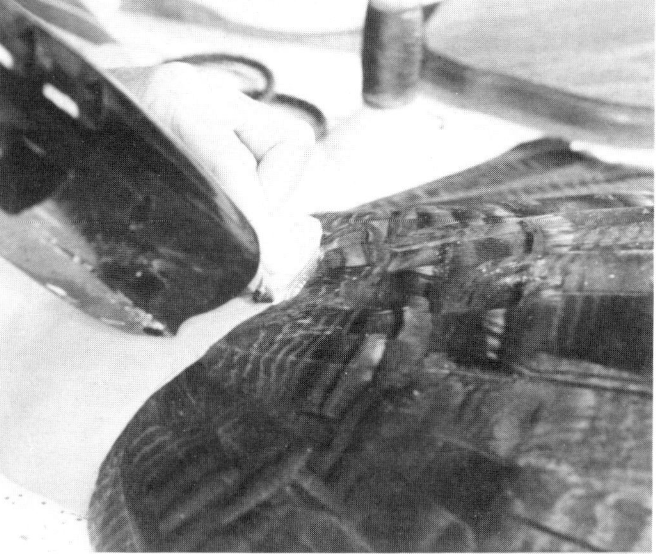

Run a bead of hot glue from one side of the tail, along the top edge of the buckskin, over to the opposite side of the tail.

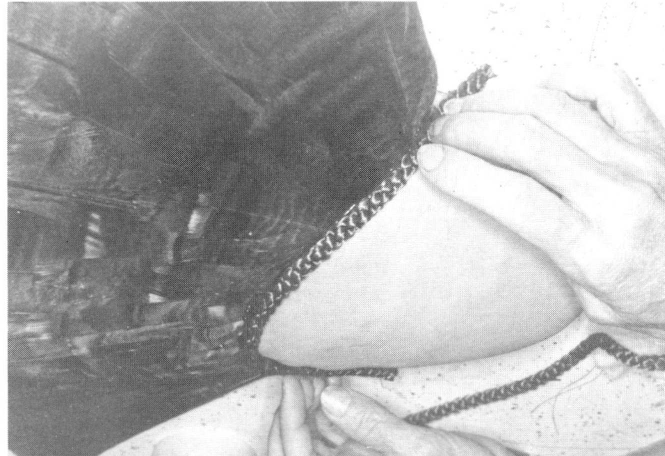

Quickly place the decorative braid on the bead of hot glue. Be very careful when placing the braid on the glue. I can personally attest to the fact that it is indeed "hot" melt glue.

When the glue and braid have set, attach the tail mount to the panel.

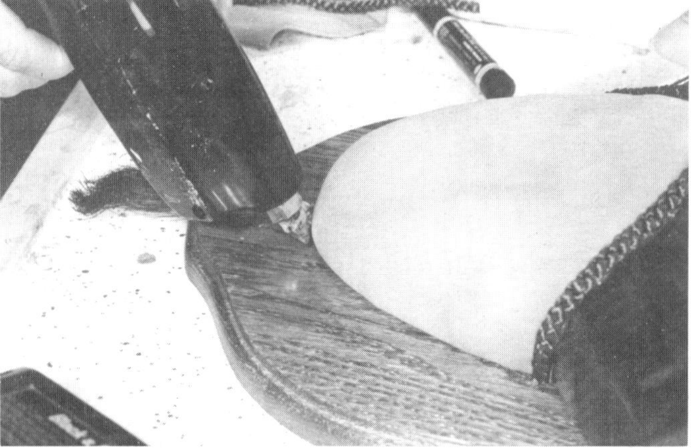

Position the turkey beard at the base of the covered antler form and hot glue the beard into place on the panel.

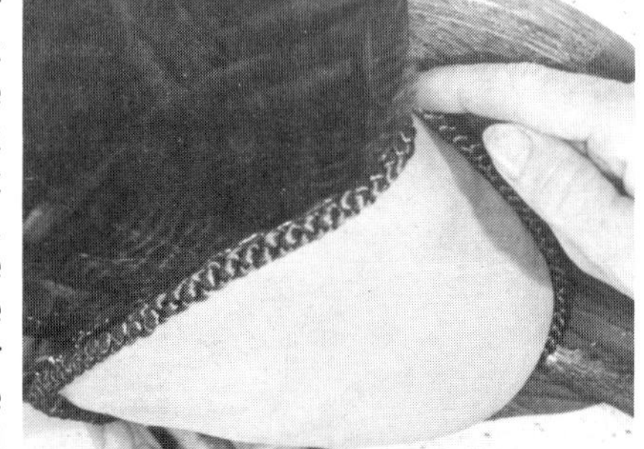

Test fit another section of braid to fit along the base of the antler form. Run another bead of hot glue on the panel from under one side of the tail feathers, along the bottom of the antler form to the opposite side under the tail feathers. Quickly set the braid in its proper position.

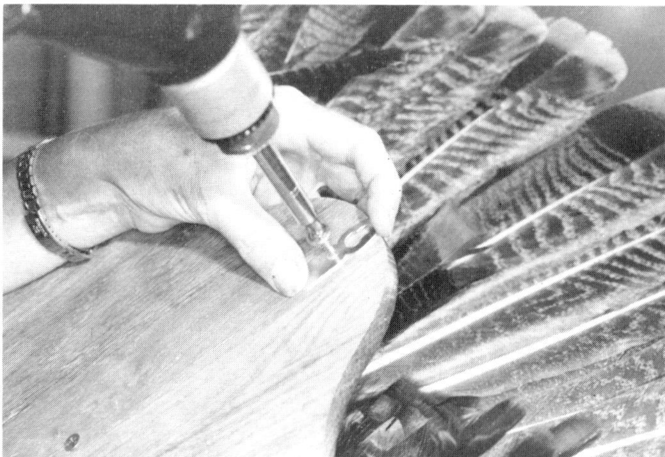

When the last application of hot-melt glue has set, the turkey tail mount is ready for a hanger, a final "once over" with a feather duster, and *one satisfied customer.*

Executive Cases

One of the best methods of doubling your profits on "bird work" is to incorporate enclosed habitat bases with the mount. Habitat bases are relatively inexpensive to produce in relation to the amount of money that they can generate. A customer will usually prefer their mount in a natural scene (unless they are concerned about spending a little extra money) and will, as a rule, decide to have more than one bird done in this style once they discover how much an executive case adds to the decor of their home or office.

Building habitat scenes is very simple, once the basic information has been obtained and tested from sources such as "The Breakthrough Habitat and Exhibit Manual" or *Breakthrough* Videotapes. It requires very little experimenting to get your "feet wet" with the newly-learned procedures, and habitat building will really become a "snap" to master.

One selling point that should be stressed to a potential customer is that glass-covered executive cases are much easier to maintain than mounts which are "exposed to the elements." A simple wipe down with glass cleaner and a paper towel will keep the mount and habitat looking as good as it did the day it was constructed. Another advantage to a covered case is the extra protection provided by the glass or plexiglass from house pets and insects. The family cat or dog can do some "serious" damage to a valuable bird mount.

In addition to the attractive executive cases, many taxidermists are placing mounts and habitat scenes into furniture such as coffee tables, end tables, and lamps. One of the most unique examples of creative habitats that I have seen recently was a habitat case made within a gun cabinet. An executive case or photograph of a beautiful habitat scene in a piece of furniture on display in your shop will definitely convince the avid sportsman that they "have" to have one for their home or office.

Why limit a glass display case to the table or floor? Incorporating a mount and habitat to hang on the wall would truly be a unique idea that would stimulate extra business. The case would have to be well constructed, but it would save on table or floor space.

Another simple idea that adds that "extra" touch of class to an executive case is a customized name plate with your company name engraved upon it, along with the date the case was completed. The customer's name can also be added at his request. A new method now being used for personalizing cases is acid etching the glass. This has become very popular among taxidermists and creates a very pleasant image without overpowering the mount and habitat.

A perfect example of this is the case constructed by Ralph Lehrman illustrated in the photo above. Ralph donated this particular case to Ducks Unlimited for their annual fund raising auction and it is obvious that the acid etched DU logo added a very nice touch to the composition of the piece.

All of these ideas are potential moneymakers and should be taken into consideration by the aggressive taxidermist. Familiarize your clientele (especially avid bird hunting customers) with the concept of executive cases by placing a few in your showroom or having a few good photographs on hand. These high profit cases will not only mean satisfied customers but extra revenue for your business.

All of these ideas are means of marketing the skills you possess as a taxidermist, and should be taken into consideration. The smart taxidermist will advertise to attract customers, do excellent work to satisfy them, and produce innovative and creative mounts to keep them interested. If these simple guidelines are followed, good profits, customer loyalty, and satisfaction is assured!

CHAPTER 15

From The Field Into The Frying Pan

Waterfowl and upland game, like venison, require proper handling in the field if the meat is to be not only palatable but fit for human consumption.

There is no marinade in the world that can mask the taste of game that has not been eviscerated (field dressed) shortly after the kill. No other step in the entire process "from the field to the frying pan" is as important as field dressing. As soon as it is feasibly possible, birds should be field dressed and placed in a spot to adequately cool before they are placed in a container or game bag for transporting home. If birds have been "shot-up" and are not suitable for roasting whole, simply breast out the bird and allow the individual breasts to cool. However, be sure to check hunting regulations in your area before removing the breast and disposing of the carcass as federal agents sometimes frown on this practice.

After returning home, attention should be given to plucking the birds that will be roasted or grilled and breasting any birds that are not suitable for roasting.

If you don't have access to a plucking machine or to someone who *really* enjoys cleaning birds, the best method of removing feathers is to boil a big pot of water (enough to completely submerge a whole duck) and add a pound or more of paraffin wax to the boiling water. When the wax melts, simply submerge the bird in the wax and water for moment or two and then remove the bird. Once the wax coagulates on the feathers, start removing the feathers and down. Stubborn ducks may require more than one dunking.

Once all the feathers and down have been removed, check the chest cavity to make sure all organs and fascia are removed. Then allow the birds to soak in cold water for approximately one hour. This removes blood from the meat which is primarily responsible for the "gamy" taste that is often associated with wild game.

To prepare birds for the freezer, wash the bird again in cool water, again checking for shot pellets and pin feathers. Place the clean bird in a heavy duty Ziplock freezer bag, cover with water and seal. Birds prepared this way will keep easily "till next season." Be sure to mark the date and type of bird on the bag as this will eliminate the problem of serving "mystery meat" for dinner.

It is best to bag and freeze whole birds individually; however, several breasts will keep nicely when frozen in a heavy duty Ziplock bag.

To thaw frozen birds for cooking, simply run cold water over the bag to remove the bird, then place the whole birds or breasts in a container of water to soak for several hours. It is best to give the bird a quick once over to make sure all down, pin feathers and shot pellets have been removed.

A bird is now ready for one of thousands of delicious recipes such as sauted doves with cherry sauce, turkey breast marsala, roast duck, duck or turkey gumbo, duck a la orange, duck or pheasant stuffed with wild rice, turkey au povire, smoked turkey, sauted quail breast in white wine sauce, southern fried quail, or medallion of turkey breast with sauce bearnaise.

As you can see, just as with venison, if the bird is handled properly, it may be substituted in a number of recipes for beef or chicken.

Duck and dove will have a more distinctive flavor and cannot be substituted as easily as quail, woodcock, turkey or pheasant which have a more delicate flavor.

Ducks, probably more than any other bird, tend to vary in taste even among the same specie. This is mainly due to the type of food the bird had been consuming on the fall migration. In some localities hunters refuse to shoot scaup because of their poor table fare. Believe me, when scaup have been feeding on fish, I don't even like to skin them for mounting; *but*, if scaup have been feeding on grain, they are really a delicacy.

The following are some of my favorite recipes for ducks and upland game birds:

Duck Breast in Puff Pastry

Grand Mariner Marinade	Lawry's Garlic Salt	Duck breasts
Adolf's Meat Tenderizer	Butter, Flour	Beef Bouillon
Lawry's Seasoned Pepper	Orange Marmalade	Puff Pastry

Split duck breasts and soak in water for one hour. Remove from water, pat dry with a paper towel, and place breasts in a shallow glass dish. Cover with Grand Mariner Marinade for 3 hours.

Remove the breasts from the marinade and sprinkle with Adolf's Unseasoned Meat Tenderizer. Lightly pound the split breasts with a meat hammer. Season to taste with garlic salt and seasoned pepper. Saute the breasts slightly in butter. Save this pan with the drippings.

Follow the directions on the container for rolling out the puff pastry. Cut squares of pastry to completely surround each individual duck breast. Lay each breast on a piece of cut pastry. Place a teaspoon of orange marmalade on top of each breast. Cover the breast with pastry and press edges of the pastry tightly closed. Bake at 350 degrees till golden brown.

Sauce—Take the pan with the drippings and add one tablespoon of butter. Melt the drippings and add butter over a medium heat. Add 2 tablespoons of flour, stir until the mixture turns light brown. Slowly add approximately 3/4 cup of bouillon to the mixture and continue to stir. Add marinade to taste and orange marmalade to taste. Continue to wisk sauce, and add garlic salt and seasoned pepper to taste. Serve with wild rice.

Wild Turkey Breast with Red Eye Gravy

1/2 Turkey Breast
1/2 Cup Butter
1/2 Cup Minced Onions
1/2 Cup Coffee
Lawry's Garlic Salt
Lawry's Seasoned Pepper
1 Bouillon Cube
Adolf's Unseasoned Meat Tenderizer
Flour
Grits

Cut the turkey breast across the grain in 1/4" slices. Sprinkle sliced turkey breasts with meat tenderizer and pound with a meat hammer. Season the breasts to taste with garlic salt and seasoned pepper. Dust breasts with flour and saute in butter or margarine until light brown. Remove turkey breasts from the pan and save the drippings.

Add 1 tablespoon butter or margarine to the skillet with the drippings and melt butter or margarine over medium heat. Add 1/2 cup minced onions, and cook until the onions are tender. Add 2 tablespoons of flour and wisk until golden brown. Slowly add 1/2 cup bouillon and wisk until mixture is thoroughly mixed. Add approximately 1/2 cup coffee to taste and wisk until smooth. Salt and pepper to taste. Return the turkey breasts to sauce and heat slowly while preparing the grits.

Serve turkey and gravy over grits.

Pheasant Cordon Bleu with Hollandaise Sauce

1 Pheasant Breast (skin removed)
2 Slices Deli Boiled Ham
1 Cup Grated Gruyere Cheese
1 Egg
Progresso Italian Bread Crumbs
Peanut Oil
Lawry's Garlic Salt
Lawry's Seasoned Pepper

Split pheasant breast, lightly sprinkle with meat tenderizer, and pierce breast with a fork.

Lay the outside 1/2 breast down, then layer one slice deli boiled ham, grated Gruyere cheese, one slice deli boiled ham and top with remaining 1/2 breast.

Rub egg wash on outside of pheasant breasts.

Press Progresso Italian seasoned bread crumbs into the outside of the pheasant breast.

Saute in peanut oil until golden brown.

Remove from pan, place on baking dish and place in oven at 200 degrees to warm while preparing Hollandaise Sauce.

Hollandaise Sauce—Any one of the commercially packaged sauces will compliment this dish nicely.

The following chart lists the eye color and suggested glass eye size (in millimeters) for most mature North American birds. The "exposed" column lists the actual iris diameter live birds generally display.

SPECIES	MALE	FEMALE	EXPOSED
Baldpate Duck (Widgeon)	10 hazel/brown	10 hazel/brown	8 mm
Bittern	11 yellow	11 yellow	10 mm
Black Duck	11 black	11 black	9 mm
Black-Bellied Whistling Duck	11 brown	11 brown	9–10 mm
Blackbird			
Redwing Blackbird	5 brown	5 brown	4½–5 mm
Rusty Blackbird	5 brown	5 brown	4–4½ mm
Brewers Blackbird	6 yellow	6 brown	5–5½ mm
Yellow Headed Blackbird	6 brown	6 brown	5 mm
Blue Bird	4 brown	4 brown	3½–4 mm
Blue Jay	7 hazel	7 hazel	6 mm
Bobolink	5 brown	5 brown	4 mm
Bobwhite Quail	8 hazel	8 hazel	6 mm
Bufflehead Duck	10 brown	10 brown	7–8 mm
Bullfinch	6 brown	6 brown	5½ mm
Buzzard	12 brown	12 brown	12 mm
Canary	3 brown	3 brown	2½ mm
Canvasback Duck	10 red	10 brown	8–9 mm
Cardinal	5 brown	5 brown	4 mm
Cat Bird	5 brown	5 brown	4½–5 mm
Chapparel Cock			
(Road Runner)	10 yellow	10 yellow	8–9 mm
Chickadee	4 brown	4 brown	3½–4 mm
Chicken, Domestic	10 hazel	10 hazel	8½ mm
Coot, American	10 red	10 red	7½ mm
Cormorant	8–12 green	8–12 green	8–9 mm
Cowbird	6 brown	6 brown	5 mm
Crane, Sandhill	12 straw	12 straw	10 mm
Crossbill	5 brown	5 brown	4½–5 mm
Crow	11 brown	11 brown	9 mm
Cuckoo	7 brown	7 brown	6 mm
Dove			
Mourning Dove	7 brown	7 brown	6–7 mm
Turtle Dove	7 brown	7 brown	6 mm
White-Winged Dove	8 orange	8 orange	8 mm
Eagle			
Bald Eagle (Young)	16 brown	16 brown	15 mm
Bald Eagle (Adult)	17 straw	17 straw	16 mm
Golden Eagle	18 brown	18 brown	15–16 mm
Eider Ducks			
Common Eider	11 brown	11 brown	10 mm
King Eider	11 brown	11 brown	9 mm
Spectacled Eider	11 brown	11 brown	9 mm
Steller's Eider	11 brown	11 brown	8–9 mm
Finches (Most)	4 brown	4 brown	3 mm
Flicker	8 brown	8 brown	7 mm
Florida Duck (Mottled Duck)	11 brown	11 brown	9–10 mm
Fulvous Whistling Duck	11 brown	11 brown	9–10 mm
Gadwall Duck	11 brown	11 brown	8 mm
Gallinule			
Common Gallinule	9 red	9 red	7½ mm
Purple Gallinule	9 red	9 red	7½ mm
Geese			
American Brant	11–12 brown	11–12 brown	8–9 mm
Black Brant	11–12 brown	11–12 brown	8–9 mm
Blue Goose	12 brown	12 brown	10–11 mm
Canada Goose	12 brown	12 brown	11 mm
Emperor Goose	12 hazel/brown	12 hazel/brown	10 mm
Snow Goose	12 brown	12 brown	10–11 mm
White-Fronted Goose	12 brown	12 brown	10–11 mm
Goldeneye Duck	11 yellow	11 yellow	9 mm
Grackle	5 straw	5 straw	4 mm
Grebe			
Eared Grebe	8 red	8 red	6½–7 mm
Horned Grebe	9 red	9 red	7½ mm
Pied Billed Grebe	8 red	8 red	6½–7 mm
Grouse			
Blue Grouse	10 hazel	10 hazel	8½ mm
Ruffed Grouse	10 hazel	10 hazel	8½–9 mm

SPECIES	MALE	FEMALE	EXPOSED
Sage Grouse	11 hazel	11 hazel	8½–9 mm
Sharp-Tailed Grouse	10 hazel	10 hazel	8½–9 mm
Spruce Grouse	10 hazel	10 hazel	8–8½ mm
Gull			
Black-Backed Gull	14 yellow	14 yellow	12 mm
Bonaparte Gull	9 brown	9 brown	7–8 mm
Franklin Gull	9 brown	9 brown	7–8 mm
Glaucus Gull	14 straw	14 straw	11 mm
Herring Gull	12 yellow	12 yellow	11 mm
Laughing Gull	10 red	10 red	8 mm
Ring Bill Gull	10 yellow	10 yellow	9 mm
Harlequin Duck	10 brown	10 brown	9 mm
Hawk			
Cooper Hawk	12 red	12 straw	10–11 mm
Fish Hawk (Osprey)	14 yellow	14 yellow	13 mm
Goshawk Hawk	14 red	14 hazel	12 mm
Marsh Hawk	12 yellow	12 yellow	10 mm
Pigeon Hawk	8 brown	8 brown	7 mm
Red Tail Hawk	14 brown	14 brown	12 mm
Rough Leg Hawk	14 hazel	14 hazel	13 mm
Red Shoulder Hawk	14 hazel	14 hazel	12 mm
Sharp Shinned Hawk	10 yellow	10 yellow	7–8 mm
Sparrow Hawk	9 brown	9 brown	8 mm
Swainsons Hawk	14 hazel	14 hazel	12 mm
Shoveler Hawk	10 yellow	10 yellow	9 mm
Kingfisher, Belted	10 brown	10 brown	9–10 mm
Killdeer	7 brown	7 brown	5½–6 mm
Lark, Skylark	5 brown	5 brown	4 mm
Loon	14 red	14 red	13 mm
Magpie	8 brown	8 brown	7 mm
Mallard Duck	11 brown	11 brown	9 mm
Mandarin Duck	10 dk. brown	10 dk. brown	9 mm
Meadowlark	7 brown	7 brown	5 mm
Merganser			
American Merganser	11 red	11 red	10 mm
Hooded Merganser	9 yellow	9 yellow	8 mm
Red-Breasted Merganser	11 red	11 red	10 mm
Mockingbird	5 brown	5 brown	4 mm
Mottled Duck	11 brown	11 brown	10 mm
Mudhen	10 red	10 red	9 mm
Nighthawk	8 black	8 black	7 mm
Old Squaw Duck	10 hazel	10 hazel	8 mm
Osprey	14 yellow	14 yellow	13 mm
Owl			
Arctic Owl	20 straw	20 straw	19 mm
Barn Owl	14 brown	14 brown	11 mm
Barred Owl	18 brown/purple	18 brown/purple	17 mm
Burrowing Owl	15 yellow	15 yellow	12–14 mm
Elf Owl	10 yellow	10 yellow	6–7 mm
Great Horned Owl	20 yellow	20 yellow	19 mm
Long Eared Owl	14 yellow	14 yellow	10–11 mm
Pigmy Owl	12 yellow	12 yellow	6–7 mm
Richardson Owl	14 yellow	14 yellow	13 mm
Screech Owl	14 straw	14 straw	13 mm
Sawwhet Owl	12 straw	12 straw	6–7 mm
Snowy Owl	18 straw	18 straw	17 mm
Short Eared Owl	14 yellow	14 yellow	10–11 mm
Partridge, Gray (Hungarian)	9 brown	9 brown	7½ mm
Pelican	17 straw	17 straw	14–15 mm
Pheasant, Ring-Necked	10–11 special	10–11 brown	8–9 mm
Pigeon	8 orange	8 orange	7 mm
Pintail Duck	9 brown	9 brown	8 mm
Plover	6 brown	6 brown	5 mm
Prairie Chicken	10 hazel	10 hazel	8–9 mm
Ptarmigan	9 brown	9 brown	7 mm
Quail			
Bobwhite Quail	8 hazel	8 hazel	6 mm
California Quail	8 brown	8 brown	6 mm
Gambel's Quail	8 brown	8 brown	6 mm

SPECIES	MALE	FEMALE	EXPOSED	SPECIES	MALE	FEMALE	EXPOSED
Quail (cont.)				Shoveler Duck (Northern)	10 yellow	10 brown	8 mm
Harlequin Quail	8 brown	8 brown	6 mm	Snipe	7 brown	7 brown	6 mm
Mountain Quail	8 brown	8 brown	6 mm	Sparrow	3 brown	3 brown	2½ mm
Scaled Quail	8 brown	8 brown	6 mm	Starling	5 brown	5 brown	4 mm
Rails				Swallow	4 brown	4 brown	3 mm
Carolina Rail	8 red	8 red	7 mm	Swans	12–13 brown	12–13 brown	11 mm
Clapper Rail	10 hazel	10 hazel	8 mm	Teal			
King Rail	10 brown	10 brown	8 mm	Blue-Winged Teal	9 brown	9 brown	7 mm
Little Black Rail	5 brown	5 brown	4–5 mm	Cinnamon Teal	9 red	9 brown	7 mm
Virginia Rail	9 hazel	9 hazel	8½ mm	Green-Winged Teal	9 hazel	9 hazel	7 mm
Yellow Rail	6 brown	6 brown	4–5 mm	Tern	6 brown	6 brown	5 mm
Redhead Duck	10 yellow	10 brown	8–9 mm	Thrush	6 brown	6 brown	5–5½ mm
Ring-Necked Duck	10 straw	10 hazel	7–8 mm	Turkey, Wild	12–13 brown	12–13 brown	10–11 mm
Road Runner	10 yellow	10 yellow	8–9 mm	Widgeon Duck	10 hazel/brown	10 hazel/brown	8 mm
Robin	6 brown	6 brown	5 mm	Woodcock	10 brown	10 brown	8 mm
Ruddy Duck	11 brown	11 brown	7 mm	Wood Duck	11 red	11 brown	9 mm
Sandpiper	6–7 brown	6–7 brown	6 mm	Woodpecker			
Scaup Ducks				Downy Woodpecker	5 brown	5 brown	4 mm
Greater Scaup (Broadbill)	10 yellow	10 yellow	7½ mm	Flicker Woodpecker	8 brown	8 brown	7 mm
Lesser Scaup (Bluebill)	10 yellow	10 yellow	7 mm	Ivory Billed Woodpecker	12 yellow	12 yellow	9 mm
Scoter Ducks				Red Bellied Woodpecker	7 red	7 red	6 mm
Surf Scoter	10 white	10 brown	9 mm	Red Headed Woodpecker	6 brown	6 brown	5 mm
Black Scoter	10 brown	10 brown	8 mm	White Headed Woodpecker	6 red	6 red	5 mm
White-Winged Scoter	10 straw	10 straw	8–9 mm				

APPENDIX 2: WATERFOWL NAMES

COMMON NAME	*Latin Name*	Other Names
DABBLING DUCKS		
BLACK DUCK	*Anas rubripes*	Black Mallard, Red Leg
GADWALL	*Anas strepera*	Gray Duck, Gray Mallard
MALLARD	*Anas platyrhynchos platyrhynchos*	Greenhead (drake), Gray Duck (hen), Susie (hen)
PINTAIL	*Anas acuta acuta*	Sprig, Sprigtail, Spike, Spiketail
SHOVELER (Northern Shoveler)	*Anas clypeata*	Spoonbill, Spoony, Bootlip, Smiling Mallard
TEALS—BLUE-WINGED TEAL	*Anas discors*	Bluewing, Summer Teal, White-faced Teal
CINNAMON TEAL	*Anas cyanoptera septentrionalium*	Red Teal, Red-Breasted Teal
GREEN-WINGED TEAL (American G.W.T.)	*Anas crecca carolinensis*	Greenwing, Common Teal, T-Bird
WIDGEON (AMERICAN)	*Anas americana*	Baldpate, Gray Duck
WOOD DUCK	*Aix sponsa*	Woodie, Summer Duck, Swamp Duck, Squealer
WHISTLING DUCKS		
BLACK-BELLIED WHISTLING DUCK	*Dendrocygna autumnalis autumnalis*	Black-Bellied Tree Duck, Cornfield Duck
FULVOUS WHISTLING DUCK	*Dendrocygna bicolor helva*	Fulvous Tree Duck, Mexican Squealer, Squealer
DIVING DUCKS		
BUFFLEHEAD	*Bucephala ableola*	Butterball, Dipper
CANVASBACK	*Aythya valisineria*	Can, Grayback
EIDERS—AMERICAN EIDER	*Somateria mollissima*	Common Eider
KING EIDER	*Somateria spectabilis*	
PACIFIC EIDER	*Somateria mollissima*	Common Eider
SPECTACLED EIDER	*Somateria fischeri*	
STELLER'S EIDER	*Polysticta stelleri*	
GOLDENEYES—AMERICAN (Common)	*Bucephala clangula americana*	Whistler
BARROW'S GOLDENEYE	*Bucephala islandica*	Whistler, Ricky Mountain Whistler
HARLEQUIN	*Histrionicus histrionicus*	
MERGANSERS—AMERICAN (Common)	*Mergus merganser americanus*	Goosander, Sawbill, Fish Duck
HOODED MERGANSER	*Mergus cucullatus*	Sawbill, Fish Duck
RED-BREASTED MERGANSER	*Mergus serrator*	Sawbill, Fish Duck
OLDSQUAW	*Clangula hyemalis*	Long-Tailed Duck, Sea Pintail, Cockertail, Coween Kakawi
SCAUPS—GREATER SCAUP	*Aythya marila mariloides*	Broadbill, Bluebill
LESSER SCAUP	*Aythya affinis*	Bluebill
SCOTER—BLACK SCOTER	*Melanitta nigra americana*	American Scoter, Common Scoter, Black Coot, Sea Coot, Black Duck
SURF SCOTER	*Melanitta perspicillata*	Skunkhead, Coot, Sea Coot
WHITE-WINGED SCOTER	*Melanitta fusca deglandi*	Coot, Sea Coot, Whitewing
REDHEAD	*Aythya americana*	Pochard
RING-NECKED DUCK	*Aythya collaris*	Ring-Billed Duck, Ringbill, Blackjack
RUDDY DUCK	*Oxyura jamaicensis rubida*	Butterball, Bull-Necked Teal
GEESE AND SWANS		
ATLANTIC BRANT	*Branta bernicla hrota*	Sea Goose, Brant Goose, White-bellied Brant
BLACK BRANT	*Branta bernicla nigricans*	Pacific Brant
CANADA GOOSE	*Branta canadensis*	Honker, Canada
EMPEROR GOOSE	*Anser canagicus*	
LESSER SNOW	*Anser c. caerulescens*	Blue Goose, Waive, Brant
WHITE-FRONTED GOOSE	*Anser albifrons frontalis*	Specklebelly, Specklebelly Brant, Speck, Laughing Goose
ROSS' GOOSE	*Anser rossii*	Little Wavie, Warty-Nosed Wavie, Horned Wavie
TRUMPETER SWAN	*Cygnus buccinator*	
WHISTLING (Tundra) SWAN	*Cygnus columbianus*	
MUTE SWAN	*Cygnus olor*	

NORTH AMERICAN WATERFOWL FAMILY TREE
Order Anseriformes
Suborder Anseres
Family Anatidae

SUBFAMILY ANSERINAE

- **SWANS** — *Tribe Cygnini*
  - Tundra (Whistling) Swan
  - Trumpeter Swan
  - Mute Swan
- **GEESE** — *Tribe Anserini*
  - Greater White-Fronted Goose
  - Snow (Blue) Goose
  - Ross' Goose
  - Emperor Goose
  - Brant Goose
    - Atlantic Brant
    - Black Brant
  - Canada Goose
- **TREE DUCKS** — *Tribe Dendrocygnini*
  - Fulvous Whistling Duck
  - Black-Bellied Whistling Duck

SUBFAMILY ANATINAE

- **DABBLING DUCKS**
  - *Tribe Cairinini*
    - Wood Duck
  - *Tribe Anatini*
    - Black Duck
    - Mottled Duck
    - Mallard
    - Pintail
    - Teal
      - Green-Winged Teal
      - Blue-Winged Teal
      - Cinnamon Teal
    - Shoveler
    - Gadwall
    - Widgeon
      - American Widgeon
      - European Widgeon
- **DIVING DUCKS**
  - **BAY DUCKS** — *Tribe Aythyini*
    - Canvasback
    - Redhead
    - Ring-Necked Duck
    - Scaup
      - Greater Scaup
      - Lesser Scaup
  - **SEA DUCKS** — *Tribe Mergini*
    - Eider
      - Common Eider
      - King Eider
      - Spectacled Eider
      - Steller's Eider
    - Harlequin
    - Oldsquaw
    - Bufflehead
    - Scoter
      - Black Scoter
      - Surf Scoter
      - White-Winged Scoter
    - Goldeneye
      - Common Goldeneye
      - Barrow's Goldeneye
    - Merganser
      - Hooded Merganser
      - Common Merganser
      - Red-Breasted Merganser
- **STIFF-TAILED DUCKS** — *Tribe Oxyurini*
  - Ruddy Duck